DATE DUE

THE STORK
AND THE PLOW

*The Equity Answer to the
Human Dilemma*

Paul R. Ehrlich, Anne H. Ehrlich,
and Gretchen C. Daily

Yale University Press
New Haven & London

First published in 1995 by G. P. Putnam's Sons. Reissued in paperback in 1997 by Yale University Press. Reprinted by arrangement with G. P. Putnam's Sons and the Authors.

Printed in the United States of America by Thomson-Shore, Dexter, Michigan.

Library of Congress Catalog Number 97-60446.
ISBN 0-300-07124-8 (pbk.)

A catalogue record for this book is available from the British Library.

The paper in this book meets the guidelines for permanence and durability of the Committee on Production Guidelines for Book Longevity of the Council on Library Resources.

10 9 8 7 6 5 4 3 2 1

TO:

Our friend Larry Condon, who carries on the work the much-missed LuEsther could not finish

AND TO:

Helen and Peter Bing and Wren and Tim Wirth, with love, respect, and thanks

ACKNOWLEDGMENTS

We are deeply grateful to Ginger Barber (Virginia Barber Literary Agency), Sam Hurst (IPAT Productions), Sally Mallam (Institute for the Study of Human Knowledge), and Peter Warshall (Zoology Department, University of Arizona), for taking the time and trouble to criticize the entire manuscript. We are also indebted to Heidi Albers (Food Research Institute, Stanford), Lisa and Timothy Daniel (Bureau of Economic Research, Federal Trade Commission), Partha Dasgupta (Department of Economics, Cambridge), Lawrence Goulder (Department of Economics, Stanford), Walter Falcon and Rosamond Naylor (Institute of International Studies, Stanford), Marcus Feldman and Jonathan Roughgarden (Department of Biological Sciences, Stanford), Frédéric Lelièvre (École Supérieure de Physique et de Chimie Industrielles de la Ville de Paris), Harold Wagner (Golden Bear Expeditions), and David Western (Director, Kenya Wildlife Service) for assistance with various aspects of the project. Heidi Albers, Wally Falcon, and Roz Naylor were kind enough to review several chapters. Needless to say, the residual errors are ours.

Our editor at Putnam, Jane Isay, once again helped us improve the manuscript and generally kept us in line. The staff of the Falconer Biology Library of Stanford's Department of Biological Sciences did their usual fantastic job of helping us deal with a massive literature, and Pat Browne and Steve Masley took care of piles of xeroxing promptly and with good cheer. Our friend Peggy

VasDias helped us overcome many last-minute problems in connection with preparation of the manuscript. Paul and Anne's youngest granddaughter, Melissa, and Frédéric Lelièvre provided inspiration.

This work has been supported in part by grants from the W. Alton Jones Foundation, the Winslow Foundation, the Heinz Foundation, the Beatrice Bliss Fund, and an anonymous grant for core support to the Center for Conservation Biology, Stanford University. We all remain indebted to LuEsther, whose support over decades has made our work possible.

CONTENTS

Contents

A Personal Preface

The three authors of this book represent three different life experiences and two different generations. That has made the process of writing it all the more interesting. Paul started his scientific career studying the ecology and evolution of insects, plants, fishes, and snakes. For him, the transition from studying populations of butterflies to those of human beings was natural. Anne and Paul first worked together on butterfly taxonomy and evolution, and then developed joint interests in policy areas—especially in population, agriculture, and the environment, and later in the environmental effects of war and preparing for war. Anne also has extensive experience in working with nongovernmental organizations, increasingly important forces in the world.

Gretchen, in contrast, grew up in a period of environmental consciousness and started out her career as an interdisciplinary scientist. She was the first of Paul's students to have done both straight biology and policy work as part of her doctoral dissertation, and now she works simultaneously on a diversity of scientific and policy problems relevant to the maintenance of human and environmental well-being.

Anne and Paul enjoy one of the very few advantages of being in their sixties: they have over three decades of intellectual and emotional experience in attempting to understand the human predicament and in trying to persuade people to solve it. Their understanding has evolved over all those years, as has the predicament. In the late 1960s and early 1970s, around the time *The Population Bomb* was published,[1] they were well aware of the unprecedented population growth in poor countries and the increasing overconsumption and use of environmentally dangerous technologies by the rich. A horrific taxi ride they had taken in New Delhi in 1966 opened the *Bomb,* describing how their intellectual understanding of the press of human numbers and profound poverty was transformed into an emotional one. They already had an emotional understanding of runaway consumerism. America's role as the world's largest consumer is emphasized in the Prologue of the *Bomb.*[2] The book also presented statistics on the lifetime of consumption ahead of the average American baby, concluding: "It's not a baby, it's Superconsumer!"[3] Soon thereafter, Paul published a scenario pointing out that every American baby will have the same negative impact on the global environment in its lifetime as about 25 Indian babies.[4]

A substantial portion of *The Population Bomb* was devoted to one of the faulty technologies that is still severely abused—synthetic organic pesticides. Paul's and Anne's understanding of technological problems had an emotional component also. Fifteen years earlier, Paul had worked as a graduate student on problems of DDT resistance in insects, and the Department of Entomology at the University of Kansas in those days was alive with stories of pesticide abuse in agriculture. Worse yet, Anne and Paul found their dates at drive-in movies in Topeka interrupted at inconvenient times by DDT being blasted directly through the car window for "mosquito control." At that time too, they were swept up in the first of many equity issues; with a colleague, Paul organized a successful series of sit-ins to desegregate the restaurants of Lawrence, Kansas.

Gradually, Anne's interests shifted from working with Paul on butterflies to finding ways to preserve a livable world for their daughter Lisa. By the late 1960s, their professional partnership was focused on the human predicament. Their solutions, back then, to producing enough food for humanity—keeping the plow up with the stork—were similar to those they recommend today. Unfortunately, the terms they used proved less than compelling: providing greater equity through "de-development" of the rich countries, "semi-development" of the poor ones, and "population control" for all.[5]

Of course, then as today, there were those who claimed that no predicament existed—that perpetual population and economic growth were both possible and desirable. Similarly, of those concerned over the years about the trajectory of social well-being, many have focused on just one or a few of an array of putative causes or cures: capitalism, communism, socialism, colonialism, religion, atheism, population growth among the poor or among people of the "wrong" color or religion, overconsumption among the rich, prejudice against women, political correctness, immigration in rich countries, migration to the planets, and on and on. One observer's cure is another's culprit.

This has often placed Paul and Anne, and their close colleague and collaborator, Professor John Holdren of the Energy and Resources Group at Berkeley, at the center of controversy. Many of their critics even today remain focused on the quarter-century-old *Population Bomb,* often basing their views of it on rumor or out-of-context quotes printed elsewhere. Few bother to look up and read the many works in which the whole complex of factors influencing environmental deterioration were analyzed.[6]

Paul and Anne have made their mistakes and taken their lumps, but hope they have also done what scientists are supposed to do: keep learning as new information comes in. They clearly overrated the difficulty of temporarily increasing food production in poor nations. They also incorrectly assumed that the public would quickly grasp that the deterioration of Earth's life-support systems (ecosystems) was the most crucial environmental issue facing civilization. And they had no idea how hard it would be to curb the expansion of consumption among both rich and poor.

The last two were big mistakes, as the 1994 elections in the United States demonstrated. In the elections, the whole focus was on individuals versus society—the latter personified and vilified as "the government." Lower taxes, fewer regulations, and job security were the issues—not how Americans could corporately make government more efficient and simultaneously save themselves by taking care of the environment and each other.

Anne has been especially distressed at related developments in a major area of her concern, the environment and the military. As this is written, Department of Defense and Department of Energy programs to prevent Russian nuclear materials from going to terrorists are under assault by the reinvigorated Republican right, and the prospects for cleaning up the colossal toxic and radioactive mess left by the U.S. weapons production program are fading. A president in political trouble now faces a hostile Congress, and American society may have to wait many years more before significant leadership can be expected from Washington on the critical issues of population, resources, and environment now facing the nation.

On a trip to Australia, Paul's and Anne's second home, they found things no better. There they interrupted visits with old friends and colleagues to try to help prevent key areas of the nation's precious old-growth forests from being woodchipped on the cheap for the benefit of Japan. They, and Australia, lost that battle. All of Australia's major ecosystems, including its forests, are deteriorating under the assault of 18 million superconsumers. And there was serious discussion in the press of the possibility of letting the population expand to more than 100 million! Ironically, the Labor government that made the short-sighted decision to go ahead with the woodchipping is usually more responsive to environmental imperatives than the Liberals who may soon replace them.

In India, three decades after Paul's and Anne's first visit, they found fewer but still abundant beggars in the streets of Delhi, people still defecating on public thoroughfares, and so many more automobiles and motor scooters that breathing was difficult. That nation has a rapidly growing middle class, which is quickly adopting the consumptive patterns of the West. Everywhere they saw scenes of grim poverty framed by billboards advertising computer software. Madras was less depressing than Delhi, with many signs of real caring for all people by the state's government, although also with many signs of unsustainable development.

Yet it was China that presented an emotional learning experience equivalent to that of Delhi in 1966. The Chinese are doing an amazing job of trying to handle their population problem. But there also, the poor, newly unleashed from a communist economic system, are clearly determined to repeat the mistakes of the West as fast as possible. In car-clogged Beijing, it is easy to envision millions of Chinese (and Indians, Thais, and more and more others) joining the club of superconsumers recklessly savaging the planet.

At half Paul's and Anne's age and armed with a more mathematical background, Gretchen adds a fresh perspective to the team. She spent her teenage years, the time when ideas and experiences seem to leave their greatest mark, in West Germany. There she developed a keen appreciation of European values and perspectives on what is worth living and fighting for, and on the power of historical and cultural forces shaping such pursuits. The worldviews to which she was exposed often contrasted strikingly with those prevailing in her native United States.

Gretchen's passion for understanding nature's machinery and human dependence on it is matched by a desire to apply her good fortune in life to improving the social and economic conditions endured by most of the human population. Sparked by a wonderful group of colleagues, she devotes a great deal of her time to studying the human predicament and communicating her work to the public. She has worked on a wide array of issues, ranging from analyses of the potential impacts of global warming, land degradation, and the loss of biodiversity on human well-being to the complex of social and economic forces perpetuating population growth, poverty, and environmental deterioration.

Gretchen has entered the public fray over how to create a sustainable society, giving speeches and appearing on radio and television. She often reflects on her determination to follow Anne's and Paul's efforts, which at times seems to her completely in vain—massively overwhelmed by greed, myopia, prejudice, and denial. She is appalled at how little the arguments have changed; at how many people still think that growth can and should continue for ever and ever, that Texas and Nevada could support millions—hell, billions!—more people because they have so much empty space; that population problems trace to the reproductive irresponsibility of the "wrong kinds" of people; that the gods of America, capitalism and technology, will save us.

Writing this book has been an interesting experience for all three of us as authors. It is basically a book about equity, and major issues of equity concern the treatment of people of different sexes and ages. Thus, our own differences have sometimes led to diverse approaches to the issue, and resolving them has been more educational than the usual process of turning the thoughts of co-authors into a coherent presentation.

Perhaps the reason we have been able to do it so readily is that we share much the same worldviews and philosophy. What keeps all three of us going is having good friends, which makes life really worth living. We share a friendship developed in working together both at Stanford and in the field in the United States, Latin America, Australia, New Zealand, Africa, and the Indian Ocean. We also share bonds with numerous mutual friends and colleagues around the world who have collaborated with us and supported our efforts, and whose help and encouragement over the years has been invaluable. "We" therefore is used throughout this book, even though all of us may not have participated in a particular incident or interview.

Finally, we share the conviction that our generations will now largely determine the future well-being of humanity. The issues discussed in this book require much more attention if our descendants are to have any chance to enjoy life's pleasures. In particular, our generations must struggle to create a more equitable world; not only would that be morally right, but nearly everyone's well-being will depend upon it. We have no fear of our diagnosis of the human predicament being proven wrong—what terrifies us is that it will be proven right.

Paul R. Ehrlich,
Anne H. Ehrlich,
Gretchen C. Daily
Stanford, California
January 1995

THE STORK
AND THE PLOW

Introduction

THE DILEMMA

As the twentieth century draws to a close, humanity faces the daunting prospect of supporting its population without inducing catastrophic and irreversible destruction on Earth's life-support systems. Human and agricultural fertility are on a collision course: the stork is threatening to overtake the plow. By 1999, the human population will surge past 6 billion in number and will still be skyrocketing. United Nations demographers project continuing expansion for another century or so to nearly 12 billion. While roughly a billion people in industrialized nations live in comfort undreamed of in centuries past, another billion suffer extremes of poverty and violence that the rich can hardly imagine. The rest are concentrated near the low end of the standard of living continuum.

The struggle merely to support today's population at today's standards of living is causing environmental destruction on a scale and at a pace unprecedented in human history. Accelerating degradation and deforestation of land, depletion of groundwater, toxic pollution, biodiversity loss, and massive atmospheric disruption are wrecking the planet's machinery for producing the basic material ingredients of human well-being.

Alarm in the international scientific community regarding the seriousness of this dilemma is widespread and mounting. Yet the very existence of the dilemma is largely unappreciated by the general public and the politically oriented and ecologically ignorant pundits of the TV/radio talk-show circuit. A student can get all the way through a major university and still not be aware of its existence. Presidents and prime ministers overlook it. Vatican policies and the exhortations of fundamentalists of various religions make it worse.

This set of circumstances prompted over six hundred of the world's most distinguished scientists, including a majority of the living Nobel laureates in the sciences, to issue the *World Scientists' Warning to Humanity* in 1993. It reads, in part:

> The earth is finite. Its ability to provide for growing numbers of people is finite. And we are fast approaching many of the earth's limits. Current economic practices which damage the environment, in both developed and underdeveloped nations, cannot be continued without the risk that vital global systems will be damaged beyond repair. . . .
>
> Pressures resulting from unrestrained population growth put demands on the natural world that can overwhelm any efforts to achieve a sustainable future. If we are to halt the destruction of our environment, we must accept limits to that growth. . . .
>
> No more than one or a few decades remain before the chance to avert the threats we now confront will be lost and the prospects for humanity immeasurably diminished. . . . A great change in our stewardship of the earth and the life on it is required, if vast human misery is to be avoided and our global home on this planet is not to be irretrievably mutilated.
>
> The developed nations are the largest polluters in the world today. They must greatly reduce their overconsumption, if we are to reduce pressures on resources and the global environment. The developed nations have the obligation to provide aid and support to developing nations, because only the developed nations have the financial resources and technical skills for these tasks.
>
> Acting on this recognition is not altruism, but enlightened self-interest: whether industrialized or not, we all have but one lifeboat.

This was followed a year later by a similar statement issued by the world's scientific academies.[1] The scientific evidence behind these statements is overwhelming. Yet we're like the profligate off-spring who inherits a fortune, puts it into a checking account, and spends it as if there were no tomorrow. We keep writing bigger and bigger checks, struggling to fend off ever-growing problems, while paying no attention to the increasingly rapid decline in the remaining balance.

Hundreds of millions of individual decisions are made each year about having or not having children. These individual decisions seem to have nothing to do with the size of the global population and everything to do with the lives, aspirations, and well-being of the women and men who make them. But the futures of their children and grandchildren will be shaped by the sum of all these decisions in the most profound way, as the seemingly inexorable demographic future unfolds. The likely outcome thus raises questions for the world community that are seldom considered by prospective parents.

How long can we go on this way? Can 8, 10, 12, or 14 billion human beings even be supported on our fragile planet? Under what conditions? What environmental costs would be incurred in the process? How will population growth ultimately end, as it must? What limiting factors will come into play first? How familiar will the Four Horsemen of the Apocalypse—famine, pestilence, war, and death—become as resource scarcities intensify? What social choices and trade-offs will humanity face in the struggle to support such huge numbers of people? How will women, in particular, be affected, and what crucial roles may women play in resolving the human dilemma? Can equity—between the sexes, between children and adults, between rich and poor—be increased in the face of population growth? Indeed, can rapid population growth be humanely ended and reversed *without* greater equity? Most important of all, can human beings, with their self-avowed intelligence, engineer and carry through a transition to a sustainable world? How?

Approaching these questions requires an understanding of the carrying capacity of Earth for human life.[2] *Carrying capacity* has a rather daunting ring to it, but it's basically a down-to-earth idea that's becoming more and more central to our lives—a concept we can ill afford to go on ignoring. Biologists define carrying capacity as the maximum population size of any organism that an area can support, without reducing its ability to support the same species in the future. Carrying capacity is determined by characteristics of both the area and the organism. For example, a larger or richer area will, all else being equal, have a higher carrying capacity. Similarly, an area can support more rabbits eating vegetation than it can support coyotes eating rabbits.

Although carrying capacity is easily defined, calculating it precisely—even for nonhuman organisms—can be very difficult since it does not remain constant. Carrying capacity may fluctuate with climate or other forces that affect the abundance of resources. Assuming no large-scale alterations in the environment, however, major changes in carrying capacity only occur as fast as organisms evolve different resource requirements.

For human beings, the matter is complicated by three factors. First, people vary tremendously in the types and quantities of resources they consume. For the most part, one rabbit is like any other rabbit in terms of resource use; however, an average Kenyan is nothing like an average American in this regard. Second, trade enables human populations to exceed local carrying capacities. The population of the Netherlands uses roughly seventeen times more land than there is within the country for food and energy alone.[3] The Netherlands is effectively importing carrying capacity from other parts of the globe. Third, human beings may undergo extremely rapid cultural (including technological) evolution in the array and amounts of resources consumed. Just consider how much American resource use has changed since pioneer days.

Carrying capacity therefore is not some fixed number of people; rather, it depends heavily on the cultural and economic characteristics of the population in question. Earth can support a larger population of cooperative, far-sighted, vegetarian pacifist saints than of competitive, myopic, meat-eating, war-making, typical human beings. All else being equal, Earth can hold more people if they have relatively equal access to the requisites of a decent life than if the few are able to monopolize resources and the many must largely do without.[4] The problems of population, social and economic inequity, and environmental deterioration are thus completely intertwined.

This makes it useful to distinguish between *biophysical carrying capacity*—the maximum population size that could be sustained under given technological capabilities—and *social carrying capacity*—the maximum that could be sustained under a given social system and its associated patterns of resource consumption. In the end, regardless of humanity's technological accomplishments, biophysical constraints will limit the number of people that can be supported without destroying Earth's future capacity to support people.[5] Social forces, however, will always come into play to limit a population before absolute biophysical constraints do. Human beings are prone to error and greed, making resource use both inefficient and inequitable. Social carrying capacity is smaller than biophysical carrying capacity also because the latter implies a standard of living—a battery chicken lifestyle for people—that would be universally undesirable.[6]

Human ingenuity has greatly increased both biophysical and social carrying capacities, and the potential exists for further increases. But has that ingenuity made today's population sustainable? The answer to this question is clearly no, by a simple standard. The current population is being maintained only through the exhaustion and dispersion of a one-time inheritance of natural capital—through rapid draining of the bank account. Worse yet, there is no way of replenishing critical, nonsubstitutable elements of Earth's natural capital on any time scale of interest. For instance, it typically takes thousands of years to form enough fertile soil to support agriculture; once it has eroded away, that's it. And soil is only one of several types of natural capital being depleted. In effect, human activity is diminishing Earth's biophysical carrying capacity just as the need for it is greatly intensifying.

Human activity needs to be *sustainable* in order for its population to remain within Earth's carrying capacity. A sustainable population is one whose activities and well-being can be maintained without interrupting, weakening, or losing valued qualities. A sustainable population would pass on its inheritance of natural capital, not unchanged but undiminished in potential to support future generations. The oft-cited Brundtland Report in 1987 defined "sustainable development" as ensuring that development "meets the needs of the present without compromising the ability of future generations to meet their own needs."[7]

The world population and food situation beautifully illustrates both the superficial simplicity and the underlying complexity of humanity's ties to the laws and limits of nature. No human activity causes as much direct environmental damage as agriculture, yet no other activity is more dependent for its success on environmental integrity. No lack of the material ingredients of well-being causes as much human suffering as lack of food. And no index so plainly measures failure of a population to remain within its carrying capacity as the extent of hunger-related disease and death.

Still, extensive environmental damage and hunger do not arise simply from ignoring nature's constraints. They are produced by a complex and often self-reinforcing interplay of social, economic, political, and natural forces operating at all scales—from rural villages to global trade agreements. As its population continues to expand, humanity must consider manipulation of these social, economic, and political forces as a means of commensurately expanding social carrying capacity.

In the past, enormous increases in carrying capacity have been wrought by pushing back intellectual frontiers—the boundaries of human knowledge and understanding. Sequential augmentation of carrying capacity occurred as newly developed physical and intellectual tools spread across territorial frontiers—regions not yet subject to the resource exploitation regime associated with the new tools (such as agriculture or minerals exploitation). New is not always better, though, and carrying capacity has also sometimes been reduced through misappropriation and misapplication of technologies. At this stage in the game, as the scientific community warns, we cannot prudently count on intellectual innovation to resolve the human dilemma in time.

The greatest promise for inducing a sufficiently rapid increase in carrying capacity today lies in pushing back cultural frontiers—converting the world's nations to a new regime of resource distribution. Any hope for providing the projected population of 8 to 12 billion people with decent lives lies in becoming more resource-efficient and, especially, more equitable than most human societies have been in a very long time. *Efficiency*, broadly speaking, is the amount of satisfaction derived per unit resource. *Equity* is the similarity of people's access to sociopolitical rights, adequate food and other material resources, health, education, and other ingredients of well-being. Increasing efficiency and equity seems anathema to the dominant, consumption-oriented culture. Yet it is now within the power of either the rich or the poor, independently, to bring the global life-support system to collapse. Perhaps this possibility of mutually assured destruction will finally motivate those on both sides to pursue efficiency and equity as matters of self-interest.

The dimensions of this dilemma and how we can extract ourselves from its grip are the central topics of this book. We explain the underlying basis of the scientists' statement from a broad historical and global perspective, often illustrated with the stories of local people and communities. We also explain the crucial importance of increasing equity at all levels of social organization: between the sexes and age classes within households; between households, with special attention to land tenure; between regions, particularly rural and urban; and between nations, especially "North" and "South." Our basic aim is to outline strategies for slowing down the stork's deliveries and enhancing those of the plow. We are, of course, summarizing an enormous topic covered by a vast literature. We have not attempted to give comprehensive coverage; rather, we emphasize what we think is most important and least appreciated.

Although many of the issues covered are extremely complex, recent progress along intellectual frontiers has vastly improved human understanding of how to extract well-being from Earth's life-support systems without destroying them. Moreover, one does not need to be a rocket scientist to grasp the fundamental predicament. A colleague of ours, Suzanne, recounted recently the following observations of a friend. Her friend is an illiterate Salvadoran woman who speaks no English. The woman, Maria, had been raped when she was thirteen and then forced to marry the rapist, who continued to abuse her unmercifully. In a series of horrible misadventures, she finally managed to flee to the United States, where she obtained a green card.

One day Maria told Suzanne that she was having an argument

with her sister. It seems that Maria's sister was on welfare, and Maria had asked her why. "If I weren't on welfare," the sister explained, "I couldn't feed my children." Maria had replied, "If you couldn't afford to feed your children, you shouldn't have had them."

At that point Suzanne interrupted the story. "But you're Roman Catholic—don't you believe that your sister should have as many children as God sends?" Maria answered, "I know what it says in the Bible; go forth, be fruitful, multiply, and subdue the Earth." She paused and added: *"¡Ya conquistada!"*—"It's subdued already!"

We begin with some background material in the next chapter on just how thoroughly we have subdued our planet.[8] The rest of the book explores in more detail the population and food security aspects of the human dilemma, focusing on the measures that need to be taken to keep the plow ahead of the stork in the race to maintain food security and create a sustainable world for our descendants.

Chapter One

THE HUMAN
JUGGERNAUT

The human enterprise has become a true juggernaut: an inexorable force that consumes and crushes everything in its path. It is now in the process of crushing its own life-support systems, especially those that underpin agricultural production. This represents not just a future threat. Hundreds of millions of people have perished of starvation and hunger-related disease in this century. While their misery can be attributed largely to maldistribution of food, recent trends suggest that the world may soon be faced with absolute shortages.

Serious flaws in our economic accounting system promote a popular illusion of security, concealing the irreversible and potentially devastating loss of critical natural capital. Brightening humanity's future will require radical social and economic change in each of the factors generating human impact: population size, per-capita consumption, and the damage caused by the technologies employed to provide that consumption. Each of these factors has tremendous momentum built into it, making change inherently slow and difficult. Their relative importance varies widely by region. Altogether, the developed nations, despite their smaller populations, at present account for the overwhelming majority of global environmental disruption. The United States, with the third largest population and the highest per-capita impacts in the world, bears the largest share of responsibility.

A child born today in sub-Saharan Africa faces a bleak set of statistics. The odds are one in 10 that he or she will not live more than a year; the odds are one in 20 that the child's mother will die giving birth.

The child born today in sub-Saharan Africa can expect to live for only 50 years—25 years less than children in the industrialized countries.

The newborn African child enters a world in which one person in five does not receive enough food to lead a healthy, productive life. . . .

By the time the child born today is 22, if present population growth continues, sub-Saharan Africa's population will have doubled. When the child is 45, it will have quadrupled. And all this in a region where many people are already poorer than they were 30 years ago.

Robert McNamara, 1990[1]

When we were conducting an ecotour in East Africa in 1983, a young Tanzanian farmer accompanied our group to act as an interpreter with villagers in northwestern Tanzania, where not everyone spoke Swahili.[2] When we crossed the border into Rwanda, a small, fast-growing, already crowded nation of about 5 million people, the first thing we saw was a river running brick red with eroded soil. The landscape changed completely from a gently rolling plain to rugged, mountainous terrain. Most striking was the density of the farming population, with all of the steep hillsides coated virtually to the top by a checkerboard of small farm plots. Some tiny fields were left fallow but plowed and exposed to the frequent rains; others were producing crops, often planted in wobbly rows marching straight up the hillside. Only a few were planted in horizontal rows following the mountain's contours or with crude terraces to hold water and soil. In the scattered plantings of eucalyptus, each tree was heavily pruned by people lopping off branches for use as firewood.

The Tanzanian farmer, who had had six years or so of schooling, was soberly looking at this scene, and we asked what he thought. "Erosion," was all he said, shaking his head sadly. To anyone with an ecological education, Rwanda was in a precarious state.[3] In 1991, James Gasana, Rwanda's Minister of Agriculture,

Livestock, and Forests, wrote about his nation's agricultural problems:

> Population pressure has made us intensify our agriculture and by doing that we have experienced significant soil losses. So we have a high level of population relative to food output. . . . Our problem is that we have no more new areas that we can colonize. And we have to stop land being lost. We estimate that our arable lands are diminishing each year by about 8000 hectares. . . . We can produce enough food for 5 million people—but we have 7.3 million people. . . . I am afraid that if the rate of population growth continues, we might have serious difficulties.[4]

Rwanda's population increased to about 7.5 million before ethnic hatreds, exacerbated by severe land shortages (some farm families of eight had to scrape a living from plots of less than a third of an acre) and other population-related pressures, caused a ferocious civil war that killed a million people.[5] That holocaust destroyed some of the remaining fragments of natural forest (which is essential to the flow of water for agricultural and household use and is inhabited by one of the last populations of mountain gorillas). As if the poor people of Rwanda did not have enough trouble, the nation had also become a center of AIDS infection.[6]

The situation in Rwanda is not unique. In 1995 we visited the island of Anjouan in the Comoros, northwest of Madagascar. It seemed in worse shape than Rwanda had a dozen years before. Anjouan also is farmed right to the mountaintops. Half of its people are under fifteen years old, the average family size is about seven children, unemployment is roughly 80 percent, and the three of us were continually begged by kids showing signs of malnutrition. The government appears to have no interest in either the environment or family planning. Authorities in nearby Mayotte (still part of France) maintain heavy patrols to ward off economic refugees from the Comoros and Madagascar.

Madagascar also presented us with a dismal picture of population-related environmental catastrophe, with continued deforestation and land degradation threatening the pitiful remnants of its unique flora and fauna. There is all too little recognition in the Malagasy government that survival of the island's lemurs, vangas (a group of birds as fascinating as Darwin's finches), baobab trees, and other natural treasures is crucial to attracting tourists, scientists, and international attention. For example, we visited a Malagasy offshore island famed for its abundant lemurs—its sole tourist attraction. The island, Nosy Komba, had been deforested in response to a rice shortage a few years earlier. The lemurs are now living in a small grove of trees supported by banana handouts from the locals and tourists. They are unlikely to persist in the absence of a suitable habitat and eating an unsuitable diet. When all of Madagascar's natural wealth is gone, it will be just one more desperately poor nation without the resources to support its burgeoning population.

Many of the obvious precursors of total disaster we found in Rwanda, Anjouan, and Madagascar are brewing elsewhere in Africa and in other parts of the less developed world. Whether humanity can come to grips with these and avert further tragedies will foretell a lot about the future of civilization. So we begin our examination of population problems in poor nations (and how to deal with them) where conditions are worst—in sub-Saharan Africa.

Sub-Saharan Africa's problems are rooted in the slave trade, colonialism, neocolonialism, and indigenous post-colonial governmental failure; it has been a long history of exploitation and inequity. Current conditions as well, caused and controlled in no small part by the behavior of today's overpopulated and overconsuming rich nations, are a leading factor in humanity's home continent being tormented by poverty, political disasters, and warfare.[7] It is the shame of those nations that so few of their citizens know or care about what is happening to the hard-working, long-suffering people of Africa.

Sub-Saharan Africa contains great biological and mineral wealth, and its people today make only a small contribution (compared to East Asians, Europeans, or North Americans) to the destruction of Earth's life-support systems. Nevertheless, the region is mired in misery; over the past decades, per-capita GNP has fallen some 15 to 20 percent. In many African countries, political turmoil has accompanied and worsened the effects of droughts and other misfortunes.[8] Other less developed nations match African ones in some measures of misery: several Middle Eastern nations have comparably high birth rates (but lower death rates and higher incomes); Bangladesh and some other Asian nations are equally poor and hungry (but have lower birth and death rates). But none has the full panoply of burdens that citizens of most countries in sub-Saharan Africa bear, and nowhere else have conditions been growing steadily worse for more than a quarter century.

The highest population growth rates in the world prevail in sub-Saharan Africa.[9] The region's 1994 population of about 570 million is projected to expand to over 1.3 billion by 2025.[10] The average number of children borne per woman (the total fertility rate, or TFR) in sub-Saharan Africa was 5.9 (most ranging between five and seven, depending on the nation) in 1994.[11] The populations of sub-Saharan countries were thus expanding by 2 to 3.5 percent per year, rates that if sustained would double their populations in twenty to thirty-four years.

Many African populations would be growing even faster if they did not also still have relatively high death rates. In some of the poorest countries, ten to fifteen babies out of a hundred newborns die before their first birthdays (as contrasted with six in Asia, five in Latin America, and fewer than one in Europe or North America). Nearly as many more African children die before reaching age five.

African population growth rates actually rose between 1970 and 1990, because of both falling death rates and rising fertility resulting from social changes and improvements in health. The persistence of such high fertility in African societies has been attributed to cultures that put a high premium on fecundity and on the social structure, which includes extended family arrangements.[12] Since the extended family shares responsibility for the children, burdens of childbearing are felt less severely by individual parents.[13] In polygynous African societies, multiple wives and children are important status symbols. Unfortunately, being able to care for them seems not so valued, as was apparent during our visit to Anjouan. There we had the pleasure of meeting Hassan M., a bright young man with a graduate degree from a first-rate American university, who was lucky to have temporary part-time jobs.

Together we came across a group of very excited but sadly malnourished children near a river. Hassan asked what they were so happy about and was told, "We just caught a [small] fish that we can sell for a kilo of rice. Our mother will be so proud—it's the first time we'll eat rice in two weeks." Hassan later remarked to us, "There is no equivalent of child-support payments here. Guys my age are already running from their first wife and children to the next, assuming no responsibility for their welfare. The kids in these huts live on coconuts, and people go hungry here without the media attention of Somalia. One of the first things that has to change to bring down the birth rate is the attitudes of men."

Social pressures on women to have large families are enormous. When Yoruba women in Ghana were asked why they wanted children, they responded that marriage had no meaning if there were no children.[14] Being an unmarried woman is a terrible fate. "For the Akamba there is no 'proper woman,' unless she is a woman who is married. . . . Indeed, one could even say that in Kikambu language there is no word for an unmarried woman."[15] In fact, childless women are without status; in some parts of sub-Saharan Africa, they are stigmatized, pitied, and sometimes even accused of witchcraft. Furthermore, husbands and the extended family make the childbearing decisions for women. The premium on fecundity is also grounded in traditional economic roles of women and children and in the fragility of the environment. These factors help explain why Ghana, despite a quarter century of family planning activity, still has a TFR of six.

Meanwhile, increases in food production have lagged behind population growth since 1967. A widespread, persistent drought in the Sahel (the southern fringe of the Sahara) was responsible for much of the shortfall during the 1970s. But other factors were at work, including erosion and depletion of soils, desertification in arid areas, and the expansion of cultivation onto less productive land.[16] As per-capita food production has steadily fallen in sub-Saharan Africa, per-capita GNP has dropped since the mid-1970s,[17] deepening poverty and forestalling progress in development, including modernization of the agricultural sector.[18]

If the gap between population growth and food production in Africa continues to widen, as it did during the 1980s, the annual food deficit around 1990 of about 10 million metric tons (mmt) of grain, some 10 percent of food needs, could mushroom beyond 200 mmt in 2020. In that case, Africans would have to import up to half the grain needed to feed a population that by then will be more than twice as large.[19] The question then arises as to whether anything like that much grain would be available on the world market, even if Africans could pay for it. According to agricultural economist Lester Brown, China alone might by then need to import all the grain available on the international market.[20]

The head of the Chinese Academy of Sciences commented on the consequences of his country's paving over farmland and diverting scarce water resources to industrial development, while increasing consumption of animal foods: "China will have to import 400 million tons of grain from the world market. And I am afraid, in that case, that all the grain output of the United States could not meet China's needs."[21] That fear is well justified, since in the early 1990s the *entire* annual grain production of the U.S. was roughly 325 million tons. Will Africa in 2020 be able to outbid China for desperately needed food imports?

In Africa, as well as in many other developing regions, a task nearly as critical as maintaining food supplies is the finding of fuel to cook the food. Wood is the primary source of energy for the vast majority of Africans, particularly in the countryside, and many foods (especially grains) cannot be eaten without cooking. Despite nearly two decades of concern about the fuelwood problem, many more trees are still being cut down than are being planted. Not only does this cause acute shortages of wood, it is a major contributor to desertification and loss of biodiversity.[22]

Another frightening aspect of the African dilemma is the explosive spread of AIDS, especially in central and eastern Africa.[23] Eight million Africans were estimated to be infected with the HIV virus by 1994, and the disease was continuing to spread rapidly. By 2015, the United Nations has estimated, the combined populations of the fifteen hardest-hit countries will be smaller by some 20 million than they otherwise would have been, because of the impact of AIDS.[24]

In Africa, AIDS affects men and women almost equally, with infection frequently abetted by the presence of other sexually transmitted diseases. Weakened immunity in AIDS victims has led to increased incidence of tuberculosis and other contagious diseases as well. The principal victims of AIDS are young adults, generally the most enterprising and productive members of their societies. Their babies often are also victims, infected during childbirth. Life expectancies after contracting the virus are relatively short in Africa, perhaps five or six years. The tragic result is the growing number of orphaned children being raised by grandparents, who still may have children of their own at home and are struggling to carry on the necessary activities of life without help from their adult offspring.

Given prevailing attitudes in much of Africa, where marital fidelity is neither valued nor expected in men, many of whom migrate regularly between farm or village and the city to work, and given the generally poor quality of medical and health services, the rapid spread of AIDS was almost inevitable. The challenge of halting its spread among people unfamiliar with contraceptives in general and unreceptive to condoms in particular is enormous. Indeed, simply supplying condoms in sufficient numbers would be a major undertaking.

Peter Piot of the World Health Organization's Global Program on AIDS thinks the epidemic reinforces a perception that Africa's problems cannot be solved. He put it this way: "Africa is looked at as a lost cause. . . . That is affecting everything—the whole health sector. Donors are getting discouraged. It's really depressing, and yet the needs nowhere in the world are greater than in Africa."[25]

Famine and plague today, and the prospect of continued explosive population growth making the situation much worse tomorrow, make the race between the stork and the plow especially poignant in the lands south of the Sahara. As in other impoverished areas, nations in sub-Saharan Africa cannot afford to put aside capital to improve the average standard of living—their resources must be allocated to attempting to care for a flood of children that threatens to engulf them. The World Bank stated the facts baldly in 1989: "Significant improvement in living standards cannot be achieved over the long term unless population growth is slowed. On current trends Africa will increasingly be unable to feed its children or find jobs for its school leavers."[26]

Imagine how difficult it would be for the United States to maintain its standard of living if the American population were to double in 25 years, a rate that is typical of African countries today. In essence, every facility the U.S. has for the support of human life would need to be duplicated in a generation. We'd need to grow and process twice as much food and draw twice as much fresh water. The number of physicians and teachers would need to be doubled, as would the capacities of hospitals, schools, and colleges. The capacities of highways, railroads, and airlines would need to double. So would the number of homes, office buildings, stores, and theaters. So would the capacity of the economy to absorb young workers.[27]

For a sub-Saharan nation like Zaire (which is fairly typical), with a doubling time of 21 years, the task appears impossible. Some 45 percent of its population is under 15 years old,[28] and it has about one hundredth of the GNP per capita of the U.S., no effective government, an inadequate and crumbling infrastructure, no capital to speak of, one physician for about every 14,000 people; and adult women have less than a year's schooling on average.[29] Zaire is not going to be able to maintain even today's miserable standard of living in the face of its exploding population.[30]

In these countries, the average citizen inevitably faces a plummeting quality of life and probably an early death. Hidden by the unfathomable numbers and horrifying trends is the individual suffering of tens of millions of people. Starving babies, vibrant young adults fading away under the assault of AIDS, people without opportunity or hope—an enormous waste of human capital. Also hidden is the disintegration of entire societies, degrading another of humanity's most valuable capital resources, its cultural diversity and its "library" of traditional knowledge.[31]

Nevertheless, a few thin rays of hope have appeared. Birth rates, after remaining at preindustrial high levels or even having risen somewhat during the two to three decades since most African nations achieved independence, began to fall in some countries in the late 1980s. In Kenya, the former world record-holder in growth with an annual rate of increase during most of the 1980s of 4 percent and an average family size (TFR) of more than eight children, the TFR had dropped to 6.3 by 1994.[32] Botswana and Zimbabwe also reported small but encouraging declines in fertility.[33]

Fertility declines have also been detected in some other African nations, but in aggregate they are too modest and too recent to be counted as a serious trend as yet. In 1994, Côte d'Ivoire, Liberia, Niger, Togo, Kenya, Comoros, Madagascar, Tanzania, Zaire, and Namibia all had TFRs between 6.0 and 7.4, growth rates of 3.3 to 3.6, and doubling times of 19 to 21 years—hardly an indication that zero population growth is just around the corner.

The Stork's History and Future

The possibility that *Homo sapiens* might attain the status of a dominant, global force would have seemed remote throughout most of human history. The total population at the dawn of the agricultural revolution, around 8000 B.C., is thought to have numbered about 5 million people. By the time of Christ, the population is thought to have grown to roughly 250 million people, as inferred from archaeological remains and the population densities of recent agricultural societies.

By 1650, the population had increased to about 500 million. A population size of 1 billion people was reached around 1825. Whereas it took virtually all of human history to grow to 1 billion, subsequent billions have been added in stunningly shorter periods. The population doubled itself again by 1930, adding the second billion in barely a century. It took only thirty years to add the third billion in 1960, fifteen years for the fourth billion in 1975, and twelve for the fifth billion in 1987. It is projected to take a mere eleven years to add the sixth billion by around 1998.[34]

The population currently is growing at a rate of about 1.6 percent per year, adding some 90 million people to the population each year. This rate, if maintained exactly, would double the mid-1995 population of 5.7 billion in forty-three years.[35]

The good news is that this represents a significant slowdown in the population growth rate since the 1960s, when it exceeded 2 percent per year.[36] The bad news is that, despite a lower growth rate, many more people are being added to the population each year now, because the percentage is applied to a considerably larger base population. This century's enormous surge in total human numbers, far greater than any in history, will continue well into the next century.[37] Projections of future growth are not comforting, and unexpected changes in birth or death rates may well make them wrong. Suffice it to say, humanity will be very, very fortunate if it can avoid a catastrophic visit from the grim reaper and slow the stork down enough to keep its numbers from rising beyond the vicinity of 8 or 9 billion people in the middle of the coming century.

To achieve that happy result, the average family size worldwide (3.2 children per couple in 1994) would soon need to fall well below "replacement reproduction"—the level at which parents on average just replace themselves in the next generation: about 2.1 children per couple. In a few developed nations today, average family sizes are as low as 1.3 children, whereas in some poor nations the average family size ranges as high as seven or eight children. In the United States, the average family size was below replacement (1.7 to 2.0) in the early 1970s, but recently has crept back up to 2.1.[38]

Global population growth statistics conceal very large differences in rates among nations and even among subgroups within many nations. At the extremes, some populations in Africa and the Middle East are growing by 3.5 percent or more per year. If they continue at that rate, they will have twice as many people before a child born today turns twenty. In contrast, many European nations have now essentially stopped growing or even have very slightly shrinking populations.[39]

Most nations have growth rates that fall between these extremes. The United States population is growing by more than 1 percent annually, but about a quarter of the increase is due to immigration.[40] Most developing nations in Latin America, southern and western Asia, and Africa have annual growth rates above 2 percent. Recently, a few countries in East Asia, most conspicuously China, have reduced their growth rates to near or below

1 percent per year. Altogether, the developing nations currently make up about 80 percent of the world's total population, a proportion that will inevitably rise because of their faster population growth.

Social scientists acknowledge that rapid population growth plays some role in generating socioeconomic stresses and dislocations. But resource limitations are not usually considered a problem, in part because many economists believe that human ingenuity will make available substitutes for any resource. They rarely consider the possibility that a large, fast-expanding and fast-consuming population may be seriously jeopardizing its future prospects.

The Plow

The basis for agricultural production is "natural capital," humanity's endowment of natural resources. Unlike human-made capital, natural capital can neither be manufactured nor substituted for on the scale required. Natural capital is often classified according to renewal rate. Renewable resources, such as fresh water and forest products, are reconstituted after human consumption through natural processes driven by solar energy; their availability is usually limited to their rates of renewal.

In contrast, nonrenewable resources, such as metals and fossil fuels, are typically limited by the size and accessibility of their stocks and have very low or no renewal rates. Reconstitution of some resources (such as metals) is possible but prohibitively costly, although a few recyclings may be possible before they must be discarded.[41] The depletion of reserves of nonrenewable resources is generally understood by the public; but it is unavoidable if civilization is to benefit from them.[42] By definition, there is no sustainable rate of consumption of nonrenewables; the closest approximation is a rate equivalent to (or lower than) the rate that substitutes can be generated. The depletion of nonrenewables, if managed wisely, need not pose a serious threat to humanity over the coming millennium (phosphorus is a possible exception).

Less clearly understood and much more important is the progressive loss or degradation of the theoretically renewable components of our natural capital. These include resources that are absolutely critical to agricultural fertility: rich soils, fresh water, and biodiversity. The rate of destruction of these resources is now so far in excess of their renewable rates that they have effectively been turned into nonrenewables on any relevant time-scale. We are sapping the power of the plow just when we need it most.[43]

Ostensibly renewable resources that support food production are being consumed or heedlessly destroyed at accelerating rates all around the world.[44] The topsoil on agricultural land is being eroded at unprecedented rates. Aquifers are being sucked dry for a few decades of crop irrigation. Forests, which help to ameliorate the climate and often provide water to farms and towns in the same watershed, are destroyed en masse, frequently to market the timber at fire-sale prices or just to clear land. Grasslands are overgrazed and denuded. Oceans are being vacuumed of edible fishes and poisoned with pollution. Wetlands, often critical to fisheries or to maintaining groundwater supplies, are drained and filled for development. Our priceless living capital—populations and species of other organisms—is being wasted in an apparent frenzy of one-time consumption or destruction.[45]

This destruction will inevitably have serious impacts on potential agricultural productivity. Since the mid-1980s, grain production per person has been dropping, although the causes are complex and the trend is not uniform from region to region.[46] In 1982, Latin America joined Africa in a trend of declining per-capita grain production. In oceanic fisheries, harvests have been essentially constant for the last few years, resulting in diminishing per-capita yields.

Expanding food production sufficiently to feed the projected additions to the population by the middle of the next century, while also improving the diets of millions now underfed, would require about a tripling of global grain production—a task that dwarfs even the remarkable achievement of the last half century, when the population more than doubled and grain production nearly tripled. Most of the readily available opportunities for substantial increases in food production have already been taken, and agriculture is now faced with a series of hurdles and potential difficulties that will not be easily surmounted.[47] No obvious new set of technologies is at hand (as the newly developed high-yielding grain varieties were in 1950) that could be used to create another miraculous "green revolution."

The ultimate biophysical limits to increasing food production have not been reached as yet. Those limits (availability of suitable land and water supplies; biological limits to potential yield increases through more fertilizer applications and to the potential genetic improvement of crops) might not be reached for many decades—if achieving sustainability were not a priority. *On a one-time basis,* humanity might well multiply today's food production severalfold—we might manage one last planetary banquet. But substantially more natural capital would inevitably be consumed in the process, with far more serious consequences than have been witnessed so far.

A Hungry World

The consequences so far have nonetheless been grimmer than most people in industrialized nations appreciate. An estimated 250 million people have died of hunger-related causes in the past quarter century—roughly 10 million each year. The victims of untimely death have been mostly infants and small children.[48] In food-short poor families, fathers are often fed first, so women and young children lose out. Children are highly vulnerable to both malnutrition and diseases. If they survive early childhood, they may have impaired growth and development, including mental development. Undernourished women are more likely to have miscarriages or underweight infants and to discontinue breast-feeding early.

International agencies estimate that as many as a billion people in the developing world today do not obtain enough energy from their food to carry on normal activities. Estimates of the numbers of hungry people are controversial and vary among international agencies.[49] But even if only a quarter-billion people were significantly underfed, the human nutritional situation would constitute a vast tragedy with grave implications.[50] In addition to the direct suffering it causes, widespread chronic hunger has negative economic effects by, among other things, reducing the productivity of the work force. Even though the *proportion* of hungry people in the world has shrunk somewhat in the last half century, their *numbers* have risen as the population has grown.[51]

Despite the widespread prevalence of chronic hunger in many poor nations, a general impression remains that the problem of feeding the growing population has been more or less permanently solved and famine mostly banished, except for local famines traceable to political conflicts.[52] Indeed, outright starvation since 1975 has mostly been restricted to situations of political strife in countries such as Ethiopia, Sudan, and Somalia. Modern communications and transport systems make it possible to move relief supplies quickly to areas of threatened famine, and the United Nations and some private international charities maintain emergency food supplies for that purpose.

Strictly speaking, hunger today *is* the result of maldistribution of resources and food supplies, rooted primarily in economics and politics.[53] Although an absolute shortage of food does not now exist, more equitable distribution would lead to major changes in the diets of both rich and poor. An assessment by Robert Kates, Robert Chen, and colleagues of the Alan Shawn Feinstein World Hunger Program at Brown University indicated that recent world harvests, if equitably distributed and with no grain diverted to feeding livestock, could supply a vegetarian diet to about 6 billion people. A diet more typical of South America, with some 15 percent of its calories derived from animal sources, could be supplied to about 4 billion people. A "full but healthy diet" (about 30 percent of calories from animal sources) of the sort enjoyed by many people in rich countries could be supplied to only 2.6 billion people.[54]

Their estimates, of course, are only approximate and were based on an assumption that some 40 percent of the food produced is wasted or spoiled after harvest.[55] They put in perspective the notion that hunger is merely a problem of maldistribution. Even if it were possible to transform most human beings into purely vegetarian saints, the sheer size and growth of the population would still be increasingly important factors in the task of providing everyone with even a minimally adequate diet because of rising population-related pressures on the world's finite food production systems.

This is not to say that a smaller population today would necessarily be better fed; the complex economics of agriculture and the extreme maldistribution of wealth might well prevent that. "Food security," defined as sustained and secure access by all people to nutritionally adequate diets, is subject to much more than environmental limitations. Economic, political, and social factors are extremely important in determining how much and what kinds of food will be produced in a particular place at a given time.[56] But agriculturally and ecologically, it certainly would be easier to feed all people well if there were fewer of them.

Externalities

Environmental concerns separate natural scientists from many social scientists in evaluating the world food situation. The former perceive environmental deterioration as a grave threat to future food security. Why are many economists and other social scientists oblivious to this threat?

The obliviousness traces primarily to a lack of basic education in the physical and natural sciences. This was illustrated in a 1987 survey of the opinions of graduate students in economics of the importance of courses in other fields to their development as economists. The lowest score was given to physics. Only 2 percent of the students considered it very important, 6 percent important, and 27 percent moderately important, whereas 64 percent rated it entirely unimportant. The writers of the survey did not even include ecology or any other discipline in the biological sciences among the fields to be scored by the students.[57]

This lack of appreciation of the natural underpinnings of economic well-being is perpetuated in the absurd models economists use to analyze the world. Our economic system has a fatal design flaw: it conceals many of the negative consequences of our choices, labeling them "externalities." Technically, an externality is any activity that affects other people—for better or worse—without those people paying or being compensated for the activity.[58] Environmental amenities, like pure air and water or the serenity of nature, are often difficult or impossible to put an exact price tag on. Sweeping this problem aside, economists typically externalize them from cost-benefit analyses, working under the convenient assumption that their value is zero.[59] Thus, the unthinkable has become commonplace: Earth's most priceless treasures are treated as if they were utterly worthless.

As Vice President Al Gore put it:

> Our current system of economics arbitrarily draws a circle of value around those things in our civilization we have decided to keep track of and measure. Then we discover that one of the easiest ways to artificially increase the value of things inside the circle is to do so at the expense of those things left outside the circle. . . . A direct and perverse ratio emerges: the more pollution dumped into the river, the higher the short-term profits for the polluter and his shareholders. . . . Our failure to measure environmental externalities is a kind of economic blindness, and its consequences can be staggering.[60]

So staggering that today the negative externalities have grown to rival the global economy itself in size and importance. Some "internalizing of externalities" has been accomplished through conventional pollution regulation. Yet this is still often counted as a cost without being weighed against the costs that would be incurred more generally by society if the pollutants were not controlled. How high those costs might be in industrialized nations like the United States can perhaps best be appreciated by a visit to Eastern Europe, where heavy industrialization proceeded for many decades with virtually no attempt to control pollution.[61]

Another example is the saturation of our environment with synthetic chemicals (like DDT and related compounds) that disrupt the endocrine system (the system of chemical messengers that control many of the functions of the body).[62] That disruption appears to be lowering sperm counts, causing both male and female babies to be born with malformed reproductive systems, and wreaking general havoc with the development processes of many kinds of nonhuman organisms. We will not deal in detail with this issue in this book, except to reiterate the warnings of biologists for more than a quarter century.[63] When we try to control the organisms that compete with us for food by spraying broad-spectrum synthetic poisons, we are taking a simultaneous chance on "controlling" ourselves. Yet, on a global scale, the costs of such risks and of outright environmental destruction remain largely invisible in the accounting of economists, business people, and policy-makers.

Population and Environmental Impact

People in industrialized countries frequently blame global environmental problems on rampant population growth among the poor. Meanwhile, citizens of poor countries rightly point out that people in rich countries are the superconsumers and big-time polluters. Resource depletion and environmental degradation, however, are the products of three factors: population size; per capita affluence (or consumption); and the environmental impact caused by the technology used to supply each unit of consumption. This concept can be expressed as a simple equation: I (environmental impact) = P (population size) × A (affluence per person) × T (technology), or I = PAT.[64]

Thus, while population growth may be much slower, or even stopped, in rich countries, the impact of each citizen (A × T) of a rich nation like the United States or Germany may be several to dozens of times higher than that of a poor country like India, Peru, or Mali. No nation keeps statistics on the A and T factors as such, but per-capita energy use can be used as a rule-of-thumb measure of environmental impact per person of a society. By this index, the impact of the average U.S. citizen in 1991 was about 20 times that of a Costa Rican, 50 times that of a Malagasi, and 70 times that of a Bangladeshi.[65] Each increment, however small, of population growth in a rich country therefore causes far more environmental damage than the considerably more rapid growth in poor countries, because it is multiplied by such high A × T factors.

These examples contrast the extremes of energy consumption in rich and poor nations. Most European countries and Japan are more energy-efficient than the U.S., using only 40 to 60 percent as much per person while pursuing lifestyles of similar comfort. Also, most developing nations are not as poor as Madagascar or Bangladesh; some are approaching the levels of energy use per person of Japan and some European countries. Many developing nations and most nations of the former Soviet bloc use energy very inefficiently and thus make high T-factor contributions to their per-capita impacts (without enjoying a commensurate quality of life).

One should note also that every person added to the planet causes, on average, proportionately more environmental damage than the previous person. Human beings naturally first farm the best soils, mine the richest ores, take the most accessible fresh water, and so on. As the population grows, it becomes necessary to farm inferior soils (with more danger of erosion and more need for artificial fertilizers), mine lower-grade ores (which requires more energy and usually causes more land degradation), and transport water further (creating still more ecological problems).

This is one reason that the relationship of damage to life-support systems to population size is "nonlinear" (the damage increases faster than the population). Another is the existence of thresholds. Below a certain population density, for instance, people living along a river may be able to dispose of their sewage in it and still drink the water with impunity, because the river will naturally purify itself. Above a density threshold, the self-cleansing mechanisms of the river will be overwhelmed, and serious health consequences would follow from continuing to consume the water without treatment. Now that environmental problems are causing perturbations of the entire global system, complacency about continued expansion of the human enterprise seems increasingly misplaced.

Momentum

Tremendous momentum is built into the P, A, and T factors. Demographic momentum is well understood and can be quite accurately quantified. Rapidly growing populations with age structures heavily skewed toward young people will experience huge surges in population growth long after replacement fertility is reached.

27

For example, even if the TFRs of sub-Saharan African nations dropped by half in the thirty years between 1990 and 2020 and then continued downward, the region's population would still top 1.2 billion in 2020 and keep on expanding to beyond 3 billion (equal to the entire world population in 1960) before growth stopped.[66] Population momentum in poor nations is one reason that it would be very difficult to halt growth before the population reaches 8 billion, if then.[67]

Demographic momentum can be a powerful force in industrialized nations as well. For instance, to stop growth instantly in the United States without changing the 1992 level of immigration, births would have to be restricted to slightly *less than one child for every two couples.* Even with zero immigration (and continued emigration), Americans would have to reduce their fertility to one child per family.[68]

Consumption patterns also exhibit tremendous cultural momentum, although it is more difficult to quantify. Changing the superconsumerism of Americans appears a much greater challenge than slowing population growth in developing nations. The American lifestyle is so seductive that a bewildering diversity of cultures around the world is abandoning long-held values and ways of life with hardly a thought in favor of the shallow but instant gratification offered by the American way of life.

Closely related to this is the momentum behind choices of technology used to supply the consumption. These may have extremely long-lasting impacts. Just consider the consequences of having designed the United States around the automobile, rather than around human beings, over the past six decades. This colossal structural problem rivals superconsumerism as an obstacle in transforming the U.S. into a sustainable enterprise. Redesigning and rebuilding the nation's urban and suburban infrastructure to sustain human and environmental well-being will require decades.

Just bringing a new energy technology to the forefront takes around forty years; research and development, deployment, and redesigning of infrastructure are not only time-consuming but costly. Unwisely, people don't want to pay the initial investment price, even though the long-run benefits may greatly outweigh it, and even though the future cost of failing to change may be devastating.

Worse, the failure of the industrialized nations to develop more environmentally benign technologies leaves developing nations with little alternative but to repeat the mistakes of the rich. Thus China is embarking on an ambitious program to make automobiles its principal form of transportation. This would have an enormous negative impact on the resources and environment of the entire planet, as was recognized long ago.[69]

The bottom line is that rich and poor nations must share the responsibility for humanity's joint predicament, although the factors are distributed differently between them: poor nations are now contributing over 90 percent of the population increase (while aspiring to the affluence of the rich), and rich nations are causing by far the largest overall impacts because of their profligate per-capita consumption. Curbing the consumption of resources and changing the technologies used in their mobilization will be necessary to slow the buildup of greenhouse gases in the atmosphere and thereby delay global warming effects, as well as to mitigate numerous other environmental impacts. But no amount of readjustment of resource consumption or improvement in technologies will suffice in the long run if population growth cannot be halted in poor nations and in the rich ones (like the U.S.) that are still growing, and a gradual decline toward a sustainable population size initiated.

Are Ecologists Alarmists?

Could it be that we and our colleagues are too concerned about the current situation? After all, the history of our species can almost be read as a history of catastrophes; famines, plagues, wars, and the like are old stuff. Here we are, fat, happy, and prosperous despite it all. Obviously, even without addressing the issue of just who and how many of us are prosperous, we don't think we're crying wolf. Today's situation is demonstrably totally unprecedented. For the very first time in human history, our species is transgressing the carrying capacity of the entire planet. In the past, relatively isolated civilizations in places as diverse as the Yucatan peninsula, the Mediterranean Basin, Cambodia, and the Tigris and Euphrates valleys have faded or collapsed, at least in part because of their abuse of environmental systems. Their disappearance, though, had little effect on distant societies.

The story at the end of the twentieth century is fundamentally different. No vast, sparsely occupied reservoirs of carrying capacity remain to be appropriated as European nations claimed the carrying capacity of the western hemisphere several hundred years ago. The global frontier is closed. Entire subcontinents are now in severe ecological peril. Sub-Saharan Africa, far from emerging from poverty and underdevelopment, seems to be sinking in deeper, with shrinking per-capita food production and GNPs and a raging epidemic of AIDS. In the words of two of India's most distinguished ecologists, their nation "is living on borrowed time. It is eating, at an accelerating rate, into the capital stock of its renewable resources of soil, water, plant and animal life."[70] Exactly the same can be said of virtually every other region on Earth.

All the continents and subcontinents are now intimately linked to each other through trade and the atmospheric and oceanic "commons." The human epidemiological environment is deteriorating, as population growth pushes larger and larger groups into contact with animal reservoirs of viruses; as disease organisms become increasingly resistant to the chemical weapons with which humanity defends itself; as absolute numbers of hungry (and thus immune-compromised) people increase; and as aircraft permit intercontinental transport of infected people in hours. A worldwide civilization is threatened with collapse through starvation, plague, social breakdown, and war—because it is overreproducing, destroying its natural capital, and crippling global life-support systems.

It is important to remember that the concerns expressed in this chapter are shared by the global scientific community. Scientists have grown increasingly concerned as the population explosion and the environmental deterioration linked to it continue, while political leaders pay scant heed. If ecologists are alarmists, a great many well-informed people share their alarm. In 1993, the same year in which the *World Scientists' Warning to Humanity* (quoted in our Introduction) was issued, fifty-eight of the world's scientific academies released a joint statement in a similar vein, as indicated in the following excerpt:

> The magnitude of the threat to the ecosystem is linked to human population size and resource use per person. Resource use, waste production and environmental degradation are accelerated by population growth.
>
> As human numbers further increase, the potential for irreversible changes of far reaching magnitude also increases. Indicators of severe environmental stress include the growing loss of biodiversity, increasing greenhouse gas emissions, increasing deforestation worldwide, stratospheric ozone depletion, acid rain, loss of topsoil, and shortages of water, food, and fuelwood in many parts of the world.
>
> The growth of population over the last half century was for a time matched by similar world-wide increases in utilizable resources. However, in the last decade food production from both land and sea has declined relative to population growth. The area of agricultural land has shrunk, both through soil erosion and reduced possibilities for irrigation. The availability of water is already a constraint in some countries. These are warnings that the earth is finite, and that natural systems are being pushed ever closer to their limits.[71]

In the rest of this book, we will examine in more detail how these warnings might be heeded, focusing on the all-important race between the stork and the plow. We first consider how human beings have always attempted to control their population sizes, how the human population outbreak developed, and how it might be contained and reversed. Then we turn to the world food situation and describe the agricultural system, the crucial and mounting environmental costs of the agricultural enterprise, and the potential for increasing food production and improving distribution in order to meet future needs. Throughout, we will be pointing out opportunities to reduce, in various combinations, P,

A, and T to lessen total human impact on the planet. Central to all these matters are the growing inequities among individuals, groups, nations, and, in particular, between the sexes. In short, we will outline what might be done to brighten the future for all of humanity.

Chapter Two

THE ONLY ANIMAL
THAT PRACTICES BIRTH
CONTROL

Sex is fun for almost everyone; but like many other amusements, it can have both intended and unintended consequences for the participants, their families, and their societies. Sex also appears to be fun for lots of nonhuman animals. But what separates human beings from other animals, as far as we know, is that only they understand the connection of sex with reproduction, and only they intentionally try to control the consequences.

One of those unintended consequences all too often is an unwelcome visit from the stork. Many people are under the impression that preindustrial societies had little control over birth rates or death rates—that "primitive" peoples believed babies to be gifts of God. Social scientists have long known better. Much early fertility control was, of course, faute de mieux, *achieved through very risky folk abortion or, worse yet, infanticide. It was not easy in an Eskimo camp fifty years ago or in Europe two hundred years ago to practice contraception. People wanted to limit their reproduction, but they lacked a safe means of keeping the sperm from getting to the egg or of expelling an early embryo. The result was an appalling slaughter of newborn babies.*

It is true that we have come a long way from using crocodile dung suppositories, teas made from poisonous plants, sheep gut "rubbers," and the ritualized infanticide of "baby farms" in pre-Victorian England, but we haven't come far enough. Techniques of contraception have gotten easier, but some are inconvenient, take some of the fun and spontaneity out of sex, or impose small risks on the user (although far smaller, ordinarily, than those associated with childbearing). Furthermore, although medical abortion is now much safer than childbearing, it is a comparatively crude technique of fertility control whose use is offensive to a sizable number of people. And, alas, infanticide is alive and well, euphemistically discussed under the term child abuse. *Much more needs to be done before all sexually active people have access to safe, convenient contraception that, ideally, would also help to limit the spread of sexually transmitted diseases.*

In the best of all possible worlds, those contraceptives would distribute any risks and burdens equitably between the sexes, rather than load most of them on women. But, while contraceptive research continues, various social factors make progress slow. One is fear of lawsuits in pharmaceutical companies, because liability for injuries imputed to contraceptives is generally higher than for drugs used to treat illness. Another is the activity of powerful groups, including celibate old men who attempt to block the use of contraceptives because they are offended by the idea of people having sex just for fun. A prude, remember, is someone who always has the nagging fear that somewhere someone is having a good time.

> The human race has in all ages and in all geographical locations *desired* to control its own fertility . . . while women have wanted babies, they have wanted them when they wanted them. And they have wanted neither too few nor too many. . . . What is new is not the desire for prevention, but effective harmless means of achieving it on a grand scale.
>
> *Norman Himes, 1936*[1]

Deer don't use condoms; sparrows don't get vasectomies; trout don't take the Pill; butterflies don't practice abortion or infanticide. These and all other nonhuman organisms are governed by the one message geneticists know for certain is written into the genetic code of the DNA of all animals: "Maximize the number of your own offspring represented in the next generation of breeding adults." That message is at the very center of the evolutionary process. For more than 3.5 billion years, those organisms that outreproduced others of their own kind passed more of their hereditary characteristics to future generations. The essence of the process of natural selection is that there is variation within populations in the genetic makeup of individuals, and some genetic types outreproduce others.[2] Obviously, this process carried out for so very long leads to a strong genetic predilection to have the number of offspring that will produce a maximum number of surviving descendants.[3]

Sex, contrary to the views of some religious leaders, does not have reproduction as its primary purpose. Lots of organisms, from amoebas and many plants to some fishes and lizards, manage to produce offspring just fine without sex. Indeed, in the short run, *sex is often a drag on reproduction*—since in a stable environment a population can reproduce more abundantly if it doesn't need to produce sperm as well as eggs.[4] One of the enduring mysteries of biology, one still debated, is how sex managed to evolve when asexual reproduction is so much more efficient! The answer is almost certainly to be found in the relationship between sex and genetic variation.[5]

The efficacy of evolutionary pressures to outbreed other organisms is sometimes not recognized by nonbiologists, leading some social scientists to believe that without cultural encouragement people would fail to reproduce and societies would disappear. One historian wrote: "There have never been any happy savages reproducing with ease. Rather, conflict between the social need to preserve the species and the individual desire to escape the burdens of childbearing is a universal part of the human experience. Societies that survive necessarily develop pronatal values [those favoring high rates of reproduction] that support the process of reproduction."[6]

Of course, natural selection ensures that the urge to reproduce will be powerful in people as it is in all animals.[7] Individuals with a genetic tendency to practice reproductive restraint and have fewer offspring than they could successfully rear will not have their genes as well represented in the next generation as those that yield the maximum possible number of reproducers in the next generation. Biologically, any tendency toward restraint (which might benefit the group or tribe if resources are short) will in the long run be overwhelmed by natural selection favoring the maximum reproducers—if a small population survives for enough time.[8] Indeed, it has been proposed that a woman's inability to detect her own ovulation resulted from selection against a tendency to avoid conception in ancestral females who found pregnancy and child-birth unpleasant.[9]

Furthermore, in the chancy circumstances of small family and tribal groups, pronatalist cultural norms have usually developed that reinforced genetic predispositions. The largest family or tribe was likely also to be the most secure. Larger groups were more likely to have viable numbers remaining after natural disasters and were more likely to prevail in intergroup conflict. This does not mean that cultural evolution cannot produce maladaption that, at least in theory, would lead a small population to reproduce at a rate too low to balance its death rate and to die out.[10]

In human beings, however, there is a vast overlay of phenomena that have developed biologically and culturally in connection with sexual pleasures, which had their origins in the complexity of the evolutionary process.[11] Human beings differ from all other mammals in being "sexy."[12] Other mammals have an estrous cycle that leads to ovulation (and copulation) for a relatively limited period each year. Human females, in contrast, ovulate monthly and (in the parlance of mostly male anthropologists) are perpetually "sexually receptive."[13] This provides, among other things, for the formation of social bonds based on sexual pleasure (one of the bases of marriage) and for a fascination with sex that pervades society. One need only consider the themes of most literature throughout history or observe popular culture today (TV or the lyrics of pop music) to dispel the notion that "sex is just for reproduction" once and for all. If that were the only role evolution gave to "sex," we would copulate occasionally and have babies at the proper season, just as most other mammals do, without all the continual fuss!

We sexy animals do something else that no other animals do. We use condoms, vasectomies, tubal ligations, and the Pill (as well as many other tactics) *consciously* to control the number of offspring we leave in the next generation. Moreover, long before modern surgery or contraceptives were available, cultural mechanisms often discouraged reproduction in times of stress when a high birth rate would result in fewer babies growing up than would a lower birth rate.

Furthermore, despite that genetic predilection to reproduce, and long-standing cultural norms promoting large numbers of off-spring, people today do not have on average anything like the number of children they could successfully rear. Cultural evolution can rapidly overwhelm the messages in our genetic code *and* the pronatal tendencies built historically into cultures.[14] To take extreme examples, the average couple in Italy in 1994 was bearing only 1.3 children, about a ninth of the maximum average of eleven or twelve produced by healthy people such as the Hutterites who make no attempt to limit their reproduction.[15] The largest family sizes in any nation in the early 1990s probably were in Rwanda, and even there, the average was only about eight children.[16] A decade earlier, the average family size in Kenya was about the same. At the national level (thus excluding subnational groups like the Hutterites), these probably represent the all-time high in average family sizes.

The ability of cultural evolution to change human attitudes quickly and even to counteract effectively the most powerful impulses built into our genetic code is the good news upon which this book is based. The change has been possible in part because evolution packaged the "reproduce to the maximum" message primarily in the form of the intense pleasures associated with sexual activity. Only when cultural evolution reached the point where our primate ancestors could learn to separate sex and reproduction did it become possible to regulate the birth rate—at first crudely (and often inhumanely) and then with increasing ease and decreasing risk.

If you don't believe that, simply contemplate what would happen if the message of those concerned about overreproduction were, of necessity, "have less sex," not "have fewer children." We have clear evidence today of how difficult it is to restrain sexual activity in such phenomena as the patent failure of the Catholic Church to maintain chastity in its celibate clergy,[17] and of most modern societies to suppress sexual activity among teenagers.

In fact, biologists have long tried to find qualitative characteristics of *Homo sapiens* that clearly distinguished them from other animals. Attempts to do so by defining human beings on the basis of the use of tools or speech or the possession of intelligence or culture have not been fully successful. Vultures, wasps, herons, and chimpanzees (among others) use tools.[18] Vervet monkeys transmit clear ideas vocally, and chimps can learn to communicate quite effectively, if simply, with human sign language. Many primates can outscore some human beings on various intelligence tests;[19] and numerous animals, from oystercatchers to lions, transmit information from generation to generation culturally. But no animals besides human beings consciously produce many fewer descendants in future generations than they potentially could. Indeed, many human beings today clearly have chosen to have few offspring in part out of concern for society as a whole and for the environmental integrity of the entire planet. This sort of altruism is confined entirely to *Homo sapiens.*

It could even be claimed that consciously practicing birth control is the single most distinguishing biological feature of our species. As the respected human geneticist J. V. Neel noted: "I conclude . . . that perhaps the most significant of the many milestones in the transition from higher primate to man—on a par with speech and toolmaking—occurred when human social organization and parental care permitted the survival of a higher proportion of infants than the culture and economy could absorb in each generation and when population control, including abortion and infanticide, was therefore adopted as the only practical recourse available."[20]

Scientific recognition of the importance of cultural control of fertility in early human groups traces at least back to A. M. Carr-Saunders, writing in 1922.[21] Ecologist V. C. Wynne-Edwards was convinced that cultural mechanisms regulated the size of early human populations. He wrote in 1962 that "civilized man . . . has thrown off the primitive customs that prevented population increase—so long ago that their former existence is today scarcely even credible—and he draws more and more for his livelihood on the united resources of the world, many of which are rapidly diminishing. The sequence has now reached the point where, if the increase is not controlled and halted within the next two or three generations, a crash seems certain; and even if some fragments of the human species were to survive, civilization as we know it would almost certainly have been lost for good in the resulting chaos."[22] This prophetic statement was penned more than a generation ago. Unhappily, however, much of the anthropological literature has assumed that virtually all cultural phenomena must be adaptive.[23] While it is evident that human beings have always manipulated their birth and death rates to achieve desired individual and social goals, one cannot assume that all cultural devices that influence vital rates are adaptive—any more than one can assume that all features of organisms are adaptive.[24]

A Short History of Fertility Control

The idea of limiting reproduction can be traced back virtually to the beginning of civilization, and it quite likely was practiced widely by our preliterate ancestors. Ancient Egyptians wrote four thousand years ago about contraception, recommending the use of saltpeter, honey, and crocodile dung as vaginal suppositories.[25] Crocodile dung suppositories, in our view, should have been very effective in preventing pregnancy! The efficacy of the other two seems questionable. That people have long managed to control their reproduction is nonetheless well documented.[26] At one time, Roman senators worried about the economic consequences of falling birth rates, just as some politicians do today; and a shortage of military personnel was a common source of worry.[27] The strength of the official pronatalist policy in Rome can be judged by the word used for common people, *proletarii,* which meant "breeders."

In the Middle Ages, there were population declines without obvious explanation in war or plague. The mechanisms causing these trends remain controversial. Most historians have looked to infanticide for an answer, but a small amount of anecdotal evidence plus some studies of female skeletons indicate an actual decline in births, rather than the killing of infants directly or by neglect after birth. Abortion by mechanical means is another possibility, but its great dangers (before the development of modern antiseptic techniques) have been widely recognized since ancient times. Abstinence and withdrawal are other possibilities.

A more interesting explanation has emerged recently. Scholars have begun piecing together evidence for widespread use of effective oral contraceptives tracing at least to the ancient Greeks and

Romans. Modern medicine has been dominated by men, and areas related to sex and reproduction have usually been so poorly handled in instruction that it is a wonder that many male physicians know where babies come from. But women, with a substantially larger stake in reproduction, appear to have accumulated over centuries and passed along an extensive folklore on the antifertility properties of a wide variety of plants.

Both the Greeks and Romans greatly valued a plant called silphion, apparently a species of giant fennel *(Ferula,* in the carrot family).[28] It grew in North Africa near the Greek city-state of Cyrene, where harvesting it formed an important part of the economy. Attempts to cultivate silphion elsewhere failed, and the species was eventually wiped out by overharvesting. There is ample evidence that silphion sap was an antifertility drug, capable of either preventing conception or stopping the implantation of a fertilized egg.[29] Extracts from surviving relatives of silphion (other members of the genus *Ferula)* also seem to have those properties. Interestingly, the seeds of a more distant relative, Queen Anne's lace *(Daucus carota,* another member of the carrot family), are used as an oral contraceptive in India and in the Appalachian mountains of the United States today.

Extracts of many other plants have been used as oral contraceptives or abortifacients (abortion-inducing drugs) over the ages, but in most cases it is not known whether they had specific hormonal effects that modified reproductive processes in women or simply created such stress by their toxicity that implantation or pregnancy was disrupted. It is notable, however, that the key hormonal component of the contraceptive pill (estrogen) is synthesized from plant steroids, and that compounds with estrogenlike properties or the ability to stimulate estrogen production in mammals are widespread in plants.

Much of the folk knowledge of herbal contraception was lost as formal medicine replaced midwifery, but there seems little doubt that at least some of the treatments were efficacious. Pennyroyal *(Mentha pulegium* in the mint family) has been used since antiquity as an abortifacient, apparently with success. It is a dangerous drug, though, and in 1978 caused the death of a young woman in Colorado. She took a preparation in which pulegone, the aborti-

facient agent, was more concentrated than in the ancient prepara-
tions. Thus the partial loss of folk knowledge in this case led to
tragedy.

Regardless of the degree to which women have historically
managed to use oral contraception successfully, there is no doubt
that *Homo sapiens* has long exercised some degree of fertility control. Condoms and diaphragms in the modern form date back
only to the middle of the last century, but other types of barrier
contraceptives (including cervical blockage with leaves, cloth, and
cotton fibers practiced by the ancient Egyptians and primitive
condoms made from animal intestines) may well go back to prehistoric times. In addition to crude barriers, crocodile dung suppositories and similar nostrums, abstinence, withdrawal, delay of
marriage (or first coitus), noncoital heterosexual activity, prolonged breastfeeding, abortion, and infanticide and pedicide (killing of children) have all been employed to influence the number
of offspring reared.

Little, of course, is known about the degree of fertility control
in early human groups. What is known is that "primitive"
gatherer-hunters could be more restrictive than more "modern"
groups. The Montagnais-Naskapi, Algonkian Indians that lived in
the tundra of Labrador, were proselytized by French Jesuit missionaries in the seventeenth and eighteenth centuries. According
to E. B. Leacock, "The Montagnais planned their families, and
their norm of three or maybe four children contrasted with the
seven or eight of the French."[30] Some inferences about early
methods of fertility control can be drawn from the behavior of
both nonhuman primates and contemporary groups of gatherer-
hunters.[31] Among nonhuman primates such as marmosets, various
baboons, talapoin monkeys, and certain macaques, spontaneous
abortion is one response to various forms of stress, including ha-
rassment by or the mere presence of more dominant females.[32]

Abortion in Early Societies

The role of spontaneous abortion in controlling the size of human
gathering-hunting populations may never be known with certainty, but nutritional or social stress and workload all influence

estrogen levels (which, in turn, influence the chances of miscarriage).[33] So spontaneous abortion may partly explain the concentration of births in the months of March and April among nomadic Kalahari zu/oasi San bushmen. The nomadic San vary greatly in weight throughout the year, being heaviest in the dry season (June through August) when their diet is richest. The preponderance of births nine months later may indicate increased spontaneous abortions of embryos conceived in slimmer times.[34]

As anthropologists Marvin Harris and Eric Ross noted in their provocative book, *Death, Sex, and Fertility*, "the distinction between spontaneous and deliberate ('induced') abortion is not nearly as sharp as modern forensic and religious notions would have it."[35] In times of resource stress, practices such as taboos against the consumption of protein-rich foods by pregnant women may be a way of increasing the "spontaneous" abortion rate.[36] Periods of high workload and stress may increase the spontaneous abortion rate among the Venezuelan Yanomamo, a group in which one anthropologist reported a "deep commitment to regulating the entry of infants into the population."[37]

Induced abortion was certainly widely practiced in antiquity despite its manifest dangers; a wide variety of methods were described by the ancient Romans.[38] Besides the usual needles, potions, and violent exercises, riding a cart along a bumpy road was recommended—a technique much discussed among teenagers in the United States in the 1950s, with a car substituted for the cart.

We will not go into the long and complex history of abortion since Roman days.[39] Few would claim that it should be a major means of fertility control, but until safe, convenient, and reliable contraception is available and universally practiced, abortion clearly will continue to be utilized by women with unwanted pregnancies.

Infanticide

Infanticide is widespread in nonhuman mammals.[40] For instance, it often occurs when a newly dominant male takes over a mate or harem and kills the offspring of the previous male.[41] Such behavior is seen primarily in polygynous species—those where one male mates with more than one female. Young of preceding males commonly are killed when new males take over a pride of lions,[42] or a new silverback becomes leader of a gorilla family.[43] The females then come into heat and mate with the new males, who thereby advance in the game of natural selection by promoting the presence of their own genes, rather than their predecessor's, in the next generation. Similar patterns are seen in *Homo sapiens,* where children are abused or killed much more frequently by boyfriends and stepfathers than by biological fathers.[44]

Infanticide by females is also known to occur. One example is seen in lily-trotters, a bird species that is polyandrous (one female mates with many males).[45] The evolutionary explanation here is the same—a female lily-trotter wants to be sure her male consorts are raising *her* offspring, not those of another female.

Human beings, like most other mammals, are probably basically polygynous.[46] Females must be very careful in their choice of sexual partners. A human male who chooses an inferior mate may spend less than an hour and a few hundred calories in producing an offspring with her. The female's investment in an offspring necessarily stretches over years and may involve an energetic expenditure of millions of calories. So natural selection presumably has produced male mammals that are less choosy about their mates than are females, and polygyny is the usual mating system in mammals. The male puts so much less effort into each offspring that it is to his evolutionary advantage to spread his genes around.[47]

How can one explain then that mothers commit much of the infanticide among basically polygynous human beings, and that the practice seems to be more frequent among people than other mammals? We suspect that it is a culturally reinforced instance of what biologists call parent–offspring conflict—situations where the evolutionary interests of parent and offspring are not necessarily congruent. For example, female birds that lay four eggs may, on average, successfully rear 1.5 young. The stress of trying to care for six chicks may reduce the average success rate to less than one, and may sharply reduce the female's chances of surviving the winter to breed next season. Since it is *lifetime* reproduction that evolution presumably attempts to maximize, a female bird might "profitably" (from her point of view) dispense with two offspring.[48] That would be a favorable strategy for the female, but hardly for the two offspring.

Females (and males) of many species of gulls, hawks, woodpeckers, herons, and grackles, among others, in fact do this indirectly by a process known as asynchronous hatching and brood reduction. Instead of starting to incubate their eggs only when the last one is laid, they start immediately, so the first egg laid hatches before the second, and so on. The oldest hatchlings have an advantage in begging for food; and if it is scarce, they are the ones that get fed well while their younger siblings starve. Only when food is abundant will the entire brood have a chance to fledge.[49]

It seems likely that some infanticide by human mothers serves a purpose analogous to brood reduction in birds. A mother or a couple who find it difficult to care for existing children may not wish to jeopardize the entire family unit by trying to care for more children than they think they can "afford." They may opt for infanticide instead. Even the Inuit, renowned for their kind treatment of children, sometimes turn against them. When a child is not growing up properly (as determined by each particular family), "the love and caring, and respect can all too quickly turn to hatred, violence, and 'giving-up' on the child."[50] Or an unwed young woman with no immediate prospects for marriage and no way of supporting a child might also be driven to infanticide. The frequency with which abandoned newborn infants are found even today is testimony to this. Another factor may be that the mother, given her enormously greater potential investment, may not see the cost/benefit equation of raising the child in the same light as the father.

Infanticide, especially the killing of female babies, also has a long history as a fertility-control device, and may well have been common among early gatherer-hunters in times of resource scarcity.[51] Indeed, infants may even have been consumed at times, as some modern Australian aborigine and Inuit groups have done.[52] Among the Netsilik Eskimo, a large portion of female infants was killed. Ethnographer Asen Balikci found that the Inuit themselves attributed the practice to "survival reasons in a very harsh environment. The practice was an adaptive one, for it increased the chances of survival of the community by reducing the number of non-food-procuring people to be fed."[53]

But infanticide among the Inuit had functions other than promoting group persistence. For instance, in one case, a man decided to kill a child "because he had another father"—an adaptive behavior pattern (in the sense of the basic rule of natural selection) similar to that found in lions and mountain gorillas.[54] But female infanticide had the potential for being nonadaptive for the group as a whole. It could result in a severe shortage of women, leading to males killing each other over them and threatening the persistence of the group.[55] The sex ratio among children was almost 2:1 in favor of males. This was ameliorated somewhat by a much higher death rate among males from natural hazards and suicide, so the adult sex ratio was closer to parity. As Balikci also wrote: "From a social point of view [female infanticide] may be considered as a harmful practice which led to an imbalance in the sex ratio, an effort to keep the marriageable girls within the kinship unit, and a consequent division of the community into many small, mutually suspicious, unrelated kinship groups. This disharmony between the ecological and social results of female infanticide produced some powerful tensions within society."[56]

Until very recently, infanticide had two major advantages over abortion. It did not threaten the mother's life, and the condition and sex of the child were known.[57] In some present-day gatherer-hunter societies such as the !Kung bushmen, an individual is not considered a part of society at birth. Women give birth in isolation, and until a mother returns to her village with the child, it lacks full status as a human being. Killing the infant before that point in !Kung society would not be socially differentiated from abortion.[58] Indeed, J. V. Neel's claim that population control is, more than tool-making, a defining characteristic of humanity is based primarily on the killing of infants to control human numbers, a behavior that is odious to the vast majority of us today (because we have more palatable options). As Neel put it, *Homo sapiens* is the only animal "who regularly and without 'external' provocation purposely and knowingly commits infanticide."[59]

Infanticide was apparently a common method of fertility regulation in ancient Greece and Rome, and not illegal. In *The Republic*, Plato considered whether or not to dispose of a newborn to be solely the business of the parents. The Greeks had a saying, "Everyone, even if poor, raises a son, but even if rich, exposes a daughter."[60] In fourteenth-century England, sex ratios among serfs in the vicinity of 1.7 male children to each female indicated extensive female infanticide, perhaps because of "the necessity for restricting the number of children in view of the limited opportunities open to people of that class."[61] Sexism has often been found to pervade fertility control issues. In the Middle Ages in Europe, the law distinguished between infanticide and "exposure" and did not punish the latter, which was "practiced on a gigantic scale with absolute impunity."[62]

Although we find infanticide horrifying when it occurs today, it was institutionalized in Western societies not so long ago. It was widely practiced in Europe between 1750 and 1850.[63] By 1800, Germany, France, and England were experiencing a virtual epidemic of infanticide. In 1862, Dr. Lankester, one of the coroners of Middlesex, charged that even the police seemed to think no more of finding a dead child along a roadside than of finding a dead dog or cat.[64] Most of the unwanted children were born to the poverty-stricken and others lacking the resources to care for them: domestic servants, landless laborers, and even the younger sons of the wealthy. Abandoned babies were so common that foundling hospitals were set up to care for them. The care provided, however, was minimal, and their main effect was the disappearance of dead infants from streams, roadsides, and streets where they had previously been common sights. In 1818, of 4,779 infants admitted to the Paris foundling hospital, 2,370 died in the first three months.[65]

In addition to the hospitals, the practice of wet-nursing sometimes amounted to disguised infanticide. "Baby-farming," immortalized in a song in the Gilbert and Sullivan operetta, *HMS Pinafore,* consisted of turning a newborn over to a woman who made a profession of breastfeeding, but often permitted the infant to die from neglect or killed it outright. The nurses became known as "killer nurses" or "angel-makers."[66] One London pamphleteer wrote in 1757:

> . . . in and about London a prodigious Number of Infants are
> cruelly murdered unchristened, by these Infernals, called
> Nurses; these detestable Monsters throw a Spoonful of Gin,
> Spirits of Wine, or Hungary-Water down a Child's Throat,
> which instantly strangles the Babe; when the Searchers come
> to inspect the Body, and enquire what Distemper caused the
> Death, it is answered, Convulsions, this occasions the Article
> of Convulsions in the Bills of Mortality so much to exceed all
> others. The price of destroying and interring a Child is but
> Two Guineas; and these are the Causes that near a Third die
> under the Age of Two Years, and not unlikely under two
> Months.[67]

Another form of infanticide that was widely practiced and condoned by authorities in medieval Europe was called "overlaying."[68] Overlaying was performed by the mother, who simply suffocated a nursing infant by rolling over onto it while sleeping. While this was not considered homicide, it carried a penance. Around 850, St. Hubert wrote, "If anyone overlays an unbaptized baby, she shall do penance for three years. If unintentionally, two years."[69] The standard penance was one year on bread and water and two more without meat or wine—a light sentence considering that accidentally killing an adult carried five years, three on bread and water. Overlaying became such a common practice in medieval Europe that laws (unenforceable, of course) were enacted making it illegal for mothers to sleep in the same bed with their nursing babies.

Cultural devices certainly evolved in medieval England (and probably in the rest of Europe) that made infanticide less horrendous to its perpetrators. Infants were generally dehumanized—they were, in fact, viewed as parasites sucking away their mothers' vital substance. Young children, who often died by scalding, falling in wells, and other suspicious "accidents," were viewed as too assertive by parents severely constrained by a rigid caste structure. The parent–offspring hostility and beatings that all of this fostered apparently engendered the same generation-to-generation perpetuation of abuse that it does to this day.[70]

The killing of female babies because of a higher value placed on males may also be resource-related. It has long been a practice in Asia, where sex ratios were often extremely distorted as late as the last century. Some castes in Gujarat, India, had four times as many male children as females, and certain castes were reported not to raise daughters at all, with wives being obtained by upward movement of brides (and dowries) from lower ranking subcastes.[71]

Female infanticide has persisted in parts of India and China and seems to have resurged in response to efforts by the governments of both countries to persuade people to have fewer children. Skewed sex ratios as high as 1.3 boys to each girl have been reported in some regions.[72] China's one-child family policy, launched in 1979, led to an upsurge in female infanticide because of parents' traditional preference for sons.[73] The social policy that brought on the infanticide was based on a perceived shortage of resources.[74] More recently, modern medical devices such as amniocentesis and ultrasound monitors, which permit detection of a fetus's sex, have been widely used for that purpose in both India and China, leading to high rates of induced abortions of female fetuses.[75] Both nations have now outlawed prenatal sex determination.

Meanwhile, however, chickens have begun coming home to roost. Millions of young Chinese men, born during the Great Leap Forward of 1959–1961 when the birth rate, especially of girls, was sharply reduced (doubtless because of widespread infanticide), cannot find women of appropriate ages to marry.[76] That situation is extreme, but the lesson it holds for even a moderate skewing of the sex ratio may cause Chinese parents to start valuing female children more highly.

Homosexuality

Another fertility-limiting mechanism that is widespread in mammals is homosexuality, although it is not the dominant form of sexual behavior in any animal species or human society. As social scientists Clellan Ford and Frank Beach noted almost a quarter century ago, "The basic mammalian capacity for sexual inversion tends to be obscured in societies like our own which forbid such behavior and classify it as unnatural."[77] Social attitudes toward human homosexuality have been quite plastic historically and may be related to perceived demographic needs.[78] In a study of thirty-nine preindustrialized societies, homosexual behavior was disapproved of or punished in 75 percent of pronatalist societies, but approved of or encouraged in 60 percent of those groups classified as antinatalist.[79]

In at least some societies, such as the Etoro of the New Guinea highlands, extreme levels of homosexuality occur together with resource scarcity (shortage of arable land and animals to hunt) and severe constraints on heterosexual intercourse. Coitus between married partners was taboo on more than two hundred days each year. The Etoro justified the taboos through the belief that a man died when his semen store was exhausted and that the original supply could only be obtained in boyhood by having oral intercourse with male adults. As the Etoro expressed it, "semen does not occur naturally in boys and must be 'planted' in them." All attributes of maleness are thus transmitted, and boys are inseminated nightly "so that they might grow."[80]

Views on homosexuality in the ancient world varied in part with perceptions of the demographic situation. There was a strong strain of antinatalism in the early Christian Church, based on the belief that the world was "full" and procreation was unnecessary with the Messiah about to return. The early Church therefore was tolerant of homosexuality (and contraception, infanticide, and abortion), and Christ discouraged sexual relations even within marriage. Pronatalist views began to take over a few hundred years after Christ, when St. Augustine began to worry about populating heaven, and large families were viewed as a way of strengthening the faith. Probably not coincidentally, that was a period of population decline.[81]

Witchcraft

A phenomenon that may be connected with homosexuality through attitudes toward women is witchcraft. Male hostility toward the opposite sex—often the belief that women are in some way "polluting"—is one way of reducing the birth rate. In the overpopulated highlands of New Guinea, women are thought to be a threat to men, especially when they are menstruating. Sexual intercourse is discouraged among the Hagen, for example; the penis and vagina are thought to be "bad," and menstrual blood is considered to be extremely dangerous, having supernatural powers including the ability to poison men in the tiniest amounts.[82]

Such misogyny also seems to be behind the idea of witchcraft.[83] Harris and Ross point out that in New Guinea a belief in witchcraft seems related to environmental stress and may play a role in population regulation. Witches may be killed, banished, or deprived of foods containing critical animal protein, all of which lower their reproductive potential.[84] The situation in New Guinea mirrors that in Europe around the latter part of the sixteenth century, when commercialization of agriculture and enclosing of common pastures caused the destitution of hordes of peasants, and scarcity and famine replaced the higher living standards of the century before. As resource scarcity led to social disruption, the long-held belief in witches erupted in an intense persecution of those accused of witchcraft—the great preponderance of whom were women.

The direct impact on death rates from the execution of witches may have been significant. In the fourteenth century, some three hundred fifty people were killed in the small country of Switzerland alone, but these numbers merely foreshadowed the bloodbaths that followed in the next two centuries, in which over a half million people may have been killed.[85]

Throughout the history of witchcraft persecution and trials runs a thread of fear of female sexuality, which seems to be the reason that prepubertal girls were rarely accused.[86] The connection with population regulation seemed hardly conscious, however, since there was some concern about witches causing infertility, and midwives (who often performed abortions as well as deliveries) were frequently victims. On the other hand, many witches were accused of copulating with the devil or other evil spirits, which indicates a fear of undesirable or malevolent fertility. As Harris and Ross noted: "Thus the witch craze of the sixteenth century seems to ride a wave of social contradictions, reflecting no doubt the conflicting interests of the rulers and the ruled, with the latter seeking to limit fertility and the former to stimulate it [to keep labor supplies up]."[87] Whatever the motives, the demographic result was not trivial; the number of women who died in the carnage of Europe may have exceeded 500,000, at a time when Europe's total population was perhaps 80 million.

Marriage Age

A major factor affecting reproductive rates clearly is variation in the age of marriage, and this has been especially the case in Europe. Demographers E. A. Wrigley and R. S. Schofield have shown a strong inverse relationship between a woman's age at marriage and the number of children she bears; the later she weds, the fewer children she has.[88] Many societies have encouraged or discouraged high fertility by manipulating marriage-age customs, whether deliberately or not. The trend toward later marriage in Europe in the late eighteenth and nineteenth centuries was clearly an important element in the decline of fertility that began then. More recently, a minimum age at marriage for both men and women was enforced as an integral part of China's "birth planning" program; a relaxation of restrictions on marriage age around 1980 contributed to an upsurge in the Chinese birth rate in the mid to late 1980s.[89]

Migration

A final method of "population control" widely used by human beings through the ages needs to be mentioned: migration.[90] Long-distance migration was the most obvious population control device of the Polynesians, who as islanders were keenly conscious of population pressures. But they also used other measures, including withdrawal, periods of abstinence after childbirth, abortion, infanticide, and nonmarriage of landless younger sons.[91]

Perhaps the most famous population/resource-related migration in historic times was the flight of 2 million or so Irish between 1846 and 1855 in response to a famine caused by a disastrous potato blight, the effects of which were worsened by British incompetence, indifference, and rapacity.[92] Over a much longer period and under less dire circumstances, the migration of millions of settlers to the Americas (and Australia) can be seen as arising from population pressures in Europe after 1500.

The number of ecological refugees today is large and appears destined to get much larger.[93] People fleeing from Haiti in the 1980s and 1990s were trying to escape poverty and a severely degraded environment as much as from political repression (and the two often are not unrelated). While separating economic from environmental causes of migration is difficult (and perhaps not very significant), many observers have conservatively estimated the number of environmental refugees worldwide in the early 1990s as more than 10 million, roughly half of them in sub-Saharan Africa (rivaling some 17 million political refugees). The United Nations Population Fund estimated that 100 million people were living outside the country of their birth in 1992, and a several times greater number had migrated to other areas within their nations.[94] A key point to remember, of course, is that migration cannot be a viable method of population limitation on a global scale.[95] Nevertheless, mass movements of people certainly could alter the global distribution of environmental impacts and resource usage.[96]

Fertility Control

While direct intent is impossible to demonstrate in most situations, the overall pattern makes it hard to escape the conclusion that human beings have always manipulated their fertility, using a wide variety of techniques. At the individual level, that manipulation has been prompted by changes in the perceived costs and benefits of children, which in turn vary in response to fluctuating resource-environment conditions.[97] Although they change more slowly than individual perceptions, social norms influence fertility powerfully if indirectly through customs and traditions such as age at marriage, abstinence from sexual relations after childbirth, tolerance or intolerance of premarital and extramarital relations, and so forth.[98]

This picture is a very different one from that painted by most early demographers and anthropologists, who viewed "primitive" peoples as living in a state of nature, helpless to alter either birth or death rates and doomed to have both remain very high. Indeed, the fallacious idea that "primitive" societies had no control over birth or death rates is one of the roots of the theory of the demographic transition "theory."

Anthropologists in particular were properly anxious to affirm the validity of non-European cultures and, in an early fit of political correctness, loath to dwell on distasteful topics related to fertility control such as infanticide, abortion, wife-stealing, cannibalism, and female sexual mutilation. As anthropologist Mildred Dickemann observed: "Anthropology became, and to some degree remains, the twentieth century purveyor of the myth of the noble savage."[99]

A notable early exception to this was A. M. Carr-Saunders, who pointed out that all known "primitive" people practiced some form of fertility control in order to achieve an "optimum number" related to such environmental factors as the abundance of game and fertility of soil. As he wrote three quarters of a century ago, "normally in every primitive race one or more of these [population controls such as infanticide and head-hunting] customs are in use, and the degree to which they are practised is such that there is an approach to the optimum number."[100] Later, biologists such as V. C. Wynne-Edwards and anthropologists such as Dickemann and Marvin Harris brought consideration of the possible roles of unpleasant demographic customs back into scientific discourse,[101] but in a more sophisticated fashion than that of Carr-Saunders.[102] Overall, the evidence for the pervasiveness of conscious and subconscious regulation of numbers by both individuals and societies is overwhelming, and it is clear that both mating patterns and numbers of children are influenced by environmental constraints.[103] But this evidence is still barely recognized by many who are interested in the history and future of the human population.[104]

The whole situation is made more complicated because different classes in organized societies may have different demographic goals, a circumstance that is often lost in aggregate statistics on vital rates. Since the colonial expansion of European powers, for instance, there appears to have been a tendency for the upper classes to have antinatal attitudes for their own class, but to create situations that made it difficult or unwise for the poor to control their reproduction.

A principal reason, of course, was and is the desire for a supply of cheap labor to fuel expanding capitalist economies. Even in the Middle Ages, the Church adopted pronatalist positions when it became a large-scale property owner and saw an advantage in multiplying the numbers of peasant laborers.[105] The crucial point here is that it is a widely propagated myth that poor people over-reproduce because they don't know any better. If they can, poor people, like rich people, usually have the numbers of children that are in their perceived best interests (although, we repeat, the desires of fathers and mothers may not always match).[106] In some circumstances, hard work, inadequate nutrition, and poor living conditions may serve to limit reproduction, as they apparently did for the lower socioeconomic classes in Britain in the middle of the nineteenth century.[107] But, apart from factors beyond their control, human beings generally do not have large numbers of children out of ignorance about how to exercise restraint over their reproduction.

None of this, of course, weakens the case for giving all sexually active human beings access to the best possible contraception (with medically safe abortion as a backup). Few people would argue that backstreet abortions, infanticide, and mass migration are superior ways of controlling population size. Nor does the level of control already exercised over childbearing weaken the case for creating social situations in which it is more likely that people will broaden their perceived domain of self-interest in both time and space, and change their reproductive behavior accordingly.

While individuals in gatherer-hunter and traditional agricultural societies received more or less direct feedback on their population-resource-environment balance, in modern societies that feedback has been progressively weakened. People still engage in some sort of cost-benefit calculation in connection with childbearing, but they are generally much less aware of the resource-environment consequences of their reproductive acts than were our distant ancestors. Overweight, ignorant talk-show hosts can prosper while remaining perfectly clueless about overpopulation, hunger, and environmental deterioration because they are personally buffered (at least temporarily) by their wealth from many of the consequences of Earth's increasingly perilous state.

Fertility Control Technology Today

A wide range of contraceptive technologies is available today in both rich and poor nations. These include "modern methods"—the Pill, IUDs, condoms, diaphragms, spermicidal foams and jellies, hormonal injections, sterilization—and "traditional methods" such as rhythm, withdrawal, abstinence, and douching.[108] There is considerable variation in their availability and usage, both from nation to nation and among classes within nations, although modern methods generally predominate worldwide, with the Pill, IUD, female sterilization, and the condom being most popular. In some areas, including in the former Soviet Union, a lack of adequate supplies of contraceptives has put abortion into the front ranks of antifertility technologies.

Globally, fertility control methods that put the main burden on women (tubal ligation, IUD, the Pill, injectable hormones, and diaphragm) are used by about 65 percent of couples, those on men (condom, vasectomy) about 20 percent. Rhythm (which requires cooperation by both partners) is used by about 7 percent. Tubal ligation and IUDs make up some 45 percent globally, and almost 60 percent in less developed regions.[109] Tubal ligation is two to three times as common as vasectomy overall; in China, the ratio is about four to one.[110]

In Japan, there is a bizarre situation where fertility control is achieved primarily through condoms, abortion, and the rhythm method—with low-dose oral contraceptives banned. The reasons are complex and range from cultural attitudes about women's sexuality to the profits physicians make doing abortions.[111] There are very good reasons for legalizing the Pill in Japan, even if the Japanese themselves do not choose to use it very much. Japan is pledged to increase its support of population programs in poor countries; legalizing the Pill would spare the nation "the moral dilemma of promoting population control in developing countries by means that are illegal at home."[112]

In the United States in 1988, sterilization was the method of choice of nearly 40 percent of all couples (female, 27; male, 12); the Pill, 31 percent; condom, 15 percent; diaphragm, 6; periodic abstinence, withdrawal, IUD, foam, and miscellaneous methods (douche, sponge, etc.) account for the remainder. There has been a significant drop in IUD use since 1982, causing the withdrawal of IUDs from the market in 1985 and 1986 by the two largest manufacturers.[113] A significant rise in condom use may reflect concern over AIDS.

The effectiveness of the methods varies considerably. For instance, the annual probability of failure (percent of women using the method becoming pregnant) for married women in five Latin American countries varied from a low of 4.3 percent for IUD users to 7.9 percent for the Pill and 19.4 percent for rhythm. Between nations, Pill failures ranged from 4.0 percent in Costa Rica to 11.5 percent in Panama, and rhythm failure from 16.2 percent in Panama to 38.9 percent in the Dominican Republic.[114]

Contraceptive failure rates are becoming increasingly significant as desired family sizes drop over much of the world. For example, according to a model developed by demographers John Bongaarts and Germán Rodríguez, if the target family size is two children, an increase in contraceptive effectiveness from 85 to 95 percent would reduce the average family size from 3.73 to 2.68—by more than one birth. Thus contraceptive failure can contribute significantly to population growth rates in nations where desired family sizes are small to moderate but where contraceptive effectiveness is low.

The current state of contraceptive technology and research, especially in the United States, is nothing short of a disgrace.[115] Development and deployment of new contraceptive technologies has all but stopped in the last two decades, and there are large disincentives to reviving the enterprise. It took a quarter century of review for the U.S. Food and Drug Administration (FDA) to approve the injectable synthetic hormone Depo-Provera for use in the United States, although by then it had been used in several developing countries for years. It is better tolerated than the Pill and is virtually fail-safe, although questions persist about its safety.[116] In part, the delay was founded in a reasonable concern over any method that involves tinkering with the hormonal balance of healthy human beings. Approval of the female condom came faster, presumably because, as a barrier method with no drug or hormonal component, it seemed less likely to carry unrecognized risks, and because its use can be promptly discontinued if problems arise. Also, its capacity to protect against sexually transmitted diseases, especially AIDS, may have been a factor spurring its approval.[117]

Several new contraceptive technologies are under development or on the horizon. These include everything from new and improved condoms, diaphragms, and Pills to hormones delivered by vaginal rings or via transdermal patches, as are some sea-sickness remedies. The Chinese, partly in response to the perceived need to increase the acceptability of male sterilization, have pioneered improvements of the procedure, developing systems of "no scalpel" and reversible vasectomies.[118] They also have been experimenting with the use of gossypol (from cottonseed oil) and an herbal drug as potential "male pill" ingredients, but so far side effects are too severe.[119] Some work has been done by World Health Organization (WHO) researchers on an injectable testosterone derivative that eliminates sperm from a man's ejaculate. That could open the door to a hormonal contraceptive targeted, for a change, at men. There is even the possibility of a birth-control vaccine that would immunize a woman against a hormone essential for the implantation of a fertilized egg in the uterus or stimulate a woman's immune system to produce antibodies against her partner's sperm.[120]

But progress on even the most promising contraceptive technologies has been hampered by lack of funding of agencies such as the National Institutes of Health's Contraceptive Development Branch and the World Health Organization. American pharmaceutical companies are reluctant to put huge sums of money into developing new contraceptives, despite annual domestic sales of the contraceptive pill in the vicinity of $750 million. One important reason is product liability exposure. While juries deciding lawsuits based on unwanted side-effects of drugs given to sick people tend to recognize a balancing of risks, this is much less the case with contraceptives, and the risk of large judgments against drug companies is correspondingly greater.[121] Others are that the companies fear regulatory hassles and feel there is little demand for new contraceptives—they believe the market is well served by those available now.[122]

Of course, some progress has been made in developing new techniques for regulating fertility, such as Depo-Provera and Norplant, a plastic (Silastic) rod or capsule implanted under the skin that gradually releases a synthetic progestin, levonorgestrel.[123] Norplant protects well against pregnancy for five years and appears to be very safe.[124] Fertility is promptly restored when the rod or capsule is removed.

Some scientists are very enthusiastic about chemical sterilization of women using quinacrine.[125] Pellets of the compound, a drug often prescribed for the treatment of giardia infections,[126] are inserted into the uterus using an IUD insertion device, which causes an inflammation and fibrosis (abnormal formation of fibrous tissue) that block the fallopian tubes.[127] The technique seems to be safer and much easier than surgical sterilization—slightly over 30,000 insertions in Vietnam are estimated to have averted about 240 maternal deaths from various causes related to reproduction. One researcher estimated that "1300 clinicians doing 100 or so quinacrine pellet insertion sterilizations per month could meet Vietnam's unmet need for female sterilization. That could be a significant contribution to the country's family planning needs."[128] Controversy remains about the safety of the method, as well as opposition from the World Health Organization that may have its roots in Vatican opposition.[129] Nonetheless, it is clearly promising, and we hope that field trials now being conducted in eleven countries will soon answer safety questions.[130]

Perhaps the brightest star on the fertility control horizon is the abortifacient known as RU-486, developed in France. One pill taken after pregnancy is confirmed (and not later than seven weeks after the last period), followed forty-eight hours later with a single intramuscular injection or intravaginal suppository of the appropriate hormone (prostaglandin) led to the expulsion of the embryo in 96 percent of women tested. RU-486, now being tested in the United States, has the potential for ending the contentious abortion issue. It would make abortion a totally private matter between a woman and her physician. RU-486 would seem to have considerable promise for easing fertility control problems in poor nations, where an estimated 200,000 women perish annually from botched abortions, and many more are seriously injured.[131]

Possibly the best summary of the prospects for truly new, effective contraceptive technologies was given in 1991 by the "father" of the Pill, Stanford University chemist Carl Djerassi. He wrote: "During the past two decades, research and development on birth control has declined dramatically. Except for the abortifacient RU-486, for the balance of the century there are no prospects for scientifically feasible, novel contraceptives such as a male pill, a contraceptive vaccine, or a once-a-month menses inducer. A jet-age rhythm method is the best we can expect."[132]

Djerrasi recently suggested that, in view of the virtual impossibility of developing a male contraceptive in the near future, a new approach be taken. He proposed extensive testing of the feasibility of first storing sperm in liquid nitrogen, having men undergo vasectomies, and then using the sperm for artificial insemination when the men want children. He suggests testing the viability of long-term, large-scale sperm storage, and thinks using volunteers from the military (without vasectomies), for whom health and other records are normally maintained for the duration of their careers, would be an ideal way to test this.[133] Of course, this is not a birth control method that could be easily deployed in poor countries, although it would have the enormous advantage of requiring a *positive* step to produce a child. Like other techniques that focus on males, it would help balance the obvious inequity of women bearing most of the risk of contraception.

Djerassi also has insightful views of other birth control techniques. He quotes British novelist David Lodge on the rhythm method (also known as "natural family planning," "periodic abstinence," and "Vatican roulette"):

Clerical and medical apologists . . . encouraged the faithful with assurances that Science would soon make the Safe

59

Method as reliable as artificial contraception. But the greater the efforts made to achieve this goal, the more difficult it became to distinguish between the permitted and forbidden methods. There is nothing, for instance, noticeably "natural" about sticking a thermometer up your rectum every morning compared to slipping a diaphragm into your vagina at night. And if the happy day *does* ever dawn when the Safe Method is pronounced as reliable as the Pill, what possible reason, apart from medical or economic considerations, could there be for choosing one method rather than the other?[134]

But advances in analytical biochemistry make it clearly possible to develop a home test kit that would allow women to determine with ease and high accuracy the time of ovulation and reduce the present requirement of some seventeen days of abstinence to eight or nine. It would also allow increasingly health-conscious women to know whether and when ovulation occurs—which surveys indicate would be deemed desirable by many well-educated women, and which might have important health implications. But, as Djerassi notes, as a means of fertility control in poor nations, "such a 'high tech' method would be useless on financial and operational grounds alone."

In summary, even for highly motivated, well-educated couples in rich countries, there are few prospects for improved birth control. Couples in poor nations will need to make do for a long time with the methods now available.

The Failure of Fertility Control

In view of the long record of fertility control in *Homo sapiens* and the existence of a wide array of contraceptive technologies, how can the continuing population explosion of the twentieth century be explained? One part of the answer is that the development and deployment after World War II of elegant and easily deployed "death control" technologies swiftly overwhelmed long-evolved control mechanisms. Improved sanitation, followed by antibiotics used against pathogenic bacteria and potent pesticides used against disease-spreading insects, greatly reduced death rates, especially among infants and young children. By the mid-twentieth century, too, traditional mechanisms that constrained fertility were breaking down in many areas under the impact of colonialism or of "modern" influences from Western nations spread through magazines, radio, and television.

Furthermore, many of the obvious environmental cues to "optimum" population size were no longer clearly presented to people, especially to powerful and wealthy decision-makers in both rich and poor nations who usually had little contact with the land and basically no knowledge of how natural systems function. The Industrial Revolution and the modernization of agriculture had greatly increased the lag time between overshooting the carrying capacity of the local environment and recognition of the overshoot. Access to food and resources from other regions allayed concerns about shortages or depletion. The changes were so swift that traditional prestige structures were often shattered, first by colonialism and then by displacement of agrarian peoples from the land into crowded third-world cities. The views in this area of anthropologist Mary Douglas merit careful consideration:

> . . . there is a message here for the countries whose prosperity is threatened by uncontrolled population increase. In those countries we see the well-educated and well-to-do actively preaching family limitation and setting up birth-control clinics as a social service for the teeming poorer classes. They encounter resistance and apathy from the milling poor. . . . Their failure spurs them on to more enthusiastic propaganda. But if they would succeed, let them first look to their prestige structure. What hope of advance does their system of social rewards offer to those to whom they preach? . . . If the prestige structure were adjusted, propaganda would be more effective or perhaps not be necessary. For given the right incentives some kind of population control would be likely to develop among the poor as it apparently has amongst those who seek to administer demographic policy.[135]

Douglas was writing thirty years ago and about poor nations. We think her words apply to many countries today. But since she wrote, options for changing the distribution of wealth and power have become much more restricted, by, among other things, an increase of some 65 percent in the number of people in the world, and by more than a doubling of the populations of poor nations. Nevertheless, changing that distribution in the direction of increased equity is one of the key tasks now faced by humanity.

Chapter Three

SLOWING THE STORK

Of all the harmful interventions by Western powers into the affairs of less developed nations, it is ironic that the one most heavily laden with good intentions has, in retrospect, spawned the most horrifying consequences. How much human misery today traces to the Western introduction of the means to lower mortality, with hardly a thought to satisfying the need that would thus be created—to lower fertility commensurately? That intervention to save lives helped to curse the developing world with the all-engulfing, self-reinforcing problems of rapid population growth, poverty, and environmental deterioration.

How can this grim cycle of despair possibly be interrupted? What, at this point, is the moral responsibility of rich nations in helping the poor? Trial and error have illuminated an extraordinarily diverse set of ways to interrupt the cycle, which variously attack its different components. Teaching women to read might be the best tactic in an Indian state. Supplying kerosene stoves and fuel might be most critical to improving living standards and reducing fertility rates in a village in Tanzania. Educating boys about the responsibilities of parenthood and the local impacts of overpopulation might lower fertility in Costa Rica. The very act of installing a culturally sensitive family planning program with well-integrated maternal and child health care might contribute to lowering birth rates anywhere.

The developing world is too diverse culturally, economically, and politically for any single measure to remove the curse. There is, however, one generality so obvious that not seeing it may reflect willful blindness. Increasing socioeconomic equity at all levels—between genders, families, social classes, and nations—has by far the greatest potential for improving the human condition. Today, the rich nations are perpetuating and exacerbating inequity in a million often subtle but powerful ways. Improving the behavior of the rich would be much more effective than simply telling the poor how to change theirs.

The one aspect of the population situation that almost everyone in the industrialized North is aware of is the skyrocketing numbers of people in less developed nations. Billions of people in the struggling South, with their numbers doubling every few decades, are viewed by many in the North as a major threat to the future of rich and poor alike. Indeed, for some observers, this is the *entire* population problem.[1] The rich are peculiarly blind to the seriousness of their own overpopulation and its consequences. Thus it seems reasonable to begin an examination of slowing down the stork with the problems in the South, where at least governments and people are grappling with the problem, rather than in the North, where an equally serious population problem is still being ignored. For the moment, we exclude from discussion the fascinating and unique case of China, where population growth has already been substantially slowed.

An appropriate place to begin is with the notion of a "demographic transition." That idea has been central to discussions about the world population since at least the 1950s, when unprecedentedly rapid population growth in the underdeveloped world became apparent. Many demographers and social scientists have assumed that it can be counted on to end the population explosion in developing nations.[2] The claim was based on observations of demographic change that have occurred in history.

Let us assume the first human beings on Earth were the upright but small-brained australopithecines—"Lucy" and her relatives—which existed in a small population of a few tens of thousands at most. Gatherer-hunter societies, with no modern medicine or sanitation, had high death rates and slightly higher birth rates, which produced a very gradual expansion of the population.[3] It took millions of years for the early human populations to grow from a few thousand individuals to about 5 million people at the time agriculture was invented around 8000 B.C. After the development of agriculture, birth rates rose and population growth accelerated until by 1650 there were about a half-billion human beings.

During those ten thousand years, however, overall population expansion was quite slow, with many ups and downs reflecting famines, plagues, wars, alternating with times of prosperity.[4] Birth rates probably were in the vicinity of 40 to 50 per 1000 in the population, and death rates were not much lower. The average life expectancy at birth was less than forty years, and as many as half of all children failed to survive to age five. The Victorian era opened with a world population of 1 billion early in the nineteenth century. By then the use of soap and other simple sanitation measures in Europe and North America had led to slowly dropping death rates, especially among infants and children. Consequently, the rate of natural increase (the surplus of births over deaths) began to rise in those populations.

Following the gradual decline in death rates after 1750, fertility in Europe and North America also began slowly falling—but not fast enough to catch up with declining mortality. Population growth in these industrializing regions was rapid, although it usually did not exceed 1.5 percent per year.[5] In the 1930s, spurred by the hard times of the Great Depression, fertility in Europe and the United States plunged to replacement level—an average family size (TFR) of just over two children—or even lower.[6]

Around 1940, demographers were making dire predictions that the industrialized countries would soon have declining populations, but World War II and the subsequent baby boom averted that. Total fertility rates soared again far above replacement (to as high as 3.7) in the United States in the 1950s and did not fall to replacement level again until the early 1970s. Europe's baby boom was relatively modest and brief, but TFRs remained above replacement. The low fertility of the 1930s was far too transient to wring out the momentum generated by previous growth and stop population increase.

The Demographic Transition

The progression from a situation of high birth and death rates to one of falling death rates and high birth rates, and finally to one of low birth and death rates, came to be called the *demographic transition*. Since in Europe and North America, the transition apparently had occurred in step with industrialization, demographers concluded that the development process somehow caused it, although they were not sure how.[7] Curious exceptions in the record, such as fertility declines in eastern European countries when they had scarcely any industry, muddled the theory considerably. Also, a marked fertility decline had begun in France well before 1800, decades before any significant industrialization had begun. The French fertility drop occurred long before a fall in infant mortality was recognized, calling into question the notion that increasing the chances of children's survival lowers desired family sizes.[8] Still, in the 1950s and 1960s, social scientists were confident that a similar path would in due course be followed by the developing nations as they industrialized. This led to the idea that "development is the best contraceptive."

Indeed, by 1950 the phase of death-rate decline was well under way in many poor countries as post–World War II medical technology was introduced. Control of insect-borne diseases like malaria and yellow fever by spraying DDT and other insecticides, vaccination against smallpox and other diseases, and basic sanitation to control gastrointestinal ailments (to which infants are especially vulnerable) brought dramatic results. But the spectacularly plunging death rates were accompanied by equally spectacular growth rates, as there was no perceptible change in birth rates. Growth rates in countries with the most successful public health programs, mostly in parts of Asia and Latin America, soared above 3 percent per year.[9]

In the 1960s, the "population explosion" was going full blast. The world population was growing at an aggregate rate of over 2 percent per year, fueled by the surge of growth in the less developed world and helped along by the industrialized world's baby boom.[10] In 1960, the population passed the 3 billion mark, a scant thirty years after reaching 2 billion. World War II, which caused the premature deaths of some 60 million people over six years, had only slowed population growth a little—the surplus of births over deaths in each of the six years was more than sufficient to replace that year's losses.

By the end of the 1960s, the postwar baby boom in the rich nations was turning into a bust, and in the early 1970s fertility in most industrialized nations fell to replacement level or below. Growth continued, because of the momentum from past growth, but increasingly slowly in many countries.[11] This confirmed demographers' expectations about the demographic transition. By the 1980s, some 90 percent of the world's population growth was taking place in less developed nations.

Meanwhile, whatever happened to the demographic transition in developing countries? Are they following the same general course blazed by today's industrialized nations? By 1970, even demographers were asking those questions. To help the transition along, family planning programs were introduced in less developed nations beginning in the 1960s to provide people with the means to reduce their fertility. The first program was actually launched in India in 1952.

When family planning was first introduced in less developed nations, little was understood about what persuaded people to have small families. It was widely believed that a major prerequisite was the introduction of modern sanitation and basic health care to reduce infant and child mortality rates enough to assure that most children would survive to reach adulthood. It seemed reasonable to think also that, once children were no longer income-producers for the family (by helping with farm work or selling or even begging in cities) but instead cost money to feed, clothe, and educate, people would find smaller families advantageous. Development would indeed be the best contraceptive.

Yet, in many less developed nations, birth rates remained high throughout the 1960s and 1970s, despite substantial reductions in infant and child mortalities and despite the existence of family planning programs. Moreover, although birth rates had fallen in some of those countries by the 1970s, no clear correlation with development, as measured by growth of the gross national product (GNP), was apparent.[12] Some countries such as South Korea and Taiwan seemed to show the expected declines in fertility, while others, such as Mexico and Brazil, had traveled some distance along the development path with little or no reduction in their birth rates; still others, such as China, Sri Lanka, India (especially

67

Kerala state) showed scarcely any increase in per-capita GNP, yet their birth rates were falling.

The failure of many family planning programs to achieve early success sparked controversy in the development community about the effectiveness of those programs in reducing birth rates and their value in furthering development goals. The controversy reached its zenith at the first United Nations Population Conference in Bucharest in 1974. A heated debate took place between demographers and family planning experts, on one hand, versus prodevelopment social scientists and third-world spokespeople (many of them Marxists) on the other as to whether family planning programs were effective or beneficial and whether reducing birth rates was even a useful goal.[13] A third group was composed of critics of both positions, including environmentalists concerned about unchecked population growth overstressing natural environments. They questioned whether family planning programs were having any significant impact and suggested that policies "beyond family planning" might be needed.

Development-as-contraceptive advocates insisted that "if you take care of the needs of the people, the population will take care of itself." Many Marxists asserted that family planning was a racist plot to keep the people of less developed regions suppressed and underdeveloped. Among the most vociferous in insisting that the focus be on development were the delegates from China, which ironically was already implementing a strong "birth planning" program at home. Meanwhile, demographers and family planning advocates at the conference maintained that the programs were valuable in enhancing maternal and child health and they helped facilitate lower birth rates when people were ready for smaller families. Both groups essentially rejected or ignored arguments for reducing population growth rates on environmental grounds.

Ultimately, both viewpoints were overtaken by events, and third-world leaders became increasingly aware that rapid population growth was a hindrance to socioeconomic progress. The emergence of environmental problems with connections to population too obvious to overlook drove the lesson home.[14]

Factors in Fertility Declines

By the end of the 1970s, it had become inescapably clear that "industrialization" alone—building airports, highways, factories, hydroelectric projects, and highrise office blocks—had no effect on birth rates (although these activities did boost GNP). But improving basic health and nutritional conditions (which led to falling infant mortality rates and rising life expectancies), educating women, and granting them a measure of independence did have direct effects, sometimes dramatic ones.[15] Another factor of apparent importance in many societies was some provision for old-age security; without that, people had a need to produce several sons to be assured of support in later years. The more widely improved living conditions and social security were distributed, the more effective they seemed to be in influencing the entire population to have smaller families.[16]

By the mid-1970s, several striking examples of "family planning success" were evident. Aside from the notable state of Kerala, India, they were mainly in East Asia, Latin America, and the Caribbean: Taiwan, South Korea, Singapore, Hong Kong, Sri Lanka, Costa Rica, Trinidad and Tobago, and Barbados. By 1980, the most outstanding—and in some ways astounding—large-nation example turned out to be China, where an indigenous program attained the unprecedented achievement of cutting its TFR by half in only ten years—from more than five children per woman in 1970 to 2.3 in 1980.

It is still not entirely understood why success in fertility control has been so spotty. Many demographers have been convinced that falling infant and child mortalities, leading to the survival of most children, have been key factors in motivating family limitation in countries that have experienced declines.[17] The conventional view has been that, as development proceeds, the perceived costs and benefits of having children change, and parents respond to opportunities to limit their reproduction.

One intriguing view, offered by anthropologist Virginia Abernethy, is that in many cases "development" tends to increase family sizes and poverty to decrease them.[18] Abernethy argues that perceived economic opportunity leads to larger family sizes, and that "fertility rates fall when limits are recognized." She claims that policies based on the demographic transition model raised expectations throughout the poor world, especially through "aid and the rhetoric of development and prosperity."[19] In Africa, the continent that has received the most foreign aid per capita, fertility rose to the highest in the world, surpassing that of Latin America, the 1950s record-holder. Abernethy links instances of declining fertility to deteriorating expectations in the absence of an emigration safety valve.

She relates the fertility control successes in places like Taiwan, South Korea, Singapore, and Barbados to those nations having "experienced poverty unrelieved by government subsidies on housing or food or a culture of enlightenment."[20] She suggests that, in Taiwan, the sudden crowding that followed the arrival of large numbers of Nationalist immigrants from the mainland after the Communist victory in 1949 may have been responsible for the success of family planning efforts. But several other important factors also were at play—including development of a strong agricultural sector, education and employment of women, and rapid urbanization. Sudden crowding did not occur in South Korea, but Koreans did experience "poverty unrelieved by government subsidies on housing or food."

Abernethy's theory evidently doesn't hold in all cases. Sri Lanka's fertility declined rapidly when food supplements were provided for the poor in the 1970s, and stopped falling when a "free market" government took over around 1980. Thailand's fertility decline has accompanied a steady rise in the standard of living, a rise (to a per-capita GNP about that of Poland) that has also been accompanied by an increase in economic concerns, leading to the feeling that children are too expensive.[21]

Costa Rica is a country with a strong family planning program, where fertility declined dramatically during a period of relative prosperity but then plateaued at a TFR of 3.5 to 3.8 between 1976 and 1985. Costa Rican women were educated and had access to jobs, but were still having three or four children on average. In the early 1980s, severe economic troubles did not cause a further decline in fertility, but a slight drop seems to have been associated with *improving* economic conditions after 1988.[22]

Several explanations are possible for the failure to reduce fertility further in Costa Rica. One often cited in the country is that the otherwise very effective family planning organization caters to women only and does nothing to enlist support of husbands and partners.[23] In Latin American societies, as in many others, the support and involvement of men in setting family-size goals is crucial. Moreover, while girls are educated along with boys in Costa Rica, the average woman has only an eighth-grade education and no particular marketable skills.

These examples suggest how complex the causation of fertility trends can be, and how difficult it is to determine what is actually behind them.[24] The Costa Rican experience might be interpreted to mean that tough conditions reduce fertility, but only after a time lag during which people gradually change their family-size goals in response. Or, alternatively, job opportunities for women may have opened up in the late 1980s, leading them to have fewer children. A decline in government support of birth control services—lack of governmental commitment to population limitation—could be a significant factor. Costa Rica is still ahead of most poor countries in terms of accessibility and acceptance of services, but their effectiveness may be waning.[25]

One thing is certain: development *cannot be the best contraceptive* if by development is meant the kinds of industrial development that have occurred in today's rich nations since 1950. All evidence indicates that the life-support systems of the planet could not support 10 billion people living the current lifestyle of the average citizen of the United States. In fact, that overconsumptive lifestyle threatens the viability of the entire planet when practiced by only a quarter-billion Americans. But even if it were biophysically possible to maintain 10 billion superconsumers, there isn't time for such extensive and costly development to be carried out and for that to induce low enough fertility to stop population growth at or below 10 billion.[26]

The factors that determine fertility rates clearly are extremely complex, sometimes contradictory, and vary from culture to culture. There are, however, common threads that may provide some background for taking action to control fertility while more refined understanding and policies are developed. These have to do with health, education (especially of women), and issues of equity.

Women's Education

During the last two decades, a strong connection of education for women with lower fertility has become increasingly apparent in nearly all developing nations, although the precise mechanisms are less clear.[27] At the most basic level, a young woman may apply a few years of schooling to better management of her family's health and well-being simply by obtaining pure water and choosing more nutritious food (unlike a young man, who uses his education for a job). Since many births are no longer necessary to ensure survival of some children to adulthood, the mother may become more receptive to using birth control. In addition, education often opens other opportunities besides motherhood, which may lead to further reductions in fertility.[28]

The case of India is particularly illuminating with respect to female education and fertility. Unlike the leadership of the United States and most other rich countries, those in charge in India have long been aware of the threat that population growth represented to the security of their nation. As a result, India has made more progress in lowering fertility than, for example, most African nations, having reduced its TFR from 6.0 in the 1940s to 3.6 by 1994. Although beset by severe problems of poverty, ethnic strife, and economic inequity, India is a politically unified, democratic country. More important, India's elite has come to a tacit understanding with her poor. When things get tough and famine threatens, the government will intervene to prevent it.[29] India's poor people have some sense of security, of being able to plan for a future. The moderate progress that has been made in reducing family size there, we believe, can be traced in part to that increase in nutritional security.

But India's record in some of the other measures considered important in lowering fertility has been less than sterling. In 1990, only 34 percent of adult women were literate, infant and child mortalities were high (more than 200 per 1,000 live births), and many urban dwellers and the vast majority of the rural population had no access to safe water.[30] Shri B. Shankaranand, India's Minister for Health and Family Welfare, recently commissioned an "expert group" to prepare a Draft National Population Policy. The group, chaired by the distinguished agriculturalist M. S. Swaminathan, submitted a document that could serve as a model for more backward nations such as the United States.[31] A few quotes (annotated) from the report will give a sense of its frankness and comprehensiveness:

It is high time the limits to the human carrying capacity of the supporting ecosystems are recognized.

Current global development pathways are leading to a continuous increase in the gap between the incomes of the poor and the rich and to jobless economic growth, besides damaging basic life support systems of land, water, flora, fauna and the atmosphere.

Per capita land and water availability is declining to levels where both national food and drinking water security are at great risk.

There are as many persons below the poverty line in India [360 million] today as the entire population of the country at the time of our independence [1948].

We are deeply concerned that our country makes the largest contribution to the annual net addition to the human population and that we have one of the most adverse sex ratios in the world [1,080 males for every 1,000 females of all ages, a ratio of 1.08, an indication of female infanticide and discrimination against females. A typical Western sex ratio is 0.95].

Gender equity is vital for achieving our population goals and for a better life for all. . . . Every effort [should] be made to eliminate before the end of the century all discrimination against women.

If the present trend of adding 18 million more individuals each year to India's population is not halted, economic and social justice to the poor will remain an illusory goal. Population stabilization is thus vital for safeguarding the livelihood security of the poor and the ecological security of the nation.

Central and State Governments [should establish] the goal of achieving a national average Total Fertility Rate (TFR) of 2.1 [replacement level] by the year 2010.

The history of Kerala State in southern India underscores the need to improve conditions of life for India's poor as a prerequisite to further reductions in fertility. There, early governments and missionaries introduced effective educational and health care systems, and women traditionally have been treated relatively equitably. Kerala's death rate has long been lower than those elsewhere in India, and it fell even faster than those of the rest of the nation when declines accelerated in the 1960s and 1970s. As early as the 1930s, a sexist government report complained, "the great majority of girls . . . regard their education, not as something of cultural value in itself, but as a direct means of securing employment and competing with men in the open markets."[32] Education apparently provided the basis for dramatic demographic change. Birth rates in Kerala were on their way down in the late 1960s and early 1970s, and the TFR plunged rapidly in recent decades from 3.0 in 1979 to 1.8 in 1991, even though the area has remained very poor.[33]

Careful analysis leads to the conclusion that the key factor in the fertility drop (and an important one in lowering death rates among infants and children) was that high level of female literacy. Although there was little or no structural change in the society, once a few innovators began to use contraception in the 1960s, knowledge and use spread quickly. That is because there was a highly literate society in which the idea could propagate, one where most households depended upon paid employment and where female education permitted a rapid change in the perceived costs and benefits of children.

Robin Jeffrey described a fisherwoman walking along a road in Kerala, "barefoot, thin as bamboo, a smelly, empty basket on her head. . . . As she bustled down Mahatma Gandhi Road . . . she spotted on the road a handbill advertising a political meeting. Altering direction slightly but scarcely breaking stride, she clutched

the handbill between her toes. Without setting down her basket, she bent her knee and passed the note into her hand. Then
reading intently, she hurried on down the hill."[34] Commenting
about her and other women, Jeffrey said, ". . . all read, all no
doubt had been to school and all undoubtedly expected their children and grandchildren to go to school and learn to read. Any
government that failed to provide the opportunities would do so
at its peril. Kerala's culture gave women a remarkable independence, and women have made Kerala remarkably literate."[35]

Sadly, since female literacy and status were at the heart of the
rapid fertility decline in Kerala, the prospects for similar declines in
the rest of India may not be as bright. India as a whole trails
Kerala by about forty years in female adult literacy and still has
higher infant and child death rates. Nonetheless, the process of
diffusing the use of contraception and changing perceptions of the
value of children might be speeded by improving both formal and
informal communication channels.[36]

Success in reducing birth rates has been seen in several other
Asian nations as well. In Thailand, extremely high growth rates increased the size of the population from 19 million in 1950 to 47
million in 1980 and almost 60 million in 1994. But the government started promoting family planning around 1970, and by the
mid-70s fertility was dropping rapidly.[37] While much of the credit
has gone to the unusually vigorous and imaginative public education campaign of the family planning program, it is doubtful that
so much success would have been realized without the relatively
high status and educational attainment of Thai women. Although
the nation is still growing, its total fertility rate in 1994 was just
above replacement, and Thailand's population is projected to climb
only to 75 million by 2025. (In contrast, Egypt, now the same
size, is projected to be 98 million in 2025.)

Literacy is clearly the gateway to modernity; not only are educated women equipped to engage in economic activity, they can
contribute positively to the education and socialization of their
children of both sexes. As a route to lower fertility, education of
women is likely to be effective in most societies, but the benefits
extend far beyond effects on fertility.[38]

Of the approximately 900 million illiterate adults in the world today, two thirds are women, a salient example of discrimination against the female sex.[39] Closing the gender gap in education, besides simply providing girls with educational opportunities, turns out to be an important factor when global comparisons of social and economic well-being are made. Countries with greater equality of access to education between the sexes have been found to have higher GNPs per capita (about 25 percent higher at the extreme).[40] A high level of female literacy is also associated with improvement in a series of social indicators of family well-being, including health and nutritional status, which translate directly into enhanced life expectancy and lower infant and maternal mortality. Family income, by contrast, proves to be entirely unimportant in these respects after taking into account women's education. Indeed, we know of no society that has successfully achieved an advanced state of development (in the conventional sense) without educating its women.

The mechanisms by which gender equality in education affects development are not well understood. In much of sub-Saharan Africa, men and women maintain separate budgets and are expected to sustain different components of the household. Women's expenditures on the farm and on the children may be strongly influenced by their educational status, which determines access to credit and technology. Another possible effect of the gender gap is simply on women's power in decision making, which, if low, may translate into poorer allocation of resources in the household.[41] Finally, education is necessary (but not sufficient on its own) for access to better-paying jobs.

Women's Health

Another factor that often distinguishes countries with low fertility from ones with high fertility is the state of women's and children's health. The severe health problems that women suffer in poor nations are just one indication of the lack of equity they endure. Women's health is also quite directly connected to their children's health and chances of survival, and through that, to birth rates. Lower infant and child mortalities have long been associated with

76

couples having fewer children, but the link to women's health has been largely neglected.

For women, from midteens until the approach of menopause, the greatest dangers have always been related to reproduction: the risks of childbirth especially, and those surrounding unsafe abortion and miscarriages. Before the twentieth century (and well into it in many regions), it was commonplace for a man to outlive his wife—sometimes a series of wives. Usually, wives died in childbirth or soon after because of an infection acquired during the delivery. The strain of numerous pregnancies doubtless contributed to their deaths, and they often left their widowers with several children. In those days before family planning, women ran the risks of pregnancy and childbirth many times and commonly had little time to recover physically between pregnancies. And, if nutrition was less than adequate, vulnerability was sure to be greater.

Today things are different—or at least they should be. In industrialized societies, a woman dying in childbirth is so rare as to be almost unheard of. In the United States, in the 1990s, only fifteen maternal deaths occur per 100,000 live births.[42] Rates are even lower in northern European countries, where the lifetime risk of dying from pregnancy-related causes is one in nearly 10,000. Only about 6 percent of maternal deaths in the U.S. are from complications of abortion, which, when legal and performed under appropriate medical circumstances, is extremely safe.[43]

But in the poorest developing nations, maternal mortality rates may range as high as 500 to 1,000 per 100,000 live births—twenty to a hundred times as high as in rich nations. A lifetime risk of death from pregnancy-related causes in Africa is as high as 1 in 21; in Asia it is 1 in 54.[44] Abortion complications account for 10 to 35 percent of maternal deaths in countries where it remains illegal—and most are countries where maternal mortalities are many times higher than in the U.S. or Europe.[45]

Clearly, the number of women who die or suffer illness in connection with childbearing is higher than it should or could be. In recent years, development agencies have focused on infant mortality rates, on which better statistics are gathered, while maternal and child (ages one to five) mortality rates have been overlooked.

Yet both are also closely related to basic health conditions. Infants (especially where families are poorest and conditions are worst) usually are protected by breastfeeding for the first year; malnutrition and disease take over after weaning. This often occurs when the next child is born and inadequate foods and unsafe water are substituted for mother's milk. Child mortality figures are hard to find and notoriously unreliable, but are reckoned by many to rival infant mortality rates in the poorest regions—as high as 100 or more per 1,000 live births.[46]

Where children are vulnerable, so are their mothers. Indeed, producing a baby every year or so for decades—particularly if the mother starts in her teens, as is usual among the poor—lowers the chances of survival for each subsequent child and further stresses the mother's body. Nutritional deficiencies such as anemia are common among women in poor countries, especially during pregnancy.[47] Anemia increases the risk of maternal death from hemorrhage in childbirth and nutritionally handicaps the infant from birth.

Other nutrients also are often undersupplied, partly because the women and girls in poor families are customarily fed last and least well. This widespread practice is worsened by traditional restrictions on diets of pregnant and nursing mothers, frequently forbidding the foods that would benefit them and their children most, such as eggs, meat, vegetables, and milk. Such food taboos are found in parts of Africa, the Middle East, and South Asia—all regions where infant, child, and maternal mortalities are high (as are birth rates) and food supplies are generally below estimated needs.[48] Thus the women are poorly nourished throughout their lives, jeopardizing their children's health from the start.

Today, women's health in the developing world is not only threatened by undernourishment and the rigors of frequent childbearing, but by a rising threat from sexually transmitted diseases (STDs; AIDS is only one of more than fifty) against which they often have little or no defense.[49] The problems and burdens of STDs add to difficulties of women in developing countries in various ways. Pain and disfigurement from even milder or earlier forms can be disabling and deleterious to marital relations. Cervical cancer, pelvic inflammatory disease, and ectopic pregnancy, which may result from some STDs, are life-threatening if not treated promptly. The high incidence of STDs has caused sterility in 10 to 30 percent of some populations in tropical Africa, where fertility is the measure of a woman's value.[50] Loss of fertility can result in abuse, abandonment, or worse. In poor countries, infected and infertile women (who often were infected by their husbands and then rejected) usually have no way to support themselves or their children except by prostitution—thus spreading the STDs.

These diseases and their attendant problems are widespread in Africa, Asia, and Latin America, and increasingly in low-income groups in developed nations.[51] In the poorest areas, medical services generally are weak or nonexistent, and diagnosis and treatment of STDs often has not been a high priority for overburdened community health programs—even when patients recognized the need for help. Illiterate women with no understanding of disease are likely to accept pain and other symptoms as part of woman's sad lot. Problems connected with sex and reproduction, moreover, are not considered something to talk about, except among women friends. These attitudes are particularly common in parts of Africa and the Middle East where female circumcision is still practiced. The resultant mutilation and scarring of the lower reproductive tract not only leads to chronic pain or painful intercourse (thus masking the discomfort of an STD), but also can increase susceptibility to infections, including AIDS.

STDs are intimately connected to both childbearing and birth control. Several contraceptive methods, especially the barrier methods (diaphragm, cervical cap, and condom) and, in some cases, oral contraceptives, can help prevent transmission of disease; IUDs and, in other cases, oral contraceptives may facilitate it. Some infections are not always acquired through sexual activity. Infections can be spread through unclean gynecological examinations, IUD insertions, abortion, or even childbirth, causing pelvic inflammatory disease with its potentially fatal outcome.

For historical reasons, family planning facilities in many developing countries have been set up independently, separate from the health care system. Many programs originated as private, volunteer-run organizations and only came under government support and management well after their infrastructure was established. Outside major cities, family planning services are delivered mostly by nurses or narrowly trained health workers, who may be unqualified or lack authority to deal with other problems. In some countries, initial attempts to graft family planning functions into existing maternal–child health programs failed because health workers were so overburdened dealing with acute medical needs that family planning was treated as an afterthought at best.[52]

Whatever the reasons, family planning workers in many countries have had little training or experience in recognizing or treating STDs, despite the close connections to successful childbearing and family health. They often do offer advice and instruction in basic hygiene, nutrition, and infant care, and sometimes give pre- and postnatal care and help for infertility.

The sad state of health for women and children in many poor countries has spurred many people concerned about development, population, and equity to promote women's health—especially reproductive health—as a central issue. They regard the continuing high rates of pregnancy-related illness and death in the developing world as a serious failure of health services and the development process, with considerable justification. They have also become alarmed by the sudden outbreak and global spread of STDs, which cause so much unnecessary misery and death. And they are critical of family planning organizations for not being involved in solving both sets of problems.[53]

At the 1994 International Conference on Population and Development (ICPD) in Cairo, women's health, well-being, and empowerment were major topics of discussion and were prominently included in the Programme of Action.[54] We think that emphasis was appropriate; no step toward empowering women could be more basic than meeting their health needs. There is, after all, considerable evidence that when they are empowered, by literacy, job opportunities, and freedom from illness, women (and their families) are better off in every way *and* will choose to have fewer children.

Fertility and Equity

An important factor in making progress in economic development is equity, particularly equity between the sexes. Equity also seems to play a potent role in shaping attitudes toward childbearing and family size. In societies where they have low status, women usually have little or no choice about when or how many children they will have, even though they have most of the responsibility for their care and upbringing. Even when women do have a measure of independence, men who have power over them very often use it to command sexual favors, and children frequently result.[55] This power imbalance is doubtless one reason for the "latent demand" for contraceptive services that appears widespread in poor nations.[56]

Even where women appear to have economic opportunities, it is frequently in the family-oriented, "informal enterprise" sector of the economy. In the village of Kerdassa, on the outskirts of Cairo, women do the unskilled tasks as they aid their husbands in producing a wide range of craft items. In Taipei, women tend to work in sales, clerking, accounting, and tool-operating in family enterprises where male members of the family are managers; married women are not salaried because contributing their labor is considered a family duty. In Madras, India, female vendors of fruit, flowers, and vegetables cannot operate without the aid of male family members, who take their earnings. In Thailand, even women who fully control their enterprises of selling sweets and other processed foods generally cannot control (or reinvest) the proceeds of their labor. Instead, family members have prior claims on it—children for education and support, husbands to support gambling or drinking habits, and so on.[57]

Men thus continue to hold power over working women, often through culturally ingrained arrangements within families. They often simply co-opt the fruits of women's labor, using it to free themselves from previous obligations such as support of ex-wives or children. Or they co-opt the proceeds of the women's labor and use it to strengthen their own positions. The women achieve no true economic mobility, do not see the fruits of their labors, and shield their husbands from the true costs of childrearing (since they carry the burden of child care and often improve the men's financial positions as well).[58] In such situations, women may not directly reap the benefits of employment, and that employment may not have the expected result on their fertility.[59]

It is important, of course, to value women's work not just in family enterprises such as farms or Mom and Pop stores, but also at home. One of the many insanities in current economic practice is not assigning value to work performed outside a labor market. This marginalizes not only most work done by women, but that of many men whose unpaid efforts may be primary sources of enrichment in their own lives and the lives of their families and their communities. The popular notion that markets capture all that is of economic value in human existence is surely one of the most pernicious myths of the "modern" age.

Discrimination against women may also lead parents to choose not to pay the full costs of raising a girl, as in Taiwan where older daughters are often not given much education and are married off early.[60] In societies where discrimination is greatest, women's health and the well-being of families suffer, as reflected by infant and child mortalities, and fertility is generally high.[61] In the worst situations, girls may be sold into marriage or prostitution and suffer considerable abuse.[62]

Recognition of the imbalance of power between the sexes is in part responsible for the emphasis put on the empowerment of women by people interested in reducing birth rates. It is surely no coincidence that nations in which women have substantial rights also tend to have lower fertility rates; examples include most Western developed nations, China, and Costa Rica. Two apparent exceptions are Indonesia and Brazil (where wives suspected of infidelity may be murdered with impunity). Fertility has declined substantially in both countries recently. Another is Japan, where women are given good educations and jobs but not independence or equality with men, yet fertility is very low.

Female empowerment can be a complicated business, though. It is now orthodox to assume that educating women and exposing them to Western mass media will give them greater equality and lead to reduced fertility.[63] Exposure to such "modern" ideas seems to lead to changed attitudes toward family sizes and personal goals in many societies.

In New Delhi, our 23-year-old cycle-rickshaw driver, Gyan Singh, told us that he and his 21-year-old wife have one baby daughter and want only two or three children total. He has four brothers and three sisters himself, but feels that children are "too expensive." He is glad that he has moved to the city and definitely would not return to the life of his village, some two hundred miles away, and lead the sort of life his father does running a small spice shop. In Madras, in Tamil Nadu, we went birding with a young ecologist, Dr. Ranjit Daniels, who had only one daughter. He and his wife had decided to stop there. They represented the trend among Indians to have smaller families, but it is a trend that is very uneven from place to place and among different classes.

In some Indian states, TFRs remain high—5.1 in Uttar Pradesh, 4.6 in Madhya Pradesh and Rajasthan, and 4.4 in Bihar.[64] They balance the good news from Kerala (1.8) and Tamil Nadu (2.2), another southern Indian state that has experienced a great decline in fertility. The crude birth rate in Tamil Nadu dropped from 34.9 in 1951 to 20.7 in 1991.[65] Various factors are responsible. Unlike Kerala, Tamil Nadu has been at the forefront of industrialization, with accompanying urbanization and modernization of agriculture. It has relatively good roads and public transportation, and extensive electrification and telecommunications. Its government has put considerable effort into increased literacy and maternal/child health care, including family planning.

In Tamil Nadu, we had a chance to visit both public and private-sector health clinics.[66] In Madras, we were welcomed at the Saidapet Zone Health Post by two female physicians, Dr. Renganayaki, the zone officer, and Dr. Banumathi, head of the hospital (30 to 40 percent of the area's physicians are women). Both had been trained at Madras University, as had most of the medical personnel we met in Tamil Nadu. The two doctors guided us through the bare-bones but plainly functional facility, one of nine clinics in the zone. The staff includes three physicians and an anesthetist who serve 100 to 150 patients a day from within a mile or so of the clinic's location in a lower-class neighborhood.

There, poor women could receive pre- and postnatal services, general maternal/child health care, and family planning education and services. About two hundred deliveries per month were performed, including some caesarean sections.

During our visit, women were being given medical tests (hemoglobin, syphilis) and hygiene instruction, and their weight was monitored. One young woman was in labor, and seven were on cots in the maternity ward with their newborns. The clinic was equipped with an ultrasound machine, and we were privileged to observe the examination of an attractive 21-year-old, Mrs. Gajalakshmi Vendivelu, who had been married eighteen months previously and was now some five months pregnant with her first child. The heartbeat of the healthy fetus could clearly be seen on the monitor as a nurse skillfully did a comprehensive examination. We asked whether Mrs. Vendivelu would be told her child's sex and were immediately told that was strictly against the rules; abortion of female fetuses and female infanticide are residual problems in Tamil Nadu. The local physicians were also well aware of the parallel problems with ultrasound in China.

We also were introduced to the communications officer and cinema operator who were key to the clinic's outreach program, and had a brief but informative interaction with Dr. S. S. Kodimari, the population project coordinator. She told us that the clinic supplied women with the Pill (one cycle at a time), copper-T IUDs, and condoms as temporary fertility controls, and tubal ligation and "no-scalpel" vasectomy (in which the holes snipped in the scrotum are so small as to require no stitches) as permanent ones. The IUDs were roughly five times as popular as the Pill.

The overall program she described was very impressive. Besides showing educational videotapes for new mothers, fathers, and mothers-in-law, they had instituted a system of women "link leaders," selected from the local population. Each leader was chosen on the basis of education and popularity and made responsible for staying in touch with the women in twenty families. They were trained by physicians and other staff of each health post to serve as links between the community and the post. They reported pregnancies, urging women to go to the clinic for prenatal care if they have missed even one period, and served as a general education extension service for the entire panoply of maternal/child care (even though they themselves were sometimes illiterate.) They also were assisting health workers in outreach work, especially with immunizations and video, film, and other programs designed to mobilize the community.[67] All in all, some 200,000 poor women in Tamil Nadu are involved as family planning workers. Dr. Kodimari was emphatic about the importance of the program's inclusion of mothers-in-law, who play a crucial role in determining family sizes.

Our visit to a private city clinic funded by the International Planned Parenthood Federation (IPPF) revealed a very similar operation, fully integrated with those supported by local governments. Once, it had one multipurpose health worker to serve every fifteen thousand people, but about two years earlier the staff had increased so that the coverage was doubled. Perhaps most interesting of all was our visit to the Chengalpattu rural health clinic and the adjacent school some twenty miles south of Madras.[68] The clinic served a population of 6,794 people, with 1,019 eligible couples. The health program offered was similar to that of the city clinics; immunizations, pre- and postnatal care, deliveries, contraceptives, and sterilizations were available. In two years (1991 to 1993), infant mortality in the population served had been reduced from 58 to 22 per 1,000. In a recent month, 38 women had opted for tubal ligation, 30 went on the Pill, 27 had IUDs inserted, and 24 men had begun using condoms—but there had been no vasectomies.

At the school, we were greeted by a class of healthy looking, uniformed, cheerful twelve-year-olds sitting in the sandy schoolyard, intently listening to a lecture. On the blackboard was a trigonometry problem of the sort seen all too rarely in junior high schools in the United States. In a bare, concrete-floored building, several dozen three- to five-year-olds, obviously thriving, were sitting in a circle shouting while one of their number jumped between rectangles painted on the floor. It quickly became apparent to us what they were doing. Each rectangle contained the picture of a flower, and next to it was written the name of the flower in Tamil, the local language. As each child took a turn, he or she shouted the names of the flowers while hopping from rectangle to rectangle, and the rest of the children shouted along. It was an exercise in associating names with printed words, the very beginning of a journey to literacy.

In 1951, fewer than 30 percent of men and 10 percent of women in Tamil Nadu were literate. By 1991, those numbers had risen to 74 percent and 51 percent respectively, and the state in 1993 adopted female literacy goals of 80 percent by 1995, 90 percent by 1998, and 100 percent by 2000.[69] The Chengalpattu operation included special nutritional supplements and vocational training for adolescent girls from poor, malnourished families, comprehensive public education programs on health and nutrition, programs for women to promote literacy and better home management, and training for village women in income-producing activities such as gem-cutting, doll-making, and making bamboo decorative pieces.[70]

Like every other schoolchild in Tamil Nadu, every one receives a hot lunch in this school. The program is the legacy of a Tamil Nadu political leader, Chief Minister Dr. M. G. Ramanchandran.[71] When it was introduced in 1982, his colleagues told him it would be too logistically difficult and expensive. He replied that while they had middle- and upper-class origins and had never been hungry, he came from a poor family and knew what it was like to go to bed longing for food. He declared that while he was in office, he didn't want any child in the state to know that feeling. He made it work. The program was subsequently extended by Chief Minister Ms. J. Jayalalitha, who launched the 15-point Program of Action for the Child in 1993.

The day we visited the Chengalpattu school, the preschool children sat politely in a circle while teachers and aides dished out an ample serving of rice nutritionally enriched with lentils, soya powder, oil, and iron-fortified salt.[72] The children daintily ate it with the fingers of one hand, as dictated by local custom. In mid-afternoon, each child is given a pea-soybean snack to enhance protein nutrition. The clinic carefully monitored the health of the schoolchildren, and supplied nutritional supplements to any children who appeared to be in need.[73]

Tamil Nadu has reduced infant mortality from 110 infant deaths per 1,000 live births in 1976 to 57 in 1991, with a target of lowering it to 30 by 2000. The government's birth rate goals are 15 per 1,000 in 2000 and 10 in 2010 (which would be well below replacement), ending growth of the population (56 million at the 1991 census) at 65 million by the latter date.[74] If the efforts we observed were typical, they just might make it!

The success of this poor state in a poor nation in caring for its women and children puts to shame the inadequate attention paid to the basic needs of people in the United States. There, South Dakota voters recently voted against the taxes needed to run their school system. Nationwide, fat-cat politicians fight to reduce programs for feeding, educating, and providing health care to poor children while battling to increase subsidies to the rich such as funding vastly expensive useless weapons systems for the military.

The Value of Children

An elegant study of fertility decline in Barbados by anthropologist W. Penn Handwerker suggest that education and exposure to the outside world will not necessarily produce a great reduction in fertility and will not increase equity much in a real sense (except perhaps in years of schooling).[75] Rather, unlike the situation in Kerala, structural changes in the economy may be needed so children no longer constitute sources of income, but are converted into dependents whose costs compete for income dollars with consumer durables such as TV sets and automobiles.[76]

In the 1950s, Barbadian women viewed childbearing as an investment. Young children constituted a claim on income from their fathers; later, children provided monetary support that could give their mother independence from men. That support was crucial for avoiding an old age of absolute poverty. Then Barbados underwent great economic changes with the decline of sugar (long the backbone of the economy) and the rise of tourism and manufacturing. This allowed women to find work independent of men and increased their status relative to men. The decline in the financial benefits derived from children and the simultaneous increase in the costs of raising them naturally resulted in lower fertility.

In sub-Saharan Africa and in South Asia, by contrast, children remain sources of benefits. Women and children invest enormous energy in obtaining water and the domestic energy for roughly 90 percent of households. The energy is supplied by fuelwood and, to a lesser extent, cow dung. In the 1980s, up to five or six hours a day were required for fuelwood gathering alone in Niger and Senegal, and eight hours in Tanzania. Women may now be forced to travel regularly as far as ten kilometers from home, returning with loads as great as 35 kilograms (76 pounds).[77]

The need for water and fuel directly influence fertility decisions. Economist Partha Dasgupta has described the complexities of interactions among population growth, poverty, and environmental deterioration.[78] Children are critical "producer goods" for families that have no access to commercial fuels or tapwater. Children are a good investment; they can fetch firewood and water cheaply. There is therefore little incentive for such families to limit their reproduction. As the population grows, fuelwood and water resources are placed under greater pressure, increasing effort is required to obtain ever more remote supplies, and children become even more valuable. But since more effort is required to maintain the same living standard, increasing family size also increases poverty, which in turn puts kerosene or tapwater even further out of financial reach.

A key element in this downward spiral is the loading of high costs upon women. They bear the heaviest costs of reproduction, and they do most of the work to support and care for the family. More than the men, they *need* the children to help lighten the burden of chores. A small family simply may not be viable in such a situation. Thus, in addition to empowering women, sensible "population policies" might include such measures as building water systems for the rural poor and providing them with fuel-efficient stoves and access to kerosene.[79]

Many marginalized African peasants count on large families to provide income from cities. In some francophone areas such as Mauretania and Senegal, sons are expected to migrate to Europe to earn money. They may send an irrigation pump back to their family. Peter Warshall, a development expert with extensive experience in Africa, explained:

Since water is the crucial element in food production and manually lifted well water requires enormous amounts of time for female crop growers, this strategy implicitly requires two sons or more. Assuming an equal sex ratio, a woman would have at least four children. If she expected death from disease or a non-cooperative child, she might prefer six. In this oversimplified story, the need for a pump to increase farm productivity encourages larger family size. The family then becomes dependent on some cash income for fuel and parts, which is supplied from Paris.[80]

Financial inequity between households has long been associated with low fertility among the relatively few rich families, and high fertility among the numerous poor. In part, this disparity was fostered by the desire of the rich to maintain an abundant supply of labor and prevent any upward pressure on wages.[81] This factor was recognized by Malthus; more recently it has motivated wealthy elites in many developing nations to oppose family planning programs—except for themselves.

Inequity between households in the distribution of agricultural land may have a large impact on a population's fertility. Demographer Mead Cain of the Population Council made an interesting comparison of four villages on the Indian subcontinent.[82] One, in Bangladesh, had undergone a sharp decrease in equity of landholding; three in India had seen a substantial improvement in the equity of land distribution. Cain learned that in Bangladesh selling land was a prime mode of adjusting to natural disasters (chiefly floods), and the smaller the holding, the greater the likelihood of selling out. In India, land sales were infrequent and did not increase after droughts or floods. In Bangladesh, male children play a crucial role as "risk insurance"—potential sources of support in case of catastrophe; in the Indian villages, children are "largely redundant as a source of risk insurance."[83] The average family size in the Bangladeshi village at the time of the study was roughly two children more than in the Indian villages.

The most significant inequities sometimes are those between relatively well-off city dwellers and those living in poverty in the countryside, who perceive very different costs and benefits of having children.[84] This, of course, is a major impetus for the extremely rapid growth of cities, which has long been a general feature of developing countries.[85] Migrating people are normally attracted toward wealth and a perceived opportunity for a better life.

The phenomenon has recently arisen in China, where for decades the government kept most of the population down on the farm. The economic liberalization of the 1980s led to privatization of farming communes and unemployment of millions of peasants, who are moving to the cities. Rural-urban inequity is rising rapidly in China, as the most enterprising people make fortunes (usually in the cities) while the poor are left behind.[86] While the one-child family has become the standard in China's relatively prosperous coastal areas, it was introduced later and has been much less successful in remote rural areas.[87] In those areas, medical care (the once highly successful "barefoot doctors" program) and educational systems are deteriorating, both trends that generally tend to raise fertility rates. Furthermore, when agricultural communes were disbanded, children were needed to work in newly privatized farm operations. This led to an increase in births, and with the declining status of women, the traditional preference for sons was reasserted.[88]

Fertility differentials between urban and rural areas are classic; they were features of the now-developed nations for generations (they still persist to some extent in the U.S.) and can be seen today in developing nations. In the early days of family planning programs, the surprise was that the differential was modest or sometimes imperceptible; everyone had big families. The answer to the mystery was that the real differential was between rich and poor, and the cities of developing countries were being overwhelmed by poor migrants from the countryside who brought their rural family-size norms and ideals with them.[89]

By the 1990s, the rush to the cities had slackened a little in many developing nations, as birth rates had fallen and development (including family planning facilities) reached the countryside. An urban attitude toward family size seems to spread among the migrants after they have been in the city for a time, restoring the urban-rural fertility differential. The scale of the differential largely depends on the comparative family systems of rural and city dwellers: how they deal with inheritance and the establishment of new households, and how married couples respond to changes in the economic value of children.[90]

Equity between nations also may have effects on reproductive aspirations, although other differences between rich and poor countries go far to explain the differences in fertility. As the case of Barbados suggests, the achievement of greater equity between nations is normally associated with a reduction in the fertility of the nation "catching up."[91] Between 1955 and 1965, Barbados underwent an economic transformation, shifting from being primarily a commodity supplier (sugar) to a more diversified economy rooted in manufacturing and tourism. That greatly improved the relative economic position of Barbados and allowed an extremely rapid fertility transition. In 1994, Barbados had raised its per-capita gross national product to $6,530 (more than double that of Russia or Hungary) and lowered its total fertility rate to 1.8 (equal to the average of northern Europe).

Cultural Differences

Not all reproductive choices can be explained on the basis of literacy, health, equity, or the costs and benefits of children, however. The influence of other cultural factors can also be important. This was made clear by an intensive study carried out by demographer Alaka Basu of the Institute of Economic Growth, Delhi, India, and her collaborators.[92] Basu studied two groups of immigrants living in a multicultural resettlement colony—"a slightly glorified slum"—in East Delhi. One group was from Tamil Nadu and the other from Uttar Pradesh in northern India.

In the slum, both groups were exposed to essentially the same economic conditions, so economic differences influencing family size were minimized. The colony dwellers had been allotted 25-square-yard parcels of land (about eighteen thousand in all) on which to erect shelters. Most had made one-room, windowless shelters on their small plots. Every block of five hundred plots was provided with a row of public toilets and one or two water taps in each narrow lane between huts. Much community activity took place in the lanes, where women gathered to gossip and do chores during the day and men chatted at night after work. Use of the often filthy toilets was far from universal, especially among children and adolescent girls (for cultural and security reasons), and many people simply squatted outside them or outside their homes. Sewage and garbage were essentially everywhere outside, although many homes were kept spotless inside. (Such a contrast between social and household hygiene is often a feature of slums).

Despite the common poverty of their living conditions, economic differences between the two groups persisted. Overall, the northern Indians from Uttar Pradesh were more prosperous than the southerners from Tamil Nadu. Their household incomes were about 10 percent higher and their incomes per person almost 50 percent higher.[93] Heads of household from the north more often had regular salaried employment than those from the south. They owed less and had divided their homes into more rooms; they were more likely to have solid cement homes and electricity and to have installed a hand pump to provide a private water supply.

There were numerous cultural differences as well in household size and age composition, the tendency of all family members to eat together, the sending of money to relatives back home, and the propensity to smoke and drink.

Against this background, the most interesting of Basu's findings was that, despite the relative prosperity of people who came from Uttar Pradesh, they had higher fertility (and higher infant mortality) than those from Tamil Nadu. These results mirrored those found in census data from the two regions themselves: the northerners had more children than the southerners, and those children were more likely to die in infancy. Moreover, the Tamil families did not display the higher mortality among female children characteristic of those from Uttar Pradesh—indeed, of southern Asia in general. These differences were consistent regardless of whether the people were living in their homelands or sharing the relatively uniform economic conditions of the Delhi slum.

This seems powerful evidence that cultural factors do influence family size choices. Basu defined culture as "a set of beliefs and practices common to a group defined by characteristics such as region or language, other than standard economic and social variables."[94] In other words, cultural differences are consistent differences in beliefs and practices that exist between groups of different regional or ethnic origin, even when their socioeconomic status is similar.

What cultural differences led the generally poorer families in Tamil Nadu to have fewer children? The immediate answer is that Tamil women stop childbearing sooner, mostly through voluntary sterilization (utilized by about half of Tamil couples ages 30 to 49, but only by a third of couples from Uttar Pradesh) but also through greater use of contraception. The discrepancy in practices appeared to lie in differences in the women's status.

The Basu team found three relevant aspects of that status. The first was the extent of women's exposure to the world outside the home; second was the amount of interaction with that world; and third was the degree to which women were able to make independent decisions. Female seclusion was less prevalent among Tamils, so the women were more exposed to India's aggressive family planning campaign and other modern ideas and thus more able to override other people's traditional pronatalist views. In addition, Tamil women were more often able to seek salaried employment than women from Uttar Pradesh. To a working woman, the appearance of another child every two years would seem less of a blessing and less as a needed source of old-age security.

Interestingly, Basu found no difference in satisfaction between the two groups of women. Furthermore, while increased education of women was found to be highly correlated with reduced fertility, as indicated by many other studies, Basu found the cultural differences in family size to be consistent even among uneducated women. Overall, cultural heritage played a powerful role in shaping family size decisions.

Cultural influences undoubtedly have effects on people's decisions about marriage and family in all societies. Anecdotes indicate that even Kenyan airline stewardesses, about as cosmopolitan in their experience as any human beings, still plan to return to their villages and have numerous children in order to fulfill their role as women. Little progress was made in fertility limitation among rural populations in central Asia under the Soviets, despite more than a half century of social change, including universal primary education, lowered mortality, and massive efforts at indoctrination.[95] Anyone who doubts the powerful role that cultural norms play in human life need only consider their own food preferences. For instance, the thought of eating a perfectly nutritious cockroach stew—a delicacy in some Southeast Asian cultures—turns our stomachs.[96]

Family Planning Programs

Whatever the root causes of decisions about family size, and regardless of the possible modernizing influence on attitudes that may be traced to family planning programs, there is little question that access to modern contraception and safe abortion can humanely help families stay within their goals. Fertility control is an ancient human activity, but the Pill, IUDs, condoms, sterilization, and the like are clearly vast improvements over infanticide, abortion, and folk contraceptives. When people do decide to limit their fertility, modern methods make it a lot easier and safer to do so.

Recognition that access to safe and dependable means of contraception was important led to the establishment of family planning programs in nearly every less-developed nation long before the 1990s.[97] In the early years, a key element in these programs was to help women in postponing pregnancies, since adequate spacing between births is critical to the health of both mothers and their children. The majority of programs today, however, are explicitly designed to help couples limit the number of children they bear. Family planning programs also often are supported and reinforced by other social policies such as maternal and child health programs, social security arrangements, and regulation of age at marriage, as well as by provision of education and work opportunities for women outside the home.

Diverse approaches have been tried. Some countries have reinforced the small-family message through housing assignments and tax policies that penalize overreproducers. India has had vasectomy carnivals. Indonesian youths who promise not to marry before a certain age and to limit their family sizes are given a special pin in recognition. In Thailand, there have been condom blowing-up contests in grade schools.[98] If only children in developed nations had similar exercises in, say, dissecting junked luxury cars and trying to figure out how much energy and resources went into their manufacture and use!

In Tamil Nadu, the array of tactics now being used in promoting maternal and child health care and small families is truly impressive, as are the results achieved. They give hope that India might be able to get its population control programs—derailed by overzealous efforts in the 1970s—back on track. The ideal of a family-size *maximum* of two children (illustrated by a logo showing the heads of father, mother, and one offspring), along with family health messages, is displayed on billboards, the backs of buses, police traffic umbrellas, road railings, posters, banners, leaflets, and a variety of electronic media channels. Other goals such as a minimum marriage age of 21 for women and 25 for men, a minimum age for bearing a first child of 22 (and ideal age of 24), a target average birth weight of 6.6 pounds, a minimum weight gain with pregnancy of 22 pounds, and a "one family, one child concept" are also widely publicized.[99] Some of the devices used include loudspeaker trucks, quiz programs for slum mothers with prizes for absorbing messages, seminars for mothers and mothers-in-law, healthy baby shows, health rallies organized around such themes as hygiene, sanitation, and women's literacy, and the giving of booklets to fifteen- and sixteen-year-old girls on the value of literacy and of delaying marriage.

Exhibitions are held everywhere, often involving service clubs such as the Rotary and Lions as well as youth clubs and women's clubs. Street plays and magic shows are part of the mix that promotes comprehensive health messages, as are police band music programs in slums, puppet shows, skit and oratorical competitions for college and school students, and the working of the messages into Kolu, a kind of doll exhibition conducted as a religious function by Hindus during the Dasara festival season in October. There are celebrations on occasions such as International Women's Day, Breast-Feeding Day (advertisement of baby formulas is illegal), World Population Day, World Environment Day, Nutrition Week, AIDS Awareness Week, and so on.

Men are not neglected. One ingenious device is to organize

hairdressing salons (more frequented by men than women in Tamil Nadu) as centers of education and contraceptive delivery. The salons are social centers, and often display large health posters. Hairdressers are recruited and trained in workshops to pass the message, and large open jars of condoms are located in the salons so that men can pick them up free and without embarassment—a striking contrast to the situation in some African countries where a condom can cost a day's wages.[100] Workshops on contraception and maternal and child health are also held for practitioners of traditional Indian medical systems. Knowledge of the program is ubiquitous, and so, judging from the reports (and our limited personal experience), is the goal of having just one or two children.

How easily (and far) the successes in Tamil Nadu (or Thailand or Indonesia) will be transportable is still far from understood. Basu's findings suggest that it may be more difficult than might appear at first glance. So do the results of recent sample surveys indicating that in the northern states of Uttar and Madhya Pradesh, Rajasthan, and Bihar, India has its work cut out.[101] There, female literacy rates, although climbing, are still less than half of those in Tamil Nadu, and a third of those in Kerala. Infant mortality rates, although dropping, are some 50 percent higher than Tamil Nadu and four to six times those of Kerala.[102]

In the north, women traditionally marry men from other villages and clans, and thus go as strangers to their husbands' homes. They end up at the bottom of the pecking order, are expected to defer to their in-laws, and traditionally live in seclusion within the home, hiding their faces from strangers and even from older male relatives of their husbands. Economic security and status come from raising sons.[103] This is in striking contrast to Kerala, where inheritance is largely through the female line, girls were traditionally educated alongside boys (even in military arts), and women enjoyed high status as poets, artists, and philosophers.[104]

One important lesson from India's experience thus far is that family planning programs run on a "target" basis don't work very well. In Tamil Nadu, there was great success in filling vasectomy targets, but birth rates did not come down. Men were getting vasectomized after fathering large families, simply to pick up the incentive payments.[105] It has long been recognized in India that there was too much tendency to recruit new contraceptive acceptors at the expense of better follow-up and the creation of comprehensive health-care services in which family planning programs are embedded. The comprehensive family health approach taken in Tamil Nadu, however, has been successful. There, a small payment is still given to sterilization acceptors, but it is given as compensation for lost time on the job rather than as an incentive.[106]

It is important to recognize that there is evidence that strong family planning programs alone can reduce fertility, even if other social conditions are not entirely favorable.[107] Rather than the empowerment of women, for instance, the great success of the family planning program in Indonesia is ascribed to the energy and organization invested in its operations. A hierarchy of volunteers, virtually all women, has been deployed in every village and subvillage to plan, implement, and evaluate the program's activities. Overall management is provided through a coordination meeting every month in every village, led by the village head (who is male).[108] In Indonesia, as in China (described in the next chapter), an authoritarian approach has brought results. In Kenya, analysis indicates that an important factor in the recent decline in fertility was the accessibility of family planning services.[109]

In Matlab, Bangladesh, a project was launched in 1977 specifically to test whether, in a rural traditional society, simply providing contraceptive services could bring down fertility rates significantly.[110] The Matlab region, just south of Dhaka, is typical of the huge delta region of the Meghna and Padma rivers. Like much in Bengal, Matlab had suffered extensively from environmental deterioration, economic decline, and political turmoil, and it has remained poor, traditional, and religiously conservative. The catastrophic Bangladesh famine of 1974–1975 had an especially severe impact on child mortality in the Matlab region. In short, the modernization often thought to be a necessary prerequisite of a decline in fertility was virtually absent. Children in the region traditionally are highly valued as insurance against social and environmental catastrophe; sons especially are viewed a direct source of income, prestige, and connections to patronage.

In this environment, a more intensive program was introduced to augment the Bangladeshi government's standard health and family planning program. All married women of childbearing age were visited every two weeks, asked about their contraceptive needs, and urged to adopt contraceptives by young married women from their own villages who were trained and employed by the project. The visits, although they included educational themes and motivational messages, were primarily aimed at delivery of contraceptive services. Women in their own homes were given a choice of condoms, the Pill, foam tablets, or an injectable contraceptive. Those who wished IUDs could go to a nearby clinic or arrange for an insertion at home. Sterilizations were provided in a clinic in the small rural town, Matlab Bazaar, from which the region derives its name. Other health services were added to the program gradually.

Contraceptive use in the district tripled almost instantly, and between 1977 and 1985, climbed from 7 percent of married women of reproductive age to 45 percent. The TFR dropped substantially in comparison with areas under the less intensive government program. The program clearly demonstrated the existence of a considerable latent demand for contraception among the women in the Matlab region, although in much of the population it was probably "weak and fragile."[111] In that group it appeared that an intensive contraceptive delivery program could directly help to bring birth rates down—partly by introducing and reinforcing the idea that reproduction is subject to personal control, partly by countering pronatalist pressures from the extended family, and partly by allaying concerns that using contraceptives carried social and health risks. The Matlab project showed that the "supply side" of the solution to reducing fertility can be as important as the "demand side."

Where To Now?

In the beginning, most family planning programs in less developed nations were supported mainly through funds donated by Western industrialized nations, either through private agencies such as the International Planned Parenthood Federation (IPPF), multilateral international programs such as the United Nations Population

Fund (UNFPA), or as "bilateral" aid, a direct grant from one nation to another. As a component of overseas aid to less developed nations, family planning assistance was always a very small fraction—only a few percent at most. But after 1980, contributions from rich nations (primarily the United States) diminished, so the developing nations gradually assumed more of the financial responsibility. By the 1990s, they were paying more than three fourths of the costs for their family planning programs.

Perhaps partly because of the rising unmet worldwide demand for birth control services during the 1980s, and because of upward shifts in fertility in several large nations (including China and the United States) as well, United Nations demographers revised their projections of population growth trends upward in 1992.[112] The UN's 1992 high projection, remember, indicated no end to growth, with the population passing 28 billion in 2150 and continuing to climb—a projection no ecologist worth his or her salt would grant any credibility. No allowance is made, of course, for humanity's increasing vulnerability to a rise in mortality through the consequences of increasing population, poverty, and environmental degradation.[113]

For our purposes, the UN's low projection is far more interesting than the high one. It shows the world population reaching a peak size of just under 8 billion around 2050 and thereafter slowly declining to below 5 billion by 2150. This projection was built on an assumption that below-replacement fertility could be reached on a global basis within a few decades. While some observers might consider this as unrealistic as the high projection, at least it is not built on unsupportable assumptions about the planet's carrying capacity. There is certainly nothing demographically impossible about it.

Clearly, if much more effort were put into reducing birth rates, the peak world population might be held well below the 10 to 12 billion that is now considered the most likely outcome. The allocation of that effort presents a considerable problem, though, considering the complexity of factors that operate on fertility. Putting aside for the moment the acute need for sensible demographic policies in developed countries, two things need to be done in less developed countries, where most of the numerical growth is due to occur. One is to change social and economic conditions so that both men and women desire fewer children. The other is to make the means of family limitation universally available to couples in those nations.

Changing family-size preferences involves different things in different societies, but achieving greater equity at all levels (from the family to the nation) is surely high on the list. Increasing the freedom and educational level of women and creating job opportunities for them is clearly crucial, not just for curbing population growth but to meet the goals of development in general. Educating men about their reproductive responsibilities and their society's need for smaller families should have high priority as well, since they play such an important role in the social milieus and family structures that strongly influence fertility. Improving employment opportunities for men obviously is also critical, since it is difficult to see how many jobs can be created for women in the face of armies of unemployed men. And providing both men and women with some form of old-age security should also be helpful in reducing births and making societies sustainable.

Making contraceptive and abortion services available in a context of better overall health services, especially for women, is theoretically easier, but in many societies the provision of these services, especially abortion, is still controversial. Obviously, we believe it should be a basic human right for *all* sexually active people in *all* societies to have access to such services. Extending birth control services to everyone is a huge social and logistic challenge. Yet it can be done, as many developing countries have successfully demonstrated. Some others have been dismally unsuccessful, however.

Pakistan and Bangladesh, once united as a single nation, are cases in point. Both are principally Moslem societies, and both have had family planning programs since the 1970s.[114] But Pakistan began by restricting its program to use of IUDs, with disastrous results. Since that failure, the government has put little effort into the program; only 12 percent of couples were using birth control in the early 1990s, and the average family size was six. In

Bangladesh, building on the Matlab experiment, thousands of village women were recruited to visit other women at home and distribute a range of contraceptive choices. The acceptance rate rose to 40 percent, and family sizes have fallen to around 4.5.

Even though more than half the world's married women were using contraceptives by the 1990s, birth control services are far from universally available. The unmet need—women lacking family planning services but wishing to avoid pregnancy—is estimated to be some 120 million.[115] The World Health Organization maintains that half of all pregnancies are unplanned and a quarter are "certainly unwanted."[116] Moreover, the number of people in their reproductive years (ages 15 to 49) is expected to rise by 23 percent during the 1990s, and rates of birth control acceptance are rising as well.[117] Given the expected increases in eligible users, the number of modern contraceptive users in developing nations (roughly 33 percent of couples of reproductive age in 1990, excluding China, where contraceptive use is nearly universal) would have to double during the 1990s to achieve a moderate increase in the rate of usage by 2000.[118] This would be an important first step, but no more than that.

In an effort to meet the rising need, delegates to the International Conference on Population and Development in Cairo in 1994 recommended that total annual funding for family planning be increased to $17 billion by the end of the decade, about two thirds of it to be supplied by developing countries themselves.[119] About a third of the funds are to be allocated to improving reproductive health programs, the rest to family planning. Even the roughly fourfold expansion of family planning assistance this would entail amounts to only about 11 percent of the total official development assistance provided by rich countries in 1991 and barely 0.036 percent of their combined GNPs.[120] Can the world afford to do any less?

We think not. Indeed, we think it should do much more. Those who are in favor of strong population reduction policies for the poor nations should also press for such policies in rich nations, where each person does so much more damage. Or perhaps the rich should institute strong consumption reduction policies. After all, both population reduction and consumption reduction will be automatically imposed on us by nature if humanity keeps on assaulting its life-support systems as it does today.

Intellectually isolating environmental damage from one of its root causes—overpopulation—by having separate United Nations conferences in Rio and Cairo was a serious mistake. That error led to the biggest failure of the International Conference on Population and Development: neglect of the steps rich nations should take to deal with their overpopulation and overconsumption. Unfortunately, as a result, the crazed notion that the sorry state of the world can legitimately be blamed on the poor was left intact.

Chapter Four

GOVERNMENT IN THE BEDROOM

In both poor and rich countries, all sorts of government policies influence fertility rates, although most do so indirectly. Tax policies, welfare policies, health care policies, immigration policies, and foreign aid policies all have an impact on family sizes. In the United States, retrograde population policies on the part of the Reagan and Bush administrations, especially in limiting family planning aid, greatly hindered progress toward a sustainable world, and greatly harmed women in the process. The Clinton administration has returned to the ethical course, but the human future may already have been badly compromised.

Various groups put pressure on governments to promote population growth. The most important and persistent has been the Vatican, going against its own best interests and the wishes of many Catholics. Catholic nations in Europe have some of the lowest birth rates in the world, and Catholic women in the United States have smaller families than Protestant or Jewish women and use abortion more. Nevertheless, Pope John Paul II has led a crusade against women's rights and the future of humanity. The good news is that Catholics themselves are not only showing fertility trends similar to adherents of other religions, they are also increasingly showing signs of being fed up with the Vatican's behavior.

The most successful government policy on population in a large poor country is that of China, which may soon have the most powerful economy in the world. Although China's one-child family policy has been highly criticized as draconian, it was designed and implemented to prevent mil-

lions of deaths from hunger, and serious steps have been taken to discourage abuses.

The world's rich nations need to follow China's example in adopting explicit policies to encourage population shrinkage and implementing them in a manner tailored to their varied cultural contexts. Simultaneously, rich societies must move to reduce their wasteful consumption and lessen the environmental impacts of their technologies. They should also do all they can to meet requests from poor countries for assistance in family planning and sustainable development. Such aid must be given with great care to minimize waste and unintended side-effects. Done properly, it is one of the best investments the rich could make in their own futures.

When Ronald Reagan took office as president of the United States in 1981, population suddenly became a taboo subject, and long-established U.S. policies of giving family planning assistance to developing nations changed dramatically. A decade later, the public learned why: Reagan had entered into a secret agreement with the Vatican to do whatever he could to eliminate U.S. funding of international family planning programs and especially to eliminate any funding connected with abortions—even where abortions were completely legal.[1]

The legitimacy of entering such an agreement is highly questionable. The ostensible reason for the agreement, to enlist Pope John Paul II's help in giving capitalism a foothold in his native Poland, seems less than compelling; the pope probably would have cooperated without exacting that high price. More to the point, it was morally outrageous to deny the rights of tens of millions of couples to control their own reproduction, in the process adding to an epidemic of women being killed and maimed by illegal abortions. We, like many others, consider abortion basically a last-ditch backup for failed contraception and can sympathize with other peoples' abhorrence of it. But for celibate old men first to attempt to deny people access to contraceptives and then try to force young women to carry unwanted babies to term is, in our eyes, even more abhorrent. Furthermore, the growing threat to the survival and well-being of future generations posed by unchecked population growth was completely ignored by both Ronald Reagan and the pope.

In 1984 at the U.N. Population Conference in Mexico City, to the mixed amazement and amusement of other delegates, Reagan's representative declared that "population is not important" and announced the U.S. government's intention to cut back further on funding for family planning assistance.[2] More than one observer has commented on the irony of the Reagan position; that the "right" economic policies would take care of any population problems was an eerie echo of the official Chinese position ten years earlier. The only difference was the economic system being promoted.[3]

Yet while Reagan's position on these matters was clear, George Bush's continuation of the policy, contrary to his actions and public statements before becoming Reagan's vice president, was puzzling.[4] One might well ask what *he* owed the pope? During his presidency, Congress several times passed new family planning assistance bills to restore funding, only to have them vetoed. Although other developed nations, especially Japan, Norway, Sweden, and Germany, picked up some of the slack in the international programs created by the U.S. default, there is no doubt that momentum was lost. In constant dollars, family planning assistance donations to international agencies fell by about 25 percent between 1985 and 1988 as the Reagan policies took hold.[5] Worse, the loss of moral support for family planning by the U.S. government seems to have been keenly felt. Fortunately, financial support of family planning programs by less developed nations' own governments and private organizations was rising substantially. By the 1990s, about 85 percent of funding was from domestic sources, and annual foreign contributions were a little under a billion dollars—less than 2 percent of all foreign assistance to developing countries.[6]

Even as donated funds for family planning were drying up in the 1980s, demand for services in developing nations was rising rapidly, partly because of increasing acceptance of birth control. And the population explosion itself was producing a boom in the numbers of eligible parents as the children born during the peak population growth years of the 1960s reached their prime reproductive ages.

Consequently, progress in slowing population growth essentially stopped during the 1980s. The world growth rate remained static at about 1.7 percent per year, and fertility changed little in most regions. It remained very low in most of Europe and the Soviet Union, crept slightly upward in the United States and China, stalled in India, and remained very high in most of South and West Asia and in Africa.[7] Only in a few Latin American and Southeast Asian nations did fertility fall significantly.

Since President Bush apparently felt obliged to honor his predecessor's agreement, reversal of this invidious policy had to await the election of President Clinton in 1992, who quickly set about changing it.[8] The Vatican soon made it clear, however, that the world would have to wait longer for it to adopt a humane policy on population.[9]

Most Americans were unaware of Reagan's policy, which had abruptly shifted the nation from a position of world leadership in providing family planning assistance to developing nations to one of callous indifference and stinginess. But people were acutely aware of the Reagan–Bush position and domestic policies on abortion, which had been a subject of increasingly impassioned debate in the United States.[10] Yet that debate has taken place only in a context of conflicting rights with no connection to broader population issues, in which abortion is important mainly as a backup for contraceptive failure and as a factor in women's health care. Abortion can have significant effects on an individual's life, particularly if the need for one arises in a society where it is illegal and dangerous.[11]

The population policies of the United States today, however—both domestic and toward other nations—will have a more profound influence on the conditions of life for our descendants than virtually any other decisions this generation might make, short of all-out nuclear war. The Reagan–Bush efforts to shove the issue under the rug in essence set back world progress in dealing with it forthrightly for twelve critical years.

People close to President Clinton openly recognize the seriousness of the population element in the human predicament.[12] U.S. leadership was firmly reasserted by Under-Secretary of State Timothy Wirth in preparing and coordinating the nation's participation in the 1994 International Conference on Population and

Development in Cairo. The U.S. delegation was headed by Vice President Al Gore, which signaled the importance attached to the issue, since the conference was generally attended "at the ministerial level" (not by heads of state).

Governmental Influences on Reproduction

Formulating population policy, like economic policy, clearly is an important and legitimate function of governments. *Population control* is a term we have used for years to signify a society's efforts to control the size of its population through various policies that influence individuals' choices about their reproduction—just as governments attempt to exert "control" over consumer spending patterns and financial investments through various indirect policies. Unfortunately, many people seem to have confused the idea of population control with "thought control" as described in George Orwell's *1984*. In the United States especially, people react rather strongly to any hint of government interference in reproductive behavior decisions, forgetting that in reality many existing laws, policies, and social norms either implicitly or explicitly influence that behavior already.

Governments, in fact, exert quite direct control over one important component of demographic change: international migration. But they mostly regulate migration *into* their countries; many don't even keep records on those who leave. Although large numbers of migrants (in either direction) can have significant effects on population growth (either increasing or decreasing it), immigration policies are seldom drafted as part of an overall population policy. The United States, traditionally a receiver of immigrants, was accepting nearly 900,000 legal immigrants per year in the early 1990s, accounting for more than a quarter of its annual population growth. Neighboring Mexico, the source of many immigrants, for years was not interested in joint action to curb illegal migration, because the departure of Mexican migrants helped relieve population pressure there.[14]

Governments also regulate the number of spouses people may have (at least at one time), minimum age at marriage, and the availability of birth control methods, all of which have fairly direct effects on fertility. Laws against sexual assault, statutory rape, incest, extramarital sexual behavior, and even certain techniques of lovemaking are common regulations of reproductive behavior that may have some effect on birth rates, in part depending on enforcement. People who argue against national population policies because it would bring the government into the bedroom conveniently ignore that it has always been there.

The legality of abortion clearly can have an impact on fertility, though not as much as antiabortion spokespeople appear to believe. The estimated annual number of abortions in the United States before the *Roe v. Wade* Supreme Court decision in 1973 was not much different from the numbers of legal abortions in subsequent years; the principal change was in the number of maternal deaths, which dropped dramatically.[15] Availability of abortion may be more important than legality; obviously, before legalization in the United States, abortion was available, although safe abortion often necessitated a trip to another country. Antiabortion activists are clearly aware of the difference; hence their focus on tightening restrictions at state level and harassing providers to discourage their operations—tactics that have been considerably more successful than their attempt to have *Roe v. Wade* overturned.

Policies that encourage or discourage childbearing outside marriage, including welfare policies that provide support for poor unmarried women with children, also may significantly influence fertility. In the 1994 debate on welfare reform in the United States, some commentators claimed that generous welfare benefits had inadvertently encouraged childbearing by single women, many of them teenagers.[16] In 1993, roughly one in four children in the U.S. lived with an unmarried mother. Among minorities, the proportions were even higher: 38 percent of Hispanic children and 57 percent of black children had unwed mothers.[17]

Not only has the social stigma of illegitimacy all but disappeared, the financial strain for a single young parent of raising and supporting a child has been considerably eased. In the last decade or so, there has been a surge in teenage pregnancies and births to single mothers, leading to extended welfare dependency in many cases, as well as associations with rising crime rates and other social ills.[18] Changing welfare policies to discourage unmarried teenage childbearing without penalizing the children will not be easy.[19] Welfare policies do, however, represent another clear, if indirect, presence of government in the bedroom.

Of course, the establishment of family planning facilities, whether private and merely tolerated by a government, or government-sponsored and supported, represent quite direct population policies, even in the absence of population size goals. Indeed, a lack of available birth control information and materials, or provision of only a limited set of choices, is distinctly pronatalist in effect. That is so even though couples have always found ways to avoid unwanted children, if not always efficiently. This was true in the United States during the first half of this century when selling or prescribing contraceptives was still illegal in most states. It is still true in many countries where birth control materials are poorly supplied.[20]

Many government policies are implicitly pronatalist, if not always very effective. Income tax laws exert some influence on reproductive decisions through deductions for unlimited numbers of dependent children. If nothing else they send a message of approval for large families. In early 1995, the newly elected Republican majority in Congress were considering changes in U.S. tax laws that would be essentially pronatalist. Some developed nations, notably Canada, France, and Germany, have family allowances that provide a small annual stipend for each child; but since these nations also have very low birth rates, the allowances are not credited with having much effect.[21] Even the availability of health insurance, especially for delivery and pediatric costs, can play a part in reproductive decisions.

Less directly still, government programs frequently support customs and social expectations that young people complete lengthy educations before marrying and raising children. Such practices have operated against large families (as does postponement of marriage and childbearing for any reason). So has the increasing trend in most developed nations for women to seek careers and participate in the labor force during their childbearing years, a trend encouraged by laws against sex discrimination. For working women, the presence or absence of satisfactory alternative child-care arrangements (which are often encouraged or discouraged and regulated by governments) also may influence reproductive decisions. And tax-supported or government-mandated retirement benefits relieve the pressure to have children for old-age support.

While the social trends toward extended educations, later marriages, and women joining the workforce developed in industrialized nations for reasons other than reducing birth rates, they nonetheless have had such effects and have sometimes been adopted as policies in developing nations partly for that purpose. Indeed, many poor nations, struggling to improve their living conditions in the face of rapid population growth, have employed various social policies in an effort to lower birth rates, ranging from those with broad social benefits to relatively draconian measures (at least by Western standards).

Americans and other citizens of developed countries are, of course, already living with the consequences of past policies and social practices that produced the population of today. Few people stop to think about how population size and density affect their daily lives and how society is organized. Some benefits accrue to larger populations, at least at local levels: larger urban areas offer a greater variety of products and services, more employment options, more opportunities for cultural activities (museums, concerts, theaters, etc.), and even simply more people among whom to choose one's friends.

People who have watched their communities become overcrowded, traffic-clogged, crime-ridden, and otherwise degraded, on the other hand, are quite aware of the downside of population growth, especially when the changes have taken place quickly as people flocked in from other parts of the country. Indeed, the social problems that are sometimes associated with population density may arise more from population mobility and the lack of connection and commitment of newcomers to the local community, many of whom also may only be temporary residents.[22]

Americans, who so cherish personal liberties, seldom recall that those liberties must be restricted when they impinge on someone else's freedom, and that the more people there are, the likelier it is that other people will indeed be affected by an exercise of freedom.[23] Libertarians, who ardently espouse maximum personal freedom and minimum government, mostly do not seem to have figured out that a level of government regulation appropriate for a nation of 3 or 4 million farmers and merchants simply won't work for a nation of more than 260 million mostly urbanized, industrialized, and ethnically diverse citizens.

More and more often today, Americans see clashes between various interests over the use of land and other resources. Consider how citizens who need and desire greenbelts around urban areas conflict with landowners who always assumed they had a right to sell their property for inflated sums to developers. Consider the families and neighborhoods that have had to move out or accept disruption so traffic congestion could be eased (usually temporarily) by the building of a new freeway. Consider the fury of people whose idea of sport is to drive recreational vehicles roughshod across the landscape on public lands when opposed by others who value the ecological integrity of those lands. Recall the animosity precipitated by a six-year drought in California between farmers who had first claim on water supplies (subsidized by taxpayers) and urban dwellers who were reduced to 90-second showers, unwashed cars, dead lawns, and unflushed toilets to conserve their small portion of water. In several western states such as Arizona, Colorado, and Nevada, farmers have been forced out of business because scarce water supplies were bought up by expanding cities, which could afford to pay more.

If the U.S. population were only 130 million now, would these conflicts have arisen? Possibly, but they probably would neither have come so soon nor have been so acute.

On a much more subtle level, it has long been observed that behavior in societies with high population density, such as some European nations or Japan, is often characterized by elaborate politeness and rather rigid codes of conduct toward people in different social classes or toward strangers. This appears to be a sort of social defense mechanism to protect oneself from being overwhelmed by the stress of interaction with too many individuals. It is even visible in the United States. People in small towns often greet one another on the street, even though they may not know each other personally. But in large cities, people greet only individuals they know. Over time, as a town grows, or as increased mobility or tourism brings in many new people, a town may lose its "friendliness," and residents often lament the loss.

Considering the vast cultural differences and variation in individual preferences, it is evident that an optimum human population size would be one in which the available options would somehow be maximized. There should be enough people to have large, exciting cities, but few enough to give backpackers extensive wilderness to enjoy, and others simply a chance to live in peaceful isolation. People need sanctuaries too! Obviously, there is no single optimum, but we and others have estimated that a global population of 1.5 to 2 billion people might be sustainable, enabling all to attain a higher quality of life than that of the average American or European today with a less destructive impact on life-support systems.[24] Even if that is a gross underestimate—if twice that number turned out to be an optimum—population shrinkage will be required to get there.

Population Policies: Laissez-Faire to Coercion

In view of all the indirect policies that can profoundly (if subtly) influence citizens' reproductive decisions, one cannot honestly say that governments have no place in the bedroom or that citizens have completely free choices—or perhaps even that they should. So the question then becomes, how much and what kinds of control are appropriate for a government to exercise?

Answering the first of those questions requires answering others: How important or necessary is it for a particular society to restrict its population size? What would be gained or lost if the population's size were not regulated in any way? To our minds, a society that is on the margin, whose resources are stretched almost to the limit and that has little scope for supporting more people without sacrificing the well-being of much of the present population or of future generations, has little choice but to exercise stringent population control. Failure to do so amounts to a policy of blind and insensitive promotion of human misery. And to our minds also, and those of many other scientists, the entire world civilization is quickly approaching that point, if it is not already there.[25]

It surely would be best if all societies recognized the dangers ahead and implemented policies that promote appropriate reproductive behavior well before they needed to resort to coercion. A full spectrum of possible policies exists between complete laissez-faire on childbearing (assuming that were feasible, given society's interests in family stability and an educated populace) and draconian enforcement toward a specific demographic goal. In different places, virtually the entire spectrum has been employed during the past few decades, although most societies have leaned more toward laissez-faire. The longer action is postponed while a population continues to grow, however, the more options at that end of the spectrum are foreclosed, and the likelier it becomes that coercive policies will be needed.

The Remarkable Case of China

Imagine that the United States had four and a half times as many people, fewer natural resources (especially petroleum deposits), much less industrial infrastructure, half as much land suitable for cultivation, and much of the land seriously degraded and deforested. This describes the present situation of China, whose leaders have been acutely aware of population pressures and resource constraints for decades. As Deputy Minister Jiang Zhenghua of the State Family Planning Commission explained to us in 1994, between 700 million and 1 billion people was considered to be the optimum population size for China, and with 1.2 billion people, the nation has already exceeded that.[26]

The awareness of resource constraints and a traumatic experience of mass starvation in the late 1950s led originally to the establishment of the Chinese population policies, which most Americans consider coercive and draconian. In the late 1960s, when the program was initiated, the population size was about 725 million and average family size was around six. Since then, the TFR has fallen to 2.0, but the population has grown to 1.2 billion in 1995. And, despite the evident success of the "birth planning" program, demographic momentum will add at least the equivalent of another United States to China's population before growth stops.

Americans have heard much about China's one-child family program, introduced in 1979. But most Americans know little about the events leading up to it. For many years before the birth planning program was even started, China's medical community conducted extensive research into the factors associated with high mortality, especially of mothers and infants.[27] Out of the findings of this research came some of the seemingly arbitrary rules for marriage and childbearing that were imposed around 1970 as part of the program: minimum legal ages for marriage for men and women of 25 and 23 respectively, early production of the first child, and a minimum space of five years before the second. A limit of two children per family was strongly promoted, though by no means perfectly carried out, especially in remote rural areas. The rules, rigid though they appear, were carefully designed to maximize the health and survival of both children and mothers. And, after all, they were intended to provide the Chinese people with better lives, free from the specter of famine. In that they have clearly succeeded.

China has a tragic history of famine. Cornelius Walford, in an 1878 chronology of 350 famines, described conditions in the North China famine of 1877–1878 as follows:

Appalling famine raging throughout four provinces [of] North China. Nine million people reported destitute, children daily sold in markets for [raising means to procure] food. . . . Total population of districts affected, 70 millions. . . . The people's faces are black with hunger; they are dying by thousands upon thousands. Women and girls and boys are openly offered for sale to any chance wayfarer. When I left the country, a respectable married woman could be easily bought for six dollars, and a little girl for two. In cases, however, where it was found impossible to dispose of their children, parents have been known to kill them sooner than witness their prolonged suffering, in many instances throwing themselves afterwards down wells, or committing suicide by arsenic.[28]

No doubt similar horrors were repeated during the great famine of 1957–1958 when upward of 20 million Chinese are believed to have perished. While few details reached the world outside, the Chinese leadership obviously was powerfully affected by that disastrous event.

Afterward, while strengthening the nation's agricultural base, China's leaders also strove to improve their people's health and well-being. A basic health program was extended to the remotest villages by employing "barefoot doctors," local citizens (often mothers themselves) who had received rudimentary medical and health training (basic sanitation and the rough equivalent of first aid and home nursing) and were responsible for their community's health. Problems beyond their skills were referred to more fully trained medical personnel at county or regional health centers. The success of this approach is attested to by a dramatic reduction in the Chinese death rate and increase in life expectancy. This was achieved in a nation that was still extremely poor and underdeveloped by Western standards.

Among the duties of barefoot doctors (who were encouraged to upgrade their skills) were dispensation of contraceptives and referrals for abortions or sterilizations. Decisions about marriage and childbearing were made by the community in response to applications by couples wishing to marry or have a child, in accordance with local quotas set within government guidelines and backed up by various incentives and disincentives. Uncooperative couples who bore children without permission were usually subjected to social ostracism and penalties such as a loss of educational privileges for their children, reduced rice rations, or assignment to less desirable housing.

Such policies seem strange and coercive to Westerners, but they apparently were compatible with Chinese traditions and customs as well as being consistent with the communist system of governance. Considerable variation nonetheless seems to have prevailed among regions and even communities in the strictness of enforcement of decisions and regulations. In many rural regions, third and fourth children are still not unheard of, although by the 1990s they were increasingly uncommon.[29]

The "one couple, one child" family program was added to the

birth planning program in 1979. The architect of the program, Jiang Zhenghua, is a brilliant systems engineer turned demographer and is now second in command at the State Family Planning Commission. He has visited several times with our group at Stanford and has always impressed us with his frankness and compassion. When we saw him in Beijing, he detailed the steps that his complex bureaucracy was taking to curb abuses in the program.[30]

The policy is actually not as rigid as its name implies. As explained by Dr. Zhu Yao-hua, a member of Dr. Jiang's staff, "In rural areas the couples who have practical difficulties may be allowed to have the second child after an interval of a few years. The minorities are also encouraged to practise family planning, and the autonomous regional and provincial governments formulate their own specific regulations and measures respectively."[31] In some regions, farmers and herdsmen may be allowed three or even four children.[32]

Nevertheless, implementation of the one-child program brought unexpected results. The traditional preference for sons suddenly reasserted itself, especially in rural areas, and couples whose first child was a girl reneged on pledges to stop at one. Some overzealous local authorities responded with forced abortions and sterilizations to enforce compliance. Worse, incidents of female infanticide were reported. The government was obliged to relax the program and allow couples a second chance for a boy. This situation was compounded by a revision in 1980 of marriage-age regulations that allowed women to marry younger. A surge of earlier marriages, combined with a large cohort of young women in their twenties and allowance of "second chances," seems to have caused a rise in the birth rate in the mid-1980s.[33]

Another important factor underlying both the birth rate rise and female infanticide was dissolution of the communes. When authority and earning potential were returned to the family, women lost much of the status they had gained under communism, while children gained value as workers contributing to family income.[34] In the 1990s, Chinese fertility fell again as the one-child program was extended further and the number of women reaching reproductive ages began to fall (reflecting the program's success in the 1970s). Female infanticide faded as a problem, replaced by a surge of abortions as access to ultrasound technology made it possible to determine the sex of a developing fetus. This practice has been outlawed, but whether illegality will prevent its widespread use remains to be seen.

In our discussion with Chinese feminists in 1994,[35] we found strong support for the national goal of rapidly controlling the size of the population and understanding of the reasons for it. There was no mention of newly minted "rights" such as the right to have as many children as one wanted without considering the needs of everyone in the society—a "right" prominently claimed by Western feminists at the Population Conference in Cairo. The Chinese women were not entirely happy with the implementation of family planning, however. "The policies are heavily influenced by targets set by the State Family Planning Commission. This can lead to overreaction at the local level and problems such as too many late abortions under pressure to meet the targets." They saw a need for better integration of the operations of the three ministries most responsible for women's health issues: Family Planning, Public Health, and the All-China Women's Federation.

The sorts of horizontal links that characterized the program we saw in Tamil Nadu were still weak in China, although "a firm vertical structure exists, with many dedicated people who are quite willing to forge those links." The women also strongly favored more preferential treatment for girls and better social security arrangements to help reduce the prevalence of female infanticide and differential abortion of female fetuses, which still occurs despite its illegality.

Another concern was the disproportionate responsibility for fertility control placed on women. As one woman commented, "if limiting births is the most critical national goal, why aren't there more vasectomies? After all, the Chinese invented the no-scalpel vasectomy!" In China as a whole in 1992, 157 million women were using sterilization and IUDs, while only 32 million men had vasectomies or were using condoms (although in Sichuan Province, five times as many husbands as wives were sterilized).[36]

The women wanted open discussion of the trade-offs between short-term ethical compromises and long-term gains. As one put it, "We understand that some morally difficult choices must be made—to judge, for example, whether pressuring a woman to have an abortion in the third trimester is a lesser or greater evil than contributing to unsupportable population growth. But we wish the choices were discussed more openly, and some idea would be given of the length of time over which demographic goals should override other considerations." They generally felt, though, that although problems persisted, the efforts of the State Family Planning Commission to deal with abuses had improved things substantially.

The women were also concerned, like Jiang, about the traditional sexism of Chinese men, and the problems of female infanticide. The new capitalist rage to accumulate wealth has not helped. "Some men have the attitude toward female infanticide and the abandonment of girl babies that 'we shouldn't worry—we'll buy wives elsewhere when we're rich.' "

It was agreed that women who had lived through the Cultural Revolution were more independent than younger women, who were less assertive. While the lives of women doing traditional female tasks was being made easier, they felt that the battle to give women more equal opportunities in employment was suffering reversals. "Personnel directors of medical schools now must bargain with hospitals to take female graduates. One will say, 'If you take three women and five men this year, I promise I will only ask you to take one woman along with five men next year.' "

It is interesting that setting contraceptive delivery and childbirth targets seems to be at the root of many of the perceived difficulties with the Chinese system. In India, there has been a strong reaction against targets and incentives in favor of the much more integrated approach to "population control" now pursued in Tamil Nadu.[37] The transferability of the Chinese or Tamil Nadu approaches to Indian states like Uttar Pradesh or to sub-Saharan Africa is very much in doubt, though. We would hope that programs can converge on approaches that use the least coercion consonant with effectiveness; we suspect that cultural and organizational differences mean that virtually all societies will have to devise approaches that are adjusted to their special needs.

China's program has been more successful nationally than India's, but its implementation varies greatly from region to region.

In a vast poor nation with fifty-five culturally diverse minority nationalities, mostly located in remote areas with poor transport facilities, the challenges of delivering integrated health care/family planning services are enormous. By the government's own admission, the service network in 10 percent of China's counties and 40 percent of its townships is still not functioning properly. Considerable disparity still exists between the professional skill of the technical staff and the needs of people of reproductive ages, especially in impoverished regions and those with minority populations.[38]

The Chinese government remains concerned about its population size relative to resources—the more so since the current switch to privatization has led to an enormous surge in production and consumption, with the economic growth rate soaring above 10 percent per year in the 1990s.[39] The goal is to hold the peak population size to no more than 1.5 or 1.6 billion around 2025 and then begin a decline.[40] The State Family Planning Commission is an extremely powerful ministry, more powerful than the Ministry of Public Health, since controlling its population size is China's top social goal. The world should be worried too, since China's plans for enormously boosting its already heavy use of coal and for switching the transport system directly from bicycles to automobiles represent a gigantic threat to the integrity of Earth's life-support systems.[41] In 1972, we were concerned that China would eventually imitate the consumptive patterns of the West.[42] Now our fears are being realized.

The Responsibility of Rich Nations

Fortunately, most countries have not reached the level of population-resource pressure that China faces, but rapid population growth can propel a population into trouble before realization dawns. The difficulty is that people in modern urbanized societies seem to have great difficulty perceiving the gradual deterioration of their life-support systems and depletion of natural resources.[43] In light of the actual trends of resource depletion and degradation everywhere, though, all societies should be instituting quite stringent population measures *now*. Both industrialized and less developed nations should be striving to lower their fertility as rapidly as

possible—in rich countries to reduce their heavy impacts on resources and the environment, and in poor ones to help break the
self-reinforcing cycle of population growth and poverty.

Needless to say, part of the responsibility of rich countries also
is *to reduce their impacts by cutting their wasteful per-capita consumption.*
This is a critical step to discourage developing nations from repeating the mistakes of the developed countries and condemning
future generations to destitution. As we wrote almost a quarter
century ago:

> The industrial nations really have two choices. They can con-
> tinue their present course of devouring more and more of the
> earth's resources while destroying the environment . . . [or
> they could] deal with their own overpopulation and overcon-
> sumption. They could face up to the ecological unity of planet
> Earth and to the ways in which their destinies are intertwined
> with those of poor nations. By establishing a high quality life
> instead of a high quantity of life they could provide a new
> kind of model for the developing nations to emulate. Simulta-
> neously they could begin a massive transfer of wealth to the
> poor nations so that poverty could be reduced substantially
> without destroying our planet in the process.[44]

Such a transfer might seem to be pie in the sky today, but it is
very hard to see how the world can hold together unless the rich
nations are willing to help the developing nations in endeavors
such as deploying solar energy technologies and preserving their
portion of Earth's living capital. Such actions would not be purely
altruistic since, if the accounting were done properly, they would
make everyone richer.

With regard to population, industrialized nations face two important tasks. First, they should officially advocate and promote
low fertility for themselves, so their populations can soon stop
growing and begin slowly shrinking. Every birth averted in a rich
country helps preserve Earth's life-support systems roughly seven
times as effectively as one averted in a less developed country. The
second obligation is to provide humane assistance to developing
nations that desire help in limiting the size of their populations.

Population shrinkage is especially important in the most over-populated nation of all, the United States. With the third largest population (more than a quarter billion people), the fastest growing population of the major industrialized nations, and with one of the highest per-capita impacts on the environment (about eleven times that of an average developing country), lowering the population size in the U.S. should have top priority, along with lowering consumption levels.[45] Neither is beyond the realm of possibility.

In the mid-1970s, the average American family size fell as low as 1.7. Had it remained there, and had there been no net immigration, U.S. population growth would have stopped a few decades into the twenty-first century, and a slow decline would have begun. But by 1994, the population topped 261 million, and the TFR had crept back up to replacement level: 2.1.[46] With the addition of nearly 900,000 immigrants each year (plus a guesstimated 300,000 illegal immigrants who stay), U.S. population growth exceeds 1 percent per year.[47]

In 1992, the U.S. Census Bureau issued new population projections for the nation, indicating that growth would continue indefinitely. With replacement fertility plus an influx of immigrants (many of whom desire larger families in the first generation than do native Americans), continued growth is inescapable. Specifically, the Census Bureau's best estimate anticipates a slight further rise in fertility, substantially reduced net immigration of about 700,000 per year, and a small increase in life expectancy, producing a population of 383 million in 2050.[48] While the 1989 projection had shown a population decline beginning in 2038, the new one showed slackening growth, but no halt and no decline. A low-fertility alternative (based on returning to a TFR of about 1.8) did show a decline after about 2025. A high one (assuming a TFR of over 2.5) showed even faster growth, adding some 50 million people to the population in each decade by midcentury. In the middle projection, some 82 million of the 130 million people added from 1991 to 2050 would be due to immigration during those years. Either of the latter two outcomes would be a disaster for both America and the world.

In Western Europe, by contrast, the situation has been much

brighter. Several nations essentially reached zero population growth in the 1980s, and a few even had negative growth, which helped to lower the overall world growth rate.[49] After 1990, further fertility drops appeared. Several nations in southern Europe, which had reduced fertility relatively late, did so spectacularly with TFRs of 1.2 in Spain and 1.3 in Italy by 1994.[50] Germany, with more than twice as many people per square mile as France, has seen a steep decline in births in the last few years (1994 TFR 1.3), in part due to hard economic conditions and the difficulties of reunification. But unfortunately, the fertility decline has not been accompanied by an increase in the intelligence of German politicians. The state government of Brandenburg announced in November 1994 that it would pay parents $650 for every new child they produced!

Other industrialized countries with low fertility in the 1980s were the Soviet Union and most of its eastern European satellites. Soon after the breakup of the Soviet Union, the birth rate in Russia took a further nosedive, while the death rate soared from a resurgence of infectious diseases, a symptom of the tremendous economic hardships the country was undergoing.[51] Male life expectancy dropped to 60 (equivalent to that in a poor country like Libya or Nicaragua), while the TFR fell to 1.4, producing a significant negative growth rate. Similar problems beset the newly independent European republics of Ukraine and Belarus. While the low birth rates are a positive sign, the high death rates are exactly the form of "population control" all humane people want to avoid.

Canada and Australia, too, have below replacement reproduction. Japan's fertility, which was slightly below replacement throughout the 1980s, plunged further to about 1.5 in the 1990s.[52] In an ironic twist, this new decline appears to be in part a protest by Japanese women against their society's failure to open career opportunities for married women with children.

In developing nations, including China, the stalled birth rates of the 1980s began falling again, and declines were seen for the first time in sub-Saharan Africa. The United Nations revised its projections slightly downward in 1994, announcing that the global growth rate of 1.57 percent in the early 1990s was the lowest since

World War II.[53] Nevertheless, the world population was projected to expand to 7.5 billion by 2015 and nearly 10 billion in 2050.

Roadblocks to Progress

With any luck, virtually all rich countries except the United States will soon have declining populations.[54] But the U.S., as the world's champion polluter and resource looter, should be leading the way in reducing both its population growth and its consumption. The need for a rational domestic population policy could hardly be more clear. Many social factors impede progress toward stopping population growth (let alone moving toward an optimum population) in both rich and poor countries. Here we examine one old factor and one new one: the Vatican and other religious groups, and hyperfeminists.

The position of the hierarchy of the Roman Catholic Church has long been a politically important barrier to sensible population policies around the world. That position puts many, if not most, Catholics in opposition to the teaching of their Church and makes their lives more difficult.[55] From the viewpoint of non–Catholics, the real question is one of political interference by the hierarchy in secular laws dealing with reproductive matters that cover all faiths—including humanity's attempts to control the population explosion before it is too late.

Since the early Middle Ages, Church policy has been primarily pronatalist, but it always contained contradictory doctrinal elements such as requiring clerical celibacy and banning polygyny, divorce, and various kinds of matings between relatives.[56] Basically, the Catholic Church has been torn and confused between views such as those of early theologians like St. Augustine and St. Paul, who viewed their sexual urges as something to be controlled rather than gratified, and the urges themselves, which are the product of billions of years of biological evolution and millions of years of cultural evolution, and which pervade virtually all aspects of human life. Official Vatican views of sexuality might be funny if they didn't have so many tragic consequences for so many people. For instance, the *Catholic Encylopedia* states the following in its article on lust: "A lustful action is a disordered use or pursuit of sex pleasure not only because it defeats the biological, social or moral purpose of sex activity, but also because in doing this it subjects the spiritual in man [sic] to values of the grossly material order, acting as a disintegrative force in the human personality."[57]

One should not be surprised to read such nonsense. Members of the Roman Catholic hierarchy, like all human beings, have had to deal with the problems and (perhaps to a lesser extent) the joys of their sexuality. They also have been obliged to do so largely from the position of profound ignorance of the origins and purposes of sex that is displayed in the *Catholic Encyclopedia*. In addition, they have labored in an organization where sexual repression was generally the norm, where they themselves were not supposed to give expression to their sexual feelings, and where the importance of sexual intercourse in promoting parental bonding (and thus family stability) has not been fully appreciated.[58]

Small wonder that the Church's doctrines on sex have at times been contradictory, confusing, and downright silly; they have been shaped to serve the needs of the institution rather than of its adherents. After all, who of us has not been ambivalent, frivolous, or selfish in response to our own hormones? Even people who are fully aware of the evolutionary origins and importance of sex are mostly not free of that supremely human problem of wrestling with one's own sexuality. And small wonder that the institution has had increasing difficulty in recruiting priests and problems with pederastic behavior on the part of its clergy.[59]

Despite the hierarchy's position, Catholic reproductive performance has long been converging on that of non-Catholics in similar socioeconomic situations.[60] There is evidence that in the early 1980s, *practicing* Catholic high-school students in the United States put more importance on having children, knew less about birth control, saw women in more traditional roles, and expected to have larger families than nonpracticing Catholics and members of other religions.[61] Nevertheless, by 1992, white Catholic women on average were having 1.64 births in a lifetime, while white Protestant women were having 1.91 births.[62]

Little difference can be seen today in overall use of contraceptives among religious groups in the United States, although Catholics are most likely to use the Pill, Protestants show a strong preference for female sterilization, and Jews are more likely to use diaphragms.[63] In a 1994 poll, 98 percent of Catholics aged 18 to 29 and 91 percent of those 30 to 44 believed that one could practice artificial birth control and be a good Catholic.[64] About half of both Catholics and Protestants surveyed in 1985 approved the 1973 Supreme Court decision that struck down state antiabortion laws. Furthermore, of women having abortions who report a religious affiliation, Catholic women are more likely to have an abortion than are Protestant or Jewish women.[65]

Catholics in other industrialized nations show similar reproductive patterns and attitudes; indeed, some of the lowest total fertility rates in the world today are found in Catholic nations. In 1994, the average completed family size in France was only 1.7; in Belgium it was 1.6, and Italy and Spain had record low TFRs.[66]

It is essential, in our opinion, that everyone realize *there is no particular Catholic demographic problem.* Bigots will have to find other excuses for prejudice against Roman Catholics. But there is a serious *political* problem, important within the Church and the world as a whole, stemming from antediluvian Vatican attitudes and inhumane Vatican actions in areas related to the population problem and human sexuality.

The most recent exercise of the Vatican's power was its campaign to destroy the effectiveness of the 1994 UN Conference on Population and Development in Cairo. Before the conference, the Pope repeatedly attacked the draft Programme of Action on the grounds that it promoted contraception for both married and unmarried women and seemed to promote legalization of abortion.[67]

In battles over preparation of the action plan, the Vatican managed to block agreement on several key points. As if arguing how many angels could dance on the head of a pin, it insisted that the word *reproduction* be replaced by *procreation.* The terms *safe motherhood, family planning,* and *reproductive rights* were objected to because they did not explicitly outlaw abortions. The word *condom* was also disputed in a section on AIDS.[68]

Part of the Pope's distress obviously traced to the shift of U.S. policy away from President Reagan's disastrous position at the 1984 UN Population Conference in Mexico City, which was followed by termination of U.S. funding for international family planning assistance.[69] President Clinton's more enlightened policies led to a showdown with the Pope in June 1994, with inconclusive results.[70]

The Vatican's anxiety was made clear by its attempts to find allies in attacking the Cairo conference among fundamentalist Muslims and evangelical Christians, which met with mixed success.[71] These ancient enemies, perhaps drawn together by fears of sexuality in general and women in particular, made common cause in attacking the UN.

Catholics, especially in developed countries, seem relatively little influenced by the Vatican's position. In fact, the Church accepted contraception for more than a thousand years and has only relatively recently shown interest in suppressing abortion. Why, then, is the hierarchy so furiously pursuing a policy that endangers both civilization and the future of the Roman Catholic Church? The basic answer appears to be a combination of ignorance and denial of the nature of the population-environment predicament, and a deep-seated fear of change.[72] Questions of *power* are also involved.[73] The strength of the women's movement at Cairo and the strength of opposition to the Vatican among Catholics were signs of a shift in power. That was one reason the conference was so contentious; as economist Gita Sen observed afterward, "the shifting of power is not a nice process."[74]

One obvious power factor in the hierarchy's attitude is the desire to have as large a constituency as possible. This, of course, is not given as a reason in Vatican public documents, but it occasionally emerges explictly, as when Jesuit Father Richard Ryscavage called immigration "the growing edge of Catholicism in the United States." In 1992, he said, "We are in the middle of a huge wave of immigration . . . and most of them are Catholics." He called the entry of 10 million immigrants since 1980 "the key to our future and the key to why the church is going to be very healthy in the 21st century."[75]

In an effort to keep the population of Costa Rica growing, the Vatican intervened in 1991 to prevent the government from distributing a sex education guide for seventh, eighth, and ninth graders because it contained information on birth-control methods (including the use of condoms against AIDS), even though three Catholic priests had been on the commission that originally approved its contents.[76] The Vatican has generally played a powerful role in Costa Rican government population policies, openly criticizing the family planning program and restricting publicity about it through the mass media.[77]

In the Philippines, the notorious Catholic archbishop of Manila, Jaime Cardinal Sin, led a gigantic rally against the government's plans to reestablish birth-control programs and participate in the Cairo conference.[78] President Fidel Ramos, elected in 1992, had wisely decided that the Philippines' skyrocketing population (growth rate 2.4 percent a year, 69 million people in 1994, projected to be 105 million in 2025) was causing a "serious imbalance" in both the environment and the economy. Catholic Dr. Marilen Danguilan, a health policy consultant to the Philippine Senate, described the rally as "Catholic politics being played out here in the Philippines. . . . The demonstration was a way of telling the Vatican, 'Look, we still have the clout.' "[79] At that point, the government still planned to participate in the Cairo conference and, in an attempt to compromise, it set up a committee to negotiate with the Vatican's surrogates the principles its delegation would adopt.

Another power element is simply the need of a group of men to keep women in their place. Frances Kissling, president of Catholics for a Free Choice, has bluntly said that the Vatican opposes birth control because it gives women more control over their bodies and their families—which threatens the male-dominated Vatican.[80]

There is another important aspect of power that influences the Vatican's position on population issues. As Frances Kissling saw it, although the hierarchy's opposition to birth control is a given, Pope John Paul II's bitter opposition to the UN's population conference was "about money."[81] Kissling claimed, "The way church officials see it, the greater the percentage of foreign aid budgets that is allocated to family planning, the less money there will be for education, health care and disaster relief." In 1992, Catholic Relief Services (CRS) got 77 percent of its $290 million budget from the U.S. Agency for International Development (US AID) and other U.S. government agencies. American Catholic bishops sponsor CRS, which provides humanitarian aid overseas. Finally, Kissling also pointed out that "The Pope needs a new enemy to keep religious Catholics revved up, and the new devil is modernism."

One way the Pope and the Vatican have kept religious Catholics in the United States "revved up" on these issues has been by orchestrating antiabortion campaigns in the United States. We might be more sympathetic with the Church's stand on abortion but for its retrograde policies on the use of contraceptives (even shamefully opposing the life-protecting use of condoms for prevention of AIDS) and its antique views of sexuality. Few humane people believe that abortion is a desirable method of birth control. Most people, including us, would like to see safe, effective methods of contraception made available to all sexually active people, along with competent instruction on their use and medical follow-up when appropriate. If that were the situation, "backup" abortion would rarely be needed and often could be accomplished with RU-486 or similar methods, and the abortion controversy might simply fade away.

The Vatican must bear a heavy moral responsibility for the terrorizing of hundreds of thousands of women seeking abortions in the United States and the staffs of clinics trying to help them. The propaganda, organizational skills, and financial resources of the Church, backing the idea that killing a fertilized human cell is equivalent to murder, sadly also helped create a social atmosphere conducive to the murders in 1994 of two doctors, two receptionists, and a volunteer at U.S. family planning clinics that provided abortions, not to mention innumerable bombings and arsons that have endangered lives.[82]

The number of women living in fear of unwanted pregnancy around the world because of Vatican political pressures almost certainly numbers in the tens of millions. The Vatican's effectiveness in manipulating Ronald Reagan and his government into limiting U.S. family planning aid to women in developing countries also unquestionably led to millions of additional abortions globally and tens of thousands of additional maternal deaths each year from bungled abortions.[83] When contraceptive services become available, the abortion rate virtually always gradually declines.[84] Today, an "epidemic" of 2.7 to 7.4 million abortions per year is raging in Latin America, almost all clandestine and dangerous (except for those performed in Barbados, Belize, and Cuba, where they are legal).[85] A sense of the human tragedy involved can be gained from the following account by a Chilean woman:

> First I had two injections of Methergin. Afterwards, for three days, I drank before breakfast red wine boiled with borage and rue, to which I added nine aspirins. My body was full of pimples but I did not abort. A few days later I drank cement water. It did not work either. Then I went to a lady who inserted a rubber catheter into me. I had to use it, after all the things I did I could not keep the child because he could have malformations.[86]

Such practices carry a high price in illness, injury, and death for women. Something on the order of a fifth to a half of all women who undergo induced or spontaneous abortion seek medical treatment afterward.[87] The complications of quack abortion can be diverse and frightening: perforated uteri, damaged bladders or intestines, lacerated cervices, shock, hemorrhage, and pelvic infections. During the late 1970s and early 1980s, hundreds of thousands of women needed hospitalization after abortions in Latin America.[88] The highest price paid by women for the illegality of abortion is death, a penalty exacted on perhaps 200,000 women worldwide—between a third and half of all maternal deaths. An additional price is paid by the surviving children of those women. Much of the moral burden for those deaths and injuries can be laid at the door of the Vatican and its opposition to giving women access to reliable birth control technology.

The Latin American abortion epidemic has continued for about thirty years. Nobel laureate Octavio Paz wrote in 1982 about two coexisting countries of Mexico, "one fictitious, another real." He declared, "Abortion is a clear example of this situation. Prohibitions against the practice fortify the unreal country—the one of frustrations—against the one of the facts—the country of reality."[89] The fiction that the continuum of human life must be artificially cut between sperm and egg and the zygote (fertilized egg), prop-

129

agated by biologically ignorant celibate old men, exacts a high toll from the world's young women.[90] A more unfair situation perpetuated in the name of piety and life is difficult to imagine.

We suspect that the brightest possibility for changing the Vatican's position and letting humanity get on with saving itself is the determination of many Catholics outside the Vatican to effectuate that change.[91] In fact, the Vatican's determined opposition to the Cairo conference was dealt a serious blow in June 1994 when a lay panel of the Pontifical Academy of Sciences urged limits on family size to avert "insoluble problems" in runaway population growth.[92] The panel noted that current conditions make it "unthinkable to sustain indefinitely a birth rate that notably exceeds the level of two children per couple." Not surprisingly, Pope John Paul II was reported to be "infuriated by the report, which emerged just as the Vatican [was] intensifying its campaign against several proposals to be discussed in Cairo."

One of the best Catholic statements we have come across is one that was nailed on the door of the Saint Francis Cathedral in Santa Fe, New Mexico, "by Eduardo de Los Alamos" at noon, October 31, 1991 and 1992.[93] The day and month, of course, correspond to those on which Martin Luther (in 1517) nailed his 95 Theses to the door of All Saints Church, Wittenberg. Addressed to Pope John Paul II, the Los Alamos text is, in part:

SEXUAL REFORMATION

Credo that for centuries consideration of human sexuality has not been a matter of open, easy discussion.

Credo that there needs to be a new, honest, and enlightened sexual moral code proclaimed by the church, a new sexual code for the twenty-first century.

Credo that persons should be taught that conception occurs, not as a gift-of-God, but as a consequence of a sexual act between a man and a woman.

Credo that it *is* sinful, immoral, unethical, and unthoughtful to beget an unwanted child. This is pro-life.

Credo that artifical contraception is neither evil nor sinful and that it is wrong for the Church to make even a married couple feel guilty if they use any form of artifical birth control.

Credo that a woman is denied the pleasure of her own sexuality if for most of her life she lives in a constant fear of becoming pregnant and that the Church causes untold suffering, anxiety, and hardship by insisting that every conception must be carried to term.

Credo that it *is* sinful and immoral for the Church to avoid addressing the problem of world overpopulation. This is pro-life.

Credo that homosexuality is a God-given exercise of freewill and homosexuals should not be persecuted.

Credo that adoption of this creed will greatly reduce the number of traumatic abortions. This is pro-life.

Considering current attitudes in the Vatican, we would not be surprised if the response to these theses were a papal bull and the burning of the document in Rome! But the vast majority of Catholics in the United States and many other nations have already affirmed, by their actions, many of these theses with the equivalent of "Here I stand. I can do no other."

Other religions are not as prominent in the population wars as the Vatican, and some are strongly on the side of limiting population growth to preserve Earth's ability to support civilization. The religious leadership of liberal Protestant and Jewish groups in the United States take this position.[94] Baptists, evangelicals, and Pentecostals, as well as most Muslims generally take no position on contraception. Mormons are, if anything, more extremely pronatalist than the Vatican.

Globally, things are more complex. For example, an American Muslim, Ibrahim Hooper, communications director of the Council on American-Islamic Relations, appeared on the *McNeil/Lehrer News Hour* as a spokesman for Islamic views on the population conference.[95] He stated that "any government or culture that oppresses women does so in spite of Islam, not because of it. It's through the cultural practices and traditions that are applied to women in many of these areas that the problems occur. When you apply real Islamic law and rights to women, it's a liberating ideology."

This is entirely correct—in theory. In practice, however, in most Islamic societies, women have very low status and essentially no power, and fertility continues to be high.[96] Even though there is no unified Islamic leadership equivalent to the Vatican, representatives of twenty-four Muslim nations and nine ulamas (members of the highest body of religious authorities in Islam) met and signed a declaration at Aceh, Indonesia, in February 1990. This far-sighted document, among other things, "acknowledges that population, resources, and the environment are inextricably linked and stresses our commitment to bringing about a balanced and sustainable relationship among them."[97] The Aceh Declaration also urged the eradication of illiteracy among women and the provision of accessible family planning, including safe contraceptives. And some relatively modernized Islamic nations, especially Tunisia, have set up strong and rather successful family planning programs. Tunisia's total fertility rate in 1994 was 3.3, as compared with 4.5 for North Africa on average.

So it is not Islam itself that bears the responsibility for the deplorable status of women and high population growth rates in many Islamic nations, but the traditional patriarchal cultures of those nations. The difference could be viewed, as Paz did for Mexico, as contrasting a fictional world (Islamic theory) with the real world as described by the words of Dr. Nawal el-Saadawi, a well-known fighter for women's rights in the Arab world:

> In the Arab world of 1990, there is little talk of the reproductive rights of women, or even their work rights. To this day, some conservative political and religious forces maintain that God created woman in order to stay at home, to serve her husband and children, that she is subject to the *law of obedience* which is the primary foundation of the institution of marriage

in our countries. A woman's duty is obedience: obedience to her husband, to her father, to the rules, and to the state. These forces . . . speak in the name of God, religion, religious law, and secular law. All of them regard a woman's body as part of their property and the children produced by a woman's body also as the property of the husband or state or both.[98]

Hyperfeminism

An extreme wing of the feminist movement, no doubt energized in part by the sorts of misogyny just discussed, has seized on the serious and real problems of abuse of women that exist in most (perhaps all) societies as justification for condemning national and international family planning programs. They view the programs as a racist and sexist plot by the rich and the white-skinned to suppress the poor and dark-skinned by forcing contraception on their women.[99]

Of course, racist motivations have long been an unfortunate element in promotion of family planning by some individuals. But such total denial of the origins of the birth control movement as a mechanism for female liberation is breathtaking to anyone who remembers the half-century-long struggle to legitimize contraceptives and family planning, which was finally won in the United States as recently as 1962. If the hyperfeminists' activities were focused on exposing and attempting to correct actual failings in family planning programs, they would be doing a valuable social service. But they try to make their point by downplaying or more often denying the importance of the population factor in the human predicament and by attempting to tar everyone concerned about the rapidly expanding scale of the human enterprise with a racist, fascist, and/or antifeminist brush.

Unfortunately, this branch of the feminist movement pits women against men, giving lip service at most to the victimization (and self-victimization) of both sexes by the out-of-control growthmania, scientism (science as the ultimate arbiter of values), and militarism of today's dominant culture. Just when cooperation between sexes and among races is more than ever essential to resolve the human predicament, Maria Mies and Vandana Shiva lump women, nature, and foreign peoples and countries together as "colonies" of white men. They assert that without that colonization—that is, subordination for purposes of predatory exploitation—Western civilization would not exist, nor its paradigm of progress, nor its natural science and technology.[100] The sad thing is that, although some of their diagnosis is correct, such broad-brush condemnation is not a helpful contribution to solving a set of global problems that threaten the futures of both men and women.

Equity and Population Policies at Home

One of the most important ways that people in rich nations can help keep the plow ahead of the stork is by working for equity (and hyperfeminists *are* doing that). Ending population growth (or maintaining population declines) in rich countries should be a much easier task than simply halting growth in less developed ones. Especially in the least developed (and usually the fastest growing) countries, it will be necessary to foster development, especially access to the material ingredients basic to a decent life—adequate food, education, health, sanitation, and employment—as well as family planning facilities.

Even within the United States, the source of much of the world's population-environment dilemma, increasing equity might help reduce fertility. Birth rates ordinarily fall as economic status rises; certainly health and well-being improve. Yet the U.S. is increasingly a nation where a relatively few rich people are getting richer and the mass of the population is getting poorer.[101] After decades of increasing economic equity in the U.S., around 1980 the trend was reversed. Labor Secretary Robert Reich has worried aloud that the nation is on its way to joining Brazil, India, the Philippines, and others as a "two-tiered society."[102]

There are many good reasons that the United States ought to put closing the income gap near the top of the priority list, not least that the current general mistrust of government and antagonism between groups in society might be lessened. Striving to provide access to high-quality education and secure employment for everyone might help to lower the birth rate, especially among unmarried teenagers.[103] Better education also would inform the public on the dimensions of the human dilemma and make people understand *why* reducing the birth rate would be a good idea.

Reducing racism in the United States also could help to narrow the increasing income gap; for example, roughly 33 percent of African-Americans are poor as opposed to 11 percent of white Americans.[104] Unhappily, just as the need is greatest to develop all our human capital to its maximum potential, Americans recently have been treated to yet another round of scientific racism.[105] Yet it is easy to see why the question of innate differences in intelligence among "races" is unimportant to educational policy. Even if one granted that there were distinct races of human beings (there are not), that IQ tests adequately evaluate intelligence (dubious), that there were a large genetic component to intelligence (how large is controversial), and that darker skin colors were associated with lower "genetic intelligence" (unlikely; evolutionary theory suggests that, if anything, people with lighter skin colors ought to be, on average, less genetically intelligent), what sensible social policy could be based on such findings?

If we were to treat people differently on the basis of their intelligence (as indeed we do), shouldn't that treatment be based on each individual's IQ score rather than on something weakly correlated with it? Even the worst racist will admit that many dark-skinned people are smarter than many light-skinned people. Using skin color as a basis for, say, educational policy would be roughly equivalent to using IQ tests as a basis for prescribing sunscreens.

Gross economic inequality, sexism, and racism all work against the sort of community feelings that are necessary for a nation to make difficult choices and coordinated efforts without undue conflict. More and more, as the population and pressure on resources both grow, Americans of different subcultures find themselves in opposition and alienated from other groups. But a spirit of cooperation and community will increasingly be needed both to resolve population-environment problems at home and to provide effective assistance to other nations that request help to do the same.

The most crucial population policy for the United States to adopt is one focused on solving its own population problem. The first step (and quite possibly the last) should be for the federal government to start an educational campaign around the slogan "Patriotic Americans stop at two." The campaign also might use another slogan—"Return to the days of a better America." That would take advantage of the general nostalgia for a smaller, simpler society and perhaps generate support for an announced goal of reducing U.S. population size to around the World War II figure of 130 million (half the 1995 size). The basic approach would be simple: lower the annual number of births and immigrants to less than the number of deaths and emigrants.

If the president and his minions in the executive branch as well as leaders in Congress got behind it, we suspect that alone would suffice to get the U.S. TFR well below replacement level. If it were not enough, then the tax laws could be used to provide tax penalties for overreproducers.[106] These could be graduated, so the rich (who cause disproportionately heavy impacts on the environment) paid the most per child. Tax penalties for poor people who have large families would need to be compensated by programs to protect their children from undue hardship. But penalties would nonetheless be required to reinforce the message that having more than two children is un-American. If we expect poor people in nations where children are economic assets (net contributors to household income and providers of social security) to stop at two, the very least Americans can do is the same.

Once the United States had a domestic population policy, Americans could then have a reasonable discussion of the ethically vexed immigration question.[107] The stark central issue would be the trade-off between births forgone and immigrants admitted. For anyone who is familiar with the population projections and understands that the U.S. cannot support an infinite number of people (still less keep offering them a good life), that of course is already a central issue.

How to manage the path to long-term shrinkage naturally will differ from one nation to another, but the basic principles are the same. Population growth in rich nations, whether from natural increase or net immigration, adds disproportionately to pressure on vital ecosystems and resources. While they struggle to reduce overconsumption, the rich must also struggle to reduce their own numbers.[108]

Chapter Five

GATHERING, HUNTING, AND FARMING

No human activity is more important than growing food. Indeed, no human activity is as important as growing grains—the nearly 2 billion tons of wheat, rice, maize, and other cereals that are produced annually provide our daily bread—they are the feeding base of a human species that is now more than a thousand times as numerous as it was when it invented agriculture. Foraging for food provided human groups with decent livelihoods for millions of years in the past,[1] and a very few small groups still manage to do that today. But larger, more specialized societies must depend on agriculture to feed their citizens.

First came subsistence agriculture, in which farmers' concerns were simply to provide sustenance for their families. But over many centuries, techniques evolved to the point where a single farm family was able to produce surpluses that could be traded or sold to others. This was a key development, as it paved the way for specialization—people making livings as merchants, tool makers, soldiers, and so on, and bartering or paying for their food. It opened the door to the development of cities, to books, to industrialization, and eventually to television, computers, and nuclear weapons.

Agriculture itself did not escape the trends it had set in motion. It too became industrialized in many parts of the world. Frequently today farmers grow little or none of their own food, which they buy instead at markets just like city dwellers. Industrial agriculture started in the rich countries, where it was based on mechanized production and the triumphs of plant genetics in developing varieties of grains that produced very high *yields when given sufficient fertilizer and water. That was the origin of the "green revolution," a technology that was introduced to less developed countries starting in the 1960s and was largely responsible for avoiding catastrophic famines as those nations' populations expanded at unprecedented rates.*

Global grain production roughly tripled between 1950 and 1990. Gains in production per person were not uniform, though, with Africa lagging seriously behind for three decades and Latin America losing ground since 1980. In many places, the green revolution favored big farmers over small ones, and thus increased inequities. The degree that industrialized agriculture still generates inequities is controversial, as are questions of its environmental impacts and sustainability. But humanity has made its choice. Sustainability and equity both must be enhanced within the framework of the current agricultural system, warts and all. An attempt to impose a wholesale substitute for present technologies would be extremely risky and no doubt politically impossible, but there is much room for evolutionary improvement.

The fisheries situation is grim by comparison. Per-capita yields have been dropping as the human population has grown, fish stock after fish stock has been overfished, and oceanic environments have been steadily degraded. A critical protein source in the human diet is fading away.

In short, intensifying the agricultural system and harvesting the riches of the sea have brought humanity a long way, but it has been a long way up a cliff face that we must continue climbing. Whether humanity will get safely over the teetering top and down the other side to a sustainable position remains to be seen.

The food supply is often the most important resource determining an area's carrying capacity for an animal population, and humanity is much like other animals in this respect. But most animals simply depend on the food that happens to be available in their environments. A few nonhuman animals, such as ants that "farm" fungi and "milk" aphids and certain woodpeckers that drill sap wells in trees and shrubs, work to create or enhance their food supplies. But no species approaches ours in the degree to which it can manipulate its sources of nourishment.

The transformation of Earth into a giant human food production system did not happen overnight. Our human ancestors were gatherers and hunters for millions of years.[2] Until the agricultural revolution some ten thousand years ago, people in small groups roamed the landscape searching for edible plants and game. Some groups, especially in Australia and southern Africa, were gatherer-hunters until well into this century; a very few still may be today.

Gathering and Hunting

The standard picture of gatherer-hunter societies has been one in which the men went off to pursue game while the women stayed in camp, caring for children, preparing skins to make clothes, and gathering some wild plants for a salad course to go with the meat.[3] But more recent research indicates that this is far from the real picture, hence the reversal of the earlier term "hunter-gatherer." The availability of game varied greatly from area to area and from time to time, and it is quite likely that foods from plants and small animals (eggs, nestlings, fish, trapped rodents and rabbits, etc.) usually made up the bulk of the diet.[4] Women, who collected most of these items, thus often provided the lion's share of the food in addition to their many other duties. Moreover, in many prehistoric groups, women probably participated in hunting, as they do in certain Eskimo groups.[5] Our forebears therefore might best be considered omnivorous foragers with a preference for meat, and women's role as providers has doubtless long been underestimated.[6]

During the Paleolithic, the human population grew only very slowly overall, with different groups flourishing at times and suffering setbacks and local extinctions at other times.[7] A possibly significant limiting factor in some groups was the impact on female fertility of the strenuous physical activity associated with their food-gathering efforts, as well as a life of constant nomadic movement, burdened with small children.[8] Considerable time was also spent in traveling, building shelters, processing and cooking foods, making clothing, tools, and implements, and participating in social activities.

On the other hand, evidence from contemporary gatherer-hunter groups suggest that at least some of our prefarming ancestors led lives blessed with considerable leisure, obtaining their food with only twenty to thirty hours per week of labor.[9] Even present-day Kalahari Desert Bushmen, living in some of the world's harshest environmental conditions, are reported to spend a mere twelve to nineteen hours per week obtaining food.[10]

The lives of preagricultural gatherer-hunters appear to have been relatively free of malnutrition, starvation, and epidemic diseases. Comparison of skeletal remains (which bear scars of many diseases, such as anemia, tuberculosis, leprosy, and osteoarthritis) in archaeological sites of preagricultural and agricultural societies reveals a striking trend. In preagricultural societies, the average height of gatherer-hunters was several inches more than prevails even in their present-day counterparts. Early skeletons are so healthy they don't make for interesting study; in contrast, skeletons from agricultural societies are riddled with signs of undernourishment and disease. Mortality apparently increased substantially in every age class with the advent of agriculture.[11]

Eminent biologist Jared Diamond explains that a variety of factors made agriculture surprisingly "bad for health." First, gatherer-hunters generally had rich and varied diets whereas today humanity derives over 50 percent of its calories from three plants alone—wheat, rice, and maize. Good nutrition was traded for cheap calories. Second, concentration on a relatively few staple crops put farming societies at great risk of crop failure and starvation, particularly before the development of modern transport. Finally, the majority of today's infectious diseases and parasites "persist only in societies of crowded, malnourished, sedentary people constantly reinfected by each other and by their own sewage."[12]

Agriculture was also bad for equity. As Diamond put it, gatherer-hunters "have little or no stored food, and no concentrated food sources such as an orchard or herd of cows. . . . Everybody except for infants, the sick, and the old joins in the search for food. Thus, there can be no kings, no full-time professionals, no class of social parasites who grow fat on food seized from others."[13]

Given the many curses that agriculture brought to humanity, Diamond and others wonder whether its development was ultimately a big mistake. As anthropologists Richard Lee and Irvan DeVore wrote: "To date, the hunting way of life has been the most successful and persistent adaptation man has ever achieved. . . . It is still an open question whether man will be able to survive the exceedingly complex conditions he has created for himself. If he fails in this task, interplanetary archeologists of the future will classify our planet as one in which a very long and stable period of small-scale hunting and gathering was followed by an apparently instantaneous efflorescence of technology and society leading rapidly to extinction."[14] But the die has been cast. Even if people were willing to return to gathering and hunting, humanity is far too numerous to do so.

There is considerable debate over exactly by what sequence of cultural changes agriculture developed.[15] Most likely, many factors interacted and varied from place to place. Population growth, slow though it was, may actually have played a role in inducing the first human groups to give up their gathering-hunting lifestyle and, little by little, take up agriculture.[16] More people can usually be fed from a given area of land by even primitive agriculture than by gathering and hunting. Gatherer-hunters who exceeded the carrying capacity of their territory could employ agriculture to expand that capacity. They could augment their food supply by encouraging the growth of desirable food plants (and/or domesticating animals).[17] Their larger numbers could then allow them to overwhelm and drive out smaller groups that had persisted as gatherer-hunters. The settled life has other potential benefits; not only does it eliminate the costs of frequent moving about, but it also allows for investment in manufactured goods too large or heavy to move significant distances.

The sequences and inventions no doubt differed among the various centers in which agriculture developed more or less simultaneously (on a geological time scale). The details need not concern us here. What is undeniable is that, about ten thousand years ago, *Homo sapiens* crossed an important threshold. People started extracting plants and animals from nature's genetic library and domesticating them by selective breeding for desirable characteristics. That is, they invented agriculture and started the race between the stork and the plow.

Subsistence Agriculture

Early agriculturalists had the same simple goal as gatherer-hunters. They were subsistence farmers, meaning that they wanted to provide food for themselves and their children as easily and dependably as possible—not to produce surpluses for others to consume. So they probably grew a mixture of crops that matured at different times and had different requirements for water or sunlight, thus ensuring against crop failures (if one crop failed, others might pull through).

The stress of women's work probably worsened in early agricultural societies. Bearing more closely spaced children seem to have become feasible then, as soft foods became available for earlier weaning and it was no longer necessary to carry small children and belongings long distances in a nomadic search for food.[18] In early agricultural societies (those dependent on root crops and starchy tree crops, supplemented by hunting), which developed mostly in moist tropical forest areas, childbearing still may have been constrained by the low nutritional quality of the foods.[19] As more intensive farming systems (based on cereals and legumes) were developed, food more suitable for early weaning became available, childbearing was more frequent, and more time was needed for processing and cooking foods. In either system, women are estimated to have spent an average of about four and a half hours a day, seven days per week, in the fields.[20] But the daily household workload was only an average of about three hours in less intensive systems, whereas it was nearly six hours in more intensive ones, thus making a total workload (inside and outside the household) of less than seven hours a day in the first and almost eleven in the latter.

Over time, people gradually learned the best times to plant and harvest crops, to breed more productive varieties, to care for soils by manuring, fallowing, rotating crops with nitrogen-fixing legumes, and eventually to supplement rainfall with water from irrigation. Farm families gradually became capable of supporting more people than were needed to do the farming, especially in the more intensive, grain-based systems. Some individuals were freed to specialize in other occupations, and later it became possible to establish towns. Commerce and trade developed, beginning with exchanges between farmers supplying food to the towns and the townspeople producing implements, clothing, other goods for farmers, and also weapons.

As agriculture gradually spread, the more numerous farmers wiped out gatherer-hunters and pastoralists, or pushed them into marginal habitats—deserts, remote mountainous areas, or the barrens of the Arctic. The relatively intensive grain- and legume-based systems—beginning with wheat, rye, and barley in the Near East, rice in the Far East, and maize and potatoes in the Americas—ultimately conquered the world of food production.

Thus the foundations were laid for the displacement of subsistence agriculture over much of the planet, a process that has taken millennia and continues today. Indeed, subsistence agriculture has tended to persist mainly in tropical regions, where dependence on starchy root and tree crops don't lend themselves to the surplus production and, importantly, storage that can lead to urbanization. The trend toward increasingly intensive agriculture further hastened the demise of gatherer-hunters and pastoralists, although in some areas such as India, Southeast Asia, and parts of Africa, preagricultural groups have managed to a degree to coexist with subsistence agriculturalists.[21]

The central role of women as food providers in early agricultural societies can be seen today in much of Africa and parts of Asia and Latin America, where subsistence agriculture is still carried on. Crops produced include cassava, taro, and bananas, or cereals like millet, sorghum, and maize where the climate is suitable. In Asia, rice predominates. Normally, land to be farmed is cleared close to the dwelling, permitting women to work on the land while also tending to children and other household chores. Land is still relatively abundant in much of Africa, but soil productivity is generally low, growing seasons often are short, and irrigation is little practiced. So, as in parts of Latin America and other areas of low population density, farming is often done by swidden agriculture or "shifting cultivation."

The highly variable practice of shifting cultivation generally involves slashing and burning forest patches to create temporary fields. Burning improves soil fertility, as the ash adds nutrients and reduces excess acidity. The fields are then harvested in a rotation between brief periods of cultivation and longer periods of fallowing.[22] In Africa, hand tools are mostly used in cultivation, principally hoes, in part because the deadly disease nagana tends to kill off livestock,[23] and perhaps also because the soils are relatively fragile. It probably is also more efficient, since plowing requires destumping, which is both energetically costly and a hindrance to natural regeneration of the land under temporary cultivation.[24] Traditionally, men may have spent some of their time hunting or fishing to supplement crops while the women did much of the hoeing and other work. Nowadays it is common in many areas for women to carry on the farming while the men seek employment in towns or as wage-laborers on plantations.

After one to three years of cropping, a combination of declining soil fertility, competition from weeds, and pest or pathogen outbreak then sharply diminishes yields, whereupon the plot is abandoned.[25] An invasion of grasses, brush, and trees gradually restores the land's fertility. Nitrogen is restored by the action of soil bacteria and transport in rain, while the deep roots of trees give them access to nutrients that have leached down far below the shallow roots of crops. These nutrients are pumped up by the trees and restored to upper layers of the soil through leaf fall. Fallow periods for sustainable shifting cultivation are typically about twenty years (ranging between five and forty years) in the humid tropics and may be considerably longer in subhumid and seasonal forests.[26]

Shifting cultivation of this sort has persisted for millennia in the tropics of Southeast Asia, Africa, and South America, relatively undisturbed until the late twentieth century when population growth began to upset the balance between people and the land they depended on for sustenance. With rising population density and shrinking land availability, fallow periods must be shortened, and some form of fertilization is then needed to prevent rapid exhaustion of soils and dramatic declines in crop yields.

The Rise of Modern Agriculture

Intensive grain-based agriculture has given rise to many civilizations, from ancient Sumer to China, the Indus Valley, Egypt,

Rome, the Aztecs and Mayans, and northwestern Europe. Agriculture as practiced in northern Europe since medieval times, and later transferred to North America, has always been land-intensive compared with Asian wetland rice production.[27] In the eleventh century, a typical English farm was twelve hectares (thirty acres), a size determined by the inherent limitations of a dry-grain farming system based on wheat, barley, and rye.[28] Because production of hold-over seed and regular fallowing were necessary (with only manure and crop residues available as fertilizer), dense populations of subsistence farmers could not be supported. Indeed, since yields were low and soils heavy, human labor alone could not produce enough food energy to sustain the farmers.

The need for animals and heavy equipment gave an advantage to larger farms. In economic terms, there were increasing returns to scale because bigger operations both yielded sufficient income to purchase more of the needed animals and implements and were able to organize their use more efficiently. Therefore, as the manorial system of the Middle Ages broke down in the sixteenth century, the advantage went to capitalist family farmers affluent enough to obtain the means to produce a dependable surplus.[29] Bray wrote: "Capitalist relations in agriculture had formed in many parts of northwestern Europe before the 15th century. Markets in land and labor were well-developed. The social relations necessary for the foundation of a modern mechanized agriculture were thus in place, but the necessary technical expertise was lacking."

For a long time after agriculture became commercialized (when crops were grown to be sold to distant communities), most farmers continued to grow a variety of crops, feed themselves with part of the production, and sell the rest. Yet, even before the Industrial Revolution, some crops, especially tropical and subtropical ones such as sugarcane, cotton, coffee, tea, and rubber, were grown in monocultures on large plantations. Cane, in particular, lent itself to plantation agriculture; it could be efficiently grown on very large tracts of land by unskilled crews directed by overseers. Production of these crops presaged the widespread practice today in which farmers specialize in a few crops grown on large tracts, but use machinery in place of workers. These farmers sell their produce and buy most of their food in commercial markets.

Modern agricultural systems with relatively specialized farms developed largely as a part of the Industrial Revolution. The invention of tractors, mowers and binders (machines that cut grain), threshers (which separate grain from straw), and combines (which do both) allowed farming to become much more efficient. At first, animal power was used for horse-drawn machines; later, inanimate energy was applied to the farming process. Soon, engine-driven pumps (and large-scale dam-building) facilitated the expansion of irrigation, and the development of inorganic fertilizers in effect eliminated the need for fallowing. The use of pesticides (especially synthetic organic compounds introduced after World War II, such as DDT, a chlorinated hydrocarbon, and parathion, an organophosphate) made protection of large-scale monocultures easier. How much pesticide use has actually reduced losses to pests is controversial, however.[30]

Gradually the agricultural enterprise became heavily dependent on fossil fuels—as the source of power for tractors, harvesters, and other machinery used in the fields, as power for irrigation pumps, drying, freezing, transport, refrigeration, and cooking, and as a feedstock and energy source for the production of inputs such as pesticides and fertilizers, and packaging materials (glass, plastic, aluminum, cardboard, etc.). About 17 percent of the energy used in the United States is used for food production, processing, distribution, and preparation.[31] Fossil fuels are heavily involved in agriculture worldwide, making food production and pricing sensitive to energy prices, especially in poor countries where mechanized agriculture was hard-hit during the energy crisis of the 1970s. Small wonder that ecologist Howard Odum wrote a quarter century ago: "The great conceit of industrial man imagined that his progress in agricultural yields was due to his new know-how in the use of the sun. A whole generation of citizens thought that the carrying capacity of the earth was proportional to the amount of land under cultivation and that higher efficiencies in using the energy of the sun had arrived. This is a sad hoax, for industrial man no longer eats potatoes made from solar energy; he now eats potatoes partly made of oil."[32]

A century of progressive mechanization and use of chemicals pushed millions of farmers and farmworkers off the land and into relatively high-paying urban factory and office jobs in the United States, leaving less than 2 percent of the population on the farm by 1990. Today, a medium-sized family farm in the U.S. is in the range of one hundred to two hundred hectares. In Europe, farms are smaller; indeed, not much bigger than in the sixteenth century.[33] At the extreme of developed countries, a Japanese family raising primarily rice farms less than a hectare on average.

More than anything else, however, modern agriculture is a triumph of plant evolutionary genetics. Beginning with the rediscovery of Mendel's principles of inheritance early in the twentieth century, the previous hit-or-miss process of selectively breeding plants for desirable qualities became a highly developed science. Plant scientists also developed a refined knowledge of the nutrient and other requirements of plants, which enabled them to provide appropriate soil conditions and amounts of water and nutrients to plant varieties "designed" to take advantage of them. These advances led to great increases in yields (harvest per unit area) in the major grains—the staples that form humanity's feeding base—especially corn, wheat, and rice.

All agriculture, of course, is driven first and foremost by the energy of the sun, which is captured by plants in the process of photosynthesis. Photosynthesis powers the growth of the crops that people eat and produces the forage and feed that fuel domestic animals. Because the sun's energy arrives dispersed, agricultural systems are necessarily extensive enterprises; that is, they occupy a great deal of land. Consequently, transport systems are a key part of modern agricultural systems. The means must be available to bring inputs (seeds, fertilizers, pesticides, fuel for machinery and drying of crops, etc.) to the fields, and later to take produce to markets.

The Green Revolution

Agriculture in nonindustrialized nations remained low-yield and labor-intensive (rather than mechanized and input-intensive) until well after World War II. But the rapid acceleration of population growth in those nations in the late 1940s and 1950s caused rising concern that population growth might outstrip food supplies. At that time, the colonial era was coming to an end, and political unrest occurred in many areas. Faced with prospects of massive starvation and revolution in poor nations, developed nations implemented various schemes of foreign aid for economic development. In Asia, for example, this was facilitated under a 1951 arrangement known as the "Colombo Plan," originally involving Australia, Ceylon (now Sri Lanka), Great Britain, India, New Zealand, and Pakistan, and subsequently joined by Japan, the United States, and other nations.[34] The basic idea was to use technological assistance for rural development to build a stable, prosperous, nonrevolutionary rural community.[35] Much of the impetus for the green revolution came from programs at the Rockefeller and Ford foundations, which promoted reorganization of agricultural research, extension, and training, and from the World Bank, which supplied the necessary credit.

Success with improved grain varieties in developed nations in the 1940s and 1950s led to their introduction in many developing nations in the 1960s and 1970s, with impressive results. Mexico shifted from being a net importer to an exporter of wheat in the 1970s. In India, nutritional security improved enormously as grain production increased by 4.5 percent per year between 1966 and 1971; by 1973, 75 percent of the wheat and 36 percent of the rice produced were so-called "miracle" or high-yielding varieties (HYVs). The Pakistani wheat harvest in 1968 was 37 percent higher than in any previous year.[36]

In many ways, foreign aid in those days was viewed as a "win-win strategy" for the United States and other rich nations. Vast numbers of deaths from famine could be avoided by converting traditional rural communities to Western industrialized patterns of agriculture. Simultaneously, markets for farm machinery, fertilizers, and pesticides would be opened up and guaranteed not to close. Also the U.S. Department of Agriculture (USDA) was keen to export food under subsidies or as outright aid. By the mid-1970s, the USDA was rewarded as many former aid recipients became paying customers. Timely assistance also was expected to lessen the chances that uprisings by starving peasants would undermine governments friendly to the West and further extend the influence of the Soviet Union and China.

Thus a mix of humanitarian and self-seeking motives gave impetus to a green revolution designed to prevent a red one: we would be doing well by doing good. It was a strategy that, in combination with great improvements in food distribution capabilities, was largely successful in Asia in the short run. More food was grown, and the ability to rush supplies into areas with crop failures meant that, outside of China, famine was greatly restricted in Asia after that in Bihar, India, in 1966–1967 in which the death rate in the state of Bihar rose by about a third.[37] While chronic hunger remains a fact of life for a substantial portion of the subcontinent's population, acute famines have not occurred there since.

Behind these successes were the International Rice Research Institute (IRRI) in the Philippines and the Center for the Improvement of Maize and Wheat (CIMMYT) in Mexico, which had been established in 1960 and 1966 with the support of the Ford and Rockefeller foundations.[38] By the early 1970s, a consortium of such international agricultural research centers, led by IRRI and CIMMYT, had become leaders in developing new high-yielding varieties of crops to boost grain production and keep food supplies increasing faster than Earth's booming human population. By the late 1980s, under optimal conditions, the three major grain crops were producing up to four or five times the yields of traditional varieties in developed nations. The new varieties also could do this in fewer days, increasing the potential for multiple cropping in areas where the climate was favorable.

In some developing nations, notably China, yields were converging on those of the developed nations by the 1990s. In most less developed nations, however, yields are generally well below those achieved in first-world agriculture, where temperate climates, good soils, high levels of inputs, and excellent agricultural infrastructures all contribute importantly to productivity. On a worldwide basis, average grain yields rose nearly two and a half times between 1950 and the early 1990s.[39]

Because HYVs only produce greatly increased yields if they are provided with abundant fertilizer and water, they have been more accurately called "fertilizer-sensitive varieties." The HYVs respond to high nitrogen inputs by increasing their growth rate and seed production. The tenfold increase in worldwide chemical fertilizer use between 1950 and 1990 underscores the importance of fertilizers in generating the tremendous increase in global grain production during that period—and in at least partially forestalling an ecologically damaging expansion of farming into marginal lands in many regions.

Thus the green revolution gave rise to a large-scale industry to produce synthetic fertilizers and stimulated a boom in irrigation works, both of which have led to serious environmental problems (which are discussed in the next chapter). In many areas, monocultures of HYVs have been created because planting, harvesting, and supplying inputs were more efficiently accomplished with machinery. Since fallowing and crop rotation are no longer essential, thanks to fertilizers and pesticides, large amounts of land are planted year after year with the same crop, with no apparent loss of productivity.

By the 1990s, the focus of the eighteen members of the Consultative Group on International Agricultural Research (CGIAR) had shifted to efforts to maintain previous gains in agricultural production. These tasks included: breeding new pest-resistant crop strains to combat insect pests that keep evolving the ability to attack older strains; developing more efficient systems of irrigation and weed control; improving livestock breeds; and trying to preserve as much as possible of the precious store of genetic variability of crops and wild crop relatives, which is essential for keeping humanity in the business of high-yield agriculture.[40]

Industrialized agriculture certainly has produced spectacular increases in food supplies; without it we would not be feeding today's population as well as we are. The green revolution spurred a production boom that allowed grain harvests to soar significantly ahead of population growth from 1950 until the mid-1980s.

The technology did not produce equally spectacular results in developing countries of all continents, however.[41] Its greatest success has been in Asia, where monsoon climates dominate the southern and eastern areas and run-off from Himalayan snows provide dependable water flows. When the continent heats up in the summer, low pressure is created, and moist, relatively cool air is drawn inland from the sea. It rises as it warms, and the moisture condenses and falls as rain.[42] Because the timing and amount of rain are somewhat variable, much of the continent's agriculture depends on irrigation. In 1988, some 32 percent of Asia's arable land was irrigated, as opposed to 12 percent in Latin America and 6 percent in Africa.[43] Rice is the principal crop, providing roughly half the calories consumed by Asians.

Colonial powers did not establish many large plantations in Asia, and today that crowded continent—with less than half as much arable land per person as Africa or Latin America—has huge numbers of very small family farms. More than half the farms are less than two hectares; 20 percent in India, 35 percent in Sri Lanka and South Korea, and 45 percent in Indonesia are less than a half hectare in extent.[44] In 1985, China abolished farm communes and reverted to small family farms—180 million of them—averaging just a half hectare in size.[45]

Because Asia's population density has long been high, it is not surprising that the major increases in food production between 1950 and 1985 were largely achieved not through colonizing new land but by intensification of agriculture—more careful weeding, more irrigation, more multiple cropping. After 1965, the adoption of new high-yielding varieties and application of the large amounts of fertilizers they require—the green revolution—played a major role. By 1973, the share of total area planted in HYV rice in the Philippines had increased from about 10 to almost 60 percent, and it rose from zero to 20 to 45 percent in India, Vietnam, Malaysia, and Pakistan. In less than a decade, HYVs had replaced traditional strains in over half of Pakistan's and India's wheatlands.[46] Asia also benefited from the green revolution in that most of the continent is outside the tropics; and of the areas that are tropical, many have relatively rich soils of recent volcanic origin.

Africa, by contrast, has been almost untouched by the green revolution; indeed, agriculture south of the Sahara and outside South Africa remains overwhelmingly subsistence-based. One reason for this is that the continent's biophysical characteristics, including its location spanning the tropics, constrain food production. Rainfall in most areas is undependable at best, and it is abundant in surprisingly few areas. Only about half of Africa is suitable for rain-fed agriculture, and climatic variability causes substantial fluctuations in yields where there is no irrigation.[47] Generally high temperatures create high rates of evaporation and diminish the soil moisture essential for crops. Only about 2 percent of Africa's arable land was irrigated in 1982, and the potential for expanding that area is limited.[48]

Africa has been more geologically stable than other continents, lacking the volcanic activity and land movements that produce fresh parent material for soils. This stability has produced many highly weathered soils, divested of much of their original nutrient endowment.[49] The traditional African practice of shifting agriculture, using long fallow periods, is necessary to farm such soils successfully. Population growth in recent decades has caused those periods to be shortened, but no truly satisfactory system to replace them has appeared. Humid zone agriculture there has been described as "basically labour-intensive cultivation of poor soils in an inhospitable environment."[50] An additional critical factor has been the near impossibility of using animal power in many areas because of the deadly disease, nagana.

A further problem in Africa is a political structure inimical to creating a rational agricultural system. Most African nations are very small and do not comprise sensible geographic units. Sub-Saharan Africa is carved into forty-eight separate nations, many of whose borders were established where the armies of opposing colonial powers ground to a halt. Infrastructures were designed to strip natural resources from the interior and move them to the coasts for shipment to Europe. Ethnic divisions were not considered, so many nations have been torn by tribal strife, which has battered rural areas and displaced populations. It contributed to the savage warfare and social breakdown in Rwanda in 1994. Ironically, a major problem is that too many of the political units have too few people (twenty-two countries had populations of less than 5 million in 1994) to create adequate markets, thus requiring inefficient duplication of agricultural research and extension services in each nation.

Latin America, like Africa, is sparsely populated compared to Asia. Like traditional African agricultural systems, those in tropical South America, where soils also tend to be poor, have featured long fallow periods. Here too, population growth has been associated with shrinkage of those periods. In Latin America, unlike the other less developed regions, expansion of food production has occurred largely through expansion of the area cultivated—by colonizing new land. Latin America also differs in that a much higher proportion of agricultural land has long been concentrated in large operations that farm a hundred hectares (about two hundred fifty acres) or more.[51] Some parts of the agricultural sector are industrialized, using inputs in the same manner as farmers do in North America and Europe, and concentrating largely on export crops. This industrialization, by displacing farmworkers, has given impetus to the press of migration to cities, to marginal uplands, and to other countries.

Peasant agriculture nevertheless is still very important in Latin America. It employs about half of the rural population and produces food largely for local consumption.[52] It has been argued that the principal reason for hunger in Latin America is poverty: insufficient income in the rural population to buy food or the land to grow it. In the cities, many people who fled the impoverished countryside are still poor and hungry because they have failed to find employment.

Agriculture, Equity, and Carrying Capacity

In many circumstances, the development and spread of agriculture, or of more productive farming technologies, has created serious inequities between groups. In the Americas, a European agricultural civilization displaced Native American gatherer-hunters and practitioners of simpler agriculture systems. More recently, nomadic Afars were pushed out of their traditional pastureland in the Awash Valley of Ethiopia by the development of commercial agriculture financed by corporations based in rich nations. The result was decimation of the Afar cattle herds and starvation for the Afars.[53]

Examples, of course, could be cited ad infinitum. Agricultural societies have classically displaced and destroyed gatherer-hunter societies, which certainly represents a vast inequity from the standpoint of those whose culture (at least) or lives (at worst) have been wiped out. Of course, some observers would argue that their loss is compensated by the gain in carrying capacity that results from establishing agriculture (assuming the land is suitable for cultivation), since that potentially will allow many more people to live an equitable life. We have our doubts, but resolution of this question may boil down to an ethical judgment—does a happy life for a larger number of people in the future justify destroying the lives of a few people today?

On a far grander scale, the Western model of input-intensive farming has been widely transplanted into developing regions since 1965 in the effort to increase their agricultural production. In the process, it has often displaced polycultural systems that might more satisfactorily supply the nutritional and economic needs of rural societies.[54] It also has often had profound effects on social systems and the levels of equity between individuals and social groups, including rural and urban populations.[55] And modern transport systems have made it possible for people to become culturally separated from their sources of subsistence.

Men and women fulfill different roles in agriculture, and changes in farming systems can have important impacts on equity between the sexes. In some parts of the world, notably in sub-Saharan Africa and South Asia where subsistence agriculture still prevails, women contribute a large share of farm labor. In the tribal economy of Orissa, India, women spend almost twice as much time in agricultural labor as do men,[56] and in the Himalayas, women contribute more time to farming than do men and bullocks combined.[57] This may seem to be simply another example of the gender inequity that pervades most of the world. But the role of women in agriculture is diminishing (and thus, often, their status) as more and more of the world's staple foods are produced in industrial agricultural systems, where men are preferred as wage laborers.

Evaluating the effect of this change on an area's carrying capacity is difficult. Women are often the vessels of traditional agricultural knowledge; sharing the tasks more with men, while lightening women's burdens, might lead to more factory-style farming and less care of the land, thus lowering the long-term carrying capacity. On the other hand, increased land degradation sometimes results if women are left alone with the burden of all farming responsibilities when men depart to cities or elsewhere for paid work. Yet, since children often share the women's heavy work, their economic value might be lessened if men did more of the farm work, leading to lower fertility and a population in a more favorable relationship to its area's carrying capacity.[58]

The changing importance of women in agriculture appears to be reflected in the sad figures on female infant and child mortality today in societies where commercialization of agriculture has put a premium on males to do heavy field labor.[59] The degree to which women have been devalued as producers of food has been correlated with the degree to which early weaning, protein limitation, or general food deprivation have driven up female child death rates. Thus female mortality in southern India, where women's labor in paddy rice production is highly valued, is less skewed than in Pakistan and northern India, where their labor is not as essential.

If equity is defined as a more or less even distribution of land and rewards among those laboring in agriculture, then we are convinced that greater equity would increase carrying capacity under most circumstances. The distribution of land ownership is perhaps the most important equity issue in agriculture. Is the best land concentrated in the hands of a few large landowners, while the majority of workers engaged in agriculture are either scratching out a living on tiny, marginal plots that they own or work as tenants, or working as laborers for big landowners? Or is agricultural land rather equitably distributed (in terms of both quantity and quality) among family farmers? The usual assumption is that the former situation is less socially desirable than the latter; farmworkers often command very poor wages because of artificially low food prices and labor surpluses. Hence governments at least talk about promoting more equitable distribution through "land reform."

Agricultural economists consider large plantation or estate systems to be generally less efficient than family-operated farms. While crops like cotton and sugarcane can be easily cultivated by unskilled crews, in most cases knowledge and care by the farmer are necessary to get the most out of a variety of crops and livestock. Differing microclimates and soils may require quite different treatments from field to field, and animals often need individual handling. The need for careful management can be very high, and monitoring the quality of operations even by unskilled laborers carrying out routine tasks becomes increasingly difficult as the plantation gets larger.[60]

Today we often associate size with success. People assume that, with all their machinery and inputs, big farms should be more efficient. The small farmer wishes to maximize yield per hectare; the incentives for the larger landowners, on the other hand, are to maximize yield per worker. This is reinforced by a desire to keep wages low by maintaining a large unemployed labor pool. A consequence of these contrasting incentives is that small farmers produce about twice as much per hectare as do large farmers, while using only one fourth or one fifth the amount of purchased inputs per hectare.[61] The inherent inefficiency of large-scale plantation farming has sometimes led to its replacement by family farming when demand for labor in cities put upward pressure on farm wages. This was an important factor in the sixteenth-century decline of the European manorial system.[62] Nonetheless, there are economies of scale in farming, so the relationship of efficiency to farm size can be quite complicated, and efficiency may generally peak at the upper end of the family farm size spectrum.

In overdeveloped countries like the United States, the trend has overwhelmingly been toward larger farms specializing in a single crop. The average American agricultural worker farms 137 hectares, near the upper end of the range in size of family farms. In the developing world, where the income disparity between farm families and urban workers is a serious issue, land reform has proven difficult to achieve. For example, Mexico's efforts after the revolution of 1910 were abandoned when only partially completed; the 1994 revolt in Chiapas was in part a result. The Far East (including Japan), by contrast, has been relatively successful in creating and maintaining equity in land tenure, and this unquestionably has been an important factor in its food production success. Should Japan's policy of sheltering rice prices from imports be reversed, however, the profitability of small farms may be lost and the land tenure situation may change drastically. Land inequities in Africa, mainly in eastern and southern Africa, are largely holdovers from colonial days; some of the best land is still owned and farmed by descendants of European colonials. And the employment of impoverished migrant farmworkers, often temporary immigrants, in both rich and poor countries is also a case of inequity, especially since migrant workers frequently include children.

On a global scale, the economic policies of rich and poor nations alike have helped push the food system into a situation of increasing disequilibrium. Since World War II, developed nations have created an incredibly complex system of national policies to protect domestic agriculture, discouraging imports while encouraging production and exports. Meanwhile, many developing nations have adopted industrial development policies that discriminate against domestic agricultural production in favor of industry. These mutually reinforcing trends have created the phenomenon of simultaneous food gluts and widespread hunger.[63] Roughly 45 percent of the world's grain is produced in the industrialized nations; and about 40 percent is also consumed there—although those nations have barely a fifth of the world's population.[64] About one third of the world's grain is used as livestock feed; the bulk of that is produced and consumed in rich countries.

The reasons for the loss of equity and a retreat from maximization of production[65] in many regions are varied and complex. Demand for many agricultural commodities can be quite inelastic, and technological innovation often cannot lower costs enough to compensate for falling prices caused by surplus production. Urban consumers benefit, but farm income is lowered, forcing less competitive farmers out of agriculture.[66] Technological progress in farming can work against equity, hurting the farmers who are less able to adopt innovations rapidly (often the smaller family farms), while favoring nonfarm consumers.

The introduction of modern HYVs into agricultural systems in less developed countries has been at best a mixed blessing from the standpoint of equity. From the start, it was controversial; many observers argued that a green revolution based on the HYVs would worsen existing inequities in rural areas.[67] They noted that the HYVs were more readily adopted by richer farmers, who had access to information and the financial resources to pay for the water systems and fertilizer and pesticide inputs required by the HYVs.

With greater profits, large farmers would be able to buy out smaller ones, sometimes then employing them as landless workers. As they expanded their holdings, it would eventually become profitable to replace the workers with machinery. With each step, the disparity between rich and poor farmers would widen. Agricultural wages would fall and be further depressed by the growing numbers of landless laborers.[68]

Economists Yujiro Hayami and Vernon Ruttan maintain that these predictions have not been borne out, however. They emphasize that not all technological advances have the same impact on equity in agricultural systems. Mechanical innovations, such as giant harvesting machines, generally benefit large operations that can take advantage of economies of scale. By favoring land over labor, these changes ordinarily increase land ownership inequities. Biological innovations, such as improved seeds or fertilizers, on the other hand, should increase the efficiency of small family operators who can make the small-scale, site-specific management decisions necessary to gain the most from these innovations.

Biological innovations theoretically reduce the need for land by intensifying farming, thus favoring labor over land and promoting a more equitable land distribution. That is not necessarily the case in practice, however, as shown by a current trend toward substituting herbicides for hand-weeding in Asian rice production, which is driven by higher labor costs resulting from economic development off the farm.[69] The effect of this shift on long-term carrying capacity is uncertain because the net environmental effects of herbicide use remain largely unknown. Social costs might also include a devaluation of the older, unskilled women who do most of the hand weeding, especially if no alternative employment is available.

Overall, since the HYVs are biological technologies, one might expect the green revolution's impact on equity to have been positive. The question is confounded, however, by the long-continued rapid population growth in developing countries where the green revolution was deployed, which has dramatically increased the ratio of people to land. In the 1960s and 1970s, the number of agricultural workers in South and Southeast Asia increased by 1 to 2.5 percent annually, while the amount of cultivated land increased by only about 1 percent, and the added land was mostly of lower quality.[70] It has been claimed that the green revolution, by introducing land-saving technologies and thus partially restoring the population-farmland balance, actually prevented an even greater trend toward inequity.[71] Here is an example of a possibly very serious "social trap," a situation where short-term individual incentives are inconsistent with long-term societal well-being. Population growth apparently promotes inequity, while inequity apparently lowers carrying capacity and possibly also promotes population growth. It is Partha Dasgupta's downward spiral writ large.

This sort of analysis of equity and the green revolution is suspect on several counts, however. One is that, to a great degree, the original intention of the green revolution was to substitute technological progress for trying to achieve rapid progress on difficult equity issues such as land reform. This was not necessarily an evil strategy—social change generally comes slowly, and the food crisis was severe. As M. S. Swaminathan, a distinguished agricultural geneticist, said recently, "[in 1966] there was a real threat of large-scale famine if India were not to make very rapid progress in improving agricultural production."[72] Another analyst seemed to be telling only part of the story when he stated, "the green revolution was to be a substitute for the red—and it is misleading to claim that there was nothing wrong with the original strategy and that the fault lies with inappropriate institutions and policies."[73]

Another problem is that in making yield comparisons between traditional and HYV-based systems, only grains are considered, and thus many of the yields from polyculture (for instance, fiber, fruits, vegetable, tubers, pulses, oil crops, fodder) are not included in the comparison.[74] This is especially problematic in hungry nations like India, where the entire system has tended to suppress the production of pulses (peas, beans, and lentils), thus denying many poor people a critical protein component of their diets.[75]

Of course, government policies can greatly influence the impact of green revolution technologies on equity. In India, the government has chosen to concentrate its subsidization of inputs in restricted areas where large grain surpluses could be generated.[76] Yet recent analysis suggests that the negative effects of the green revolution on equity in that nation have been uneven, were more severe early on, and have subsequently lessened or even reversed themselves. Recently, rising farm income traceable to the green revolution has increased off-farm employment opportunities for the landless and narrowed the rich-poor gap.[77] As with many other aspects of the population-food-equity nexus, things never seem to be simple.

Last, but far from least, an array of environmental problems attending the green revolution has cast considerable doubt on its long-term sustainability.[78] These will become apparent as we proceed.

The World Food Situation

Norman Borlaug, when receiving the Nobel Prize in 1970 as a founder of the green revolution, cautioned that, at best, the new technology could buy humanity thirty years to solve the population problem. When he spoke, there were still fewer than 4 billion people. Twenty-five years later, the human population has exceeded 5.7 billion and is still growing at 1.6 percent per year, adding some 90 million people annually.

Despite warnings by Borlaug and many others, the belief lingers that the green revolution has essentially solved the problem of feeding the growing population and that famines are simply the disastrous results of political conflicts.[79] Indeed, it has often been asserted that the persisting widespread chronic undernourishment in poor countries results from maldistribution of otherwise abundant food supplies, and that better distribution would solve the hunger problem.[80]

There is some truth in this view. Starvation today is primarily a problem of food distribution failures, and acute cases are often precipitated by political turmoil in an already vulnerable, poorly nourished population.[81] But while these tragic situations gain much public attention, they are a tiny tip of the iceberg of widespread hunger, mostly in poor nations, whose causes are far more complicated.

Even now, when much talk still is of food "gluts," hunger remains one of the most serious elements of the human predicament. Low grain prices are not an indicator of nutritional security, but of the inability of poor people to generate demand for food. The United Nations Children's Fund (UNICEF) estimates that one in three children under five years old in developing nations is malnourished, and that each year in recent decades on the order of 10 million people (the vast majority of them young children) have died of hunger or hunger-related diseases.[82] Other international agencies calculate that up to a billion people are unable to obtain sufficient energy from their food to carry on normal activities.[83]

Even if those estimates are greatly inflated,[84] the nutritional situation amounts to a human tragedy of enormous proportions. Besides the colossal waste of human resources resulting from impairing the productivity of millions of people, chronic hunger also threatens the educational potential of tens of millions of children, increases the population's vulnerability to epidemics such as AIDS, Ebola virus, new flu strains, and drug-resistant tuberculosis, and threatens the political stability of the nations most affected.

As the world population continues to grow, expansion of food supplies to meet their increasing needs, including those now undernourished, may prove more and more difficult, even if the population's size can be kept well below 10 billion. By 2040, production would have to be nearly tripled—in essence repeating the heroic achievements of the decades from 1950 to 1984. But most of the readily available opportunities for substantial expansion of world food production—opening new fertile lands, developing the first fertilizer-sensitive "miracle" strains of major crops, and applying the first doses of synthetic fertilizers—have already been taken. The enormously successful green revolution has now been deployed throughout the world in most suitable regions (some tropical forests and many areas of sub-Saharan Africa, for example, lack suitable soils), and it now seems to be approaching the limits of its capability for the three major crops.

Given recent trends in the condition of natural capital and the potential for further boosting crop yields, Lester Brown of Worldwatch Institute warned in 1988 that world grain harvests in the 1990s were unlikely to increase faster than the population and could fall behind population growth after 2000.[85] So far, history has not contradicted that prognosis. Since 1985, global grain production per person has been gradually slipping downward.[86]

The change has gone unnoticed for several reasons. Much of the decline was due to a substantial amount of relatively poor, highly erodable land being taken out of production in the United States, Europe, and some other countries. A desire to reduce surplus grain stocks was also a reason.[87] Moreover, demand for meat, especially grain-fattened beef, has been falling in the richest nations, mainly for health reasons, but also because of economic factors. The small per-capita grain production decline has thus been offset by reduced demand for the large fraction of the world's grain production that is used to feed livestock, so no global shortage in food for people has been apparent.[88]

While the per-capita decline in grain production at this point can be attributed largely to changes in demand and agricultural policies in rich nations, as well as economic factors, the problems that will constrain future production are very real. The green revolution's primary goal of a rapid increase in grain production has clearly been achieved in many poor countries, but it has made little impression in others. The long-term prospects for keeping the plow's production ahead of the stork's remain unclear, and the long-term costs already incurred may prove far greater than the benefits.

Fisheries Under Pressure

Have you heard that bluefin tuna for making sashimi now sells for $350 a pound in Tokyo restaurants? Have you noticed that various "blackened" fish dishes have become more common in your local restaurants? Both phenomena are symptoms of another global food problem. Provision of food that can be extracted from the sea is one of the most important free services that natural ecosystems perform for people. The roughly 90 to 100 million metric tons (mmt) of fish produced annually is a small factor in the human feeding base compared with about 1.8 billion metric tons of grains.[89] Seafood nevertheless provides an important protein supplement in the diets of many people; more than half of all human beings get most of their animal protein from fish, and for many poor people it is the only animal protein in their diets.[90]

Nowhere is the confrontation between human numbers and the planet's food supply more evident than in the area of oceanic fisheries. Under water, humanity harvests a natural biological resource. In principle, only the "interest" from this natural capital should be taken. That is, the harvest should be on a sustainable yield basis—the amount of fish extracted from each stock annually should not be more than needed for natural reproduction of the fishes to replace the losses and sustain a similar harvest the next year. But unhappily, in this case as in so many others, *Homo sapiens* is using up natural capital. We are overfishing many stocks—New England and Newfoundland cod and haddock, Atlantic bluefin tuna, red snapper in the Gulf of Mexico, pollock in the Bering Sea, orange roughy in New Zealand waters, and so on. The bluefin stock in the western Atlantic is now about a tenth of what it once was, and it is being considered for listing as an endangered species.[91] One of these days, sashimi restaurants may become an endangered species as well.

Since 1987, the world catch has hovered between 95 and 100 million metric tons, coincidentally very close to the predicted maximum sustainable yield (MSY) if the resources were properly managed.[92] But it is not being properly managed. Indeed, the current pattern is one of fierce competition for fish and overexploitation of stocks to the point of collapse, followed by shifts to exploitation of new stocks, generally of less desirable species.[93] According to the United Nations Food and Agriculture Organization (FAO), all major fishing areas have reached or exceeded their MSYs, and about half of them are in serious decline.[94] As a result, per-capita yields from the sea have been dropping since 1989 and probably will continue to drop. The decline would no doubt be much more dramatic without the fast-rising portion of the seafood harvest in recent years from aquaculture, accounting by 1991 for nearly 15 percent of the total (about two thirds of that from artificial tanks, ponds, lakes, and rivers, and the rest from coastal mariculture—fish and shellfish grown in the open ocean or bays).[95]

Fishing fleets scour the oceans and have seriously depleted the most valuable stocks such as bluefin, and have now moved on to "mining" less desirable ones like pollock. The growth in fish catches since 1983 has been in five low-value species, four of which are used only for animal feed.[96] That's why blackening has become so popular—so that fine restaurants can serve you what once were considered trash fish.

Overfishing continues for many reasons. A central cause is a lack of well-defined property rights (or enforcement thereof) which makes overexploitation the optimal strategy from nearly every fisherman's short-term perspective. This is reflected in a large investment in high-tech gear ranging from computer databases and 16-color sonar display units that help find fish schools, to gigantic nets that may bring in 30 tons of fish in a single haul. The industry is overcapitalized; too much money has been sunk into fishing fleets and processing plants.[97] Today there are more than 3 million fishing vessels, some 41 percent of all ships on the sea. This overcapitalization is abetted by natural fluctuations in fish populations. When they expand, boats are built, men are employed, and processing plants flourish. When stocks are depleted, governments are besieged for subsidies to protect investments and jobs.

These are perverse subsidies that encourage further decimation of stocks.[98] A measure of the perversity of those government subsidies is that in 1992 the fisheries industry around the world was able to spend $124 billion to catch $70 billion worth of fish.[99] Attempts to regulate fisheries have not been very successful; often they amount to too little action too late. When regional groups agree to set limits, fishers often change their ship's registry to another nation not party to the agreement.

Fish populations may also decline for reasons other than overfishing. One reason can be the general pollution of the oceans, which is especially critical in shallow near-shore areas where the richest fisheries are located. Of the marine fish catch, 90 percent comes from the 33 percent of the ocean areas closest to land.[100] Another is the pollution and outright destruction of coastal wetlands, which often serve as "nurseries" for important food fish or the species upon which those fish prey. A dramatic example is seen in the widespread decline of coral reef fisheries, which are often locally important sources of protein. Much of the decline is traceable to an influx of relatively minute quantities of nutrients that allow algae to overgrow and kill the corals.[101] The rich oyster beds of Chesapeake Bay, which once produced 8 million bushels annually, now produce about 300,000 because of pollution. About half of Nova Scotia's shellfish beds have been closed because of contamination.[102] And Georges Bank, off Cape Cod, once one of the world's richest fishing grounds, was all but shut down in October 1994 by a decision of the New England Fishery Management to protect the surviving remnants of several severely depleted fish species.[103]

The only bright spot in the fisheries picture has been a rapid rise in production from aquaculture, but that too has carried environmental costs and risks.[104] Pollution problems have plagued many fish-farming ventures, and, like crop monocultures, fish monocultures are vulnerable to diseases. Industrial aquaculture is still in its infancy, however, and appears to have great potential. As noted by prominent environmental scientist Norman Myers, there are two reasons for this:

> First, water dwelling creatures enjoy a distinct advantage over their terrestrial relatives in that their body density is almost the same as that of the water they inhabit, so they do not have to direct energy into supporting their body weight; this means, in turn, that they can allocate more food energy to the business of growing than is the case for land animals. Second, fishes, as cold-blooded creatures, do not consume large amounts of energy to keep themselves warm. Carp, for instance, can convert one unit of assimilated food into flesh one and a half times as quickly as can pigs or chickens, and twice as rapidly as cattle or sheep. The tiny shrimplike crustaceans called *Daphnia* can, when raised in a nutrient-nourished environment, generate almost 20 metric tons of flesh per hectare in just under five weeks, which is ten times the production rate for soybeans—and at one-tenth the cost per unit of protein produced.[105]

Renowned biologist Ed Wilson points out that it is high time to modernize the harvesting of aquatic organisms. The fishing industry is primitive—over 90 percent of the fish consumed worldwide are hunted in natural environments—and increasingly destructive.[106] Indeed, the overfishing and degradation of the marine environment may have reduced potential harvests from the sea for the foreseeable future. Experience, though limited, indicates that, once depressed to small numbers, many fish populations recover very slowly.

Overall, a more or less continuous decline of this critical portion of the supply of animal protein for human consumption seems very likely. It is critical because it is so important to the poor people of the world. While the overfed rich pay small fortunes for delicacies like sashimi, in Africa and Asia more than a fifth of the population depend on declining fisheries as their chief source of protein.[107]

Where Are We Going?

As the world food problem moves from a complex and confusing present into an even more uncertain future, there is one question that dominates the considerations of all ecologically literate scientists: At what point will environmental disruption start substantially reducing Earth's capacity to nourish humanity? In other words, what are the environmental limits to the agricultural enterprise? It's to that central question we now turn.

Chapter Six

THE ENVIRONMENTAL
CONNECTION

The differences between "optimists" and "pessimists" regarding the world food situation trace largely to whether the natural underpinnings of agricultural productivity are understood. Of all the major elements of the human enterprise, agriculture is both the most dependent upon and the most destructive of the natural environment. The world agricultural system is fueled by a tremendous wealth of natural resources. Paramount among these are vast areas of fertile land; fresh water taken not only from nearly every place it falls and collects now, but from aquifers into which the last ice-age glaciers melted away thousands of years ago; and biodiversity, the raw material from which our present array of crops was derived and upon which hopes for future improvements rest. Agriculture also depends critically upon protection of atmospheric quality: mitigation against global warming; control of ground-level toxic air pollution; and safeguarding of the essential UV-blocking stratospheric ozone layer.

Under extreme pressure to produce greater and greater quantities of food each year, the entire agricultural system and its natural underpinnings are showing serious signs of strain. Today's farming practices over much of the planet are not only unsustainable, but are actually diminishing Earth's capacity for agricultural production. This is apparent in accelerating rates of human-induced land degradation. Fully one third of Earth's vegetated surface has been degraded to some extent since World War II, and as much as two thirds could be degraded by 2020 if current trends continue. In addition, freshwater aquifers are being rapidly depleted, threatening to pull large areas out of irrigation in coming decades. The rate of human-induced

168

species extinction is now estimated to be one million times faster than the rate of evolution of new species. Atmospheric quality is deteriorating rapidly, due in part to the scale of agricultural production.

Virtually all of these changes are inherently or could soon become completely irreversible on any time scale of interest to humanity. They also have important underlying social causes whose alleviation could dramatically improve the overall outlook. These include a failed urban-oriented development strategy, inequitable distribution of land, lack of access to inputs and farm credit, widespread unemployment, and inequities inherent in the world food market.

Now that the world community is no longer transfixed by the Cold War, the severity and pervasiveness of threats to environmental and nutritional security have become more apparent.[1] These aspects of security are closely interconnected; agriculture is extremely sensitive to environmental conditions, and yet it ranks also as the single largest proximate cause of environmental disruption. Furthermore, both environmental and nutritional security are inherently international problems, as reflected in the global trade in commodities, the global environmental commons, and the mass migrations of people that can be provoked by regional food scarcity.[2]

Doubts about humanity's ability to expand food production massively in coming decades stem from two basic observations. First, even though agricultural output has grown faster than Malthus—or even some optimists—predicted, past expectations that the world food system could easily feed a population of 5 billion have not been met.[3] In fact, about 250 million people—nearly as many as now live in the United States—have died of hunger-related causes in the past quarter century.[4] As many as a billion people—one in every six—are chronically undernourished today, about half of them quite seriously so.[5] In several major developing regions, including Africa and South Asia, the numbers of hungry people have continued to increase,[6] despite the impressive gains in world food production. We explore the underlying causes and remedies of these tragic circumstances in the next chapter.

Here we focus more on the second source of doubt: that complacency about the security and abundance of the world food supply, even in the near future, is not justified if the environmental underpinnings of the agricultural enterprise are considered. The extraordinary expansion of food production in recent decades has been achieved at a heavy cost: the depletion of a one-time inheritance of natural capital crucial to agriculture. This cost now amounts to an annual loss of roughly 25 billion tons of topsoil—in fact, the largest export by weight (more than 2 billion tons) from the United States is eroded topsoil.[7] The cost also includes trillions of gallons of groundwater.[8] And half or more of Earth's biological diversity may be lost by the middle of the next century—populations, species, and ecosystems, which are involved directly and indirectly in maintaining agricultural productivity.[9] These growing losses in Earth's productive potential are all permanent on any time-scale of interest to humanity. Continuing to expand harvests is likely to prove increasingly difficult as the inherent constraints of a finite world more and more come into play.

One of the best measures of humanity's approach to ultimate limits is that we now collectively control about 40 percent of Earth's basic food supply for all animals on land. What is the basis of this estimate? Consider that the ultimate limit on food production may be the amount of energy from sunlight that can be converted to food energy by the chemical process of photosynthesis. The technical term for the photosynthetic product formed at the base of food chains (subtracting that used by green plants and other photosynthesizers to sustain their own life processes) is *net primary production,* or NPP.

Humanity consumes directly (as food, feed, and fiber, including timber) some 5.5 percent of the NPP on Earth's land surfaces. Including that seemingly modest amount, humanity also co-opts (by using it directly or diverting it into human-dominated ecosystems—farms, grazing lands, tree farms, exploited forests, etc.) about 30 percent of the total NPP. Taking also into account the loss of potential productivity due to land conversion (from natural to human-controlled systems) and land degradation, total human diversion of the world's potential NPP on land is about 40 percent.[10]

Of course, humanity naturally first exploited the most productive areas, thus taking over the NPP that was most readily accessible and easily put to human uses. No one knows how much more *Homo sapiens* could seize, trying to support a population surging toward 10 billion or more, without catastrophic damage to Earth's basic life-support systems. In practical terms, this means that feeding an additional 3 to 6 billion people will be much more difficult and risky than providing for the first 3 to 6 billion.

The principal environmental constraints to increasing food production are:

- *Losses of farmland* to human settlement and degradation[11]
- *Limits to freshwater supplies* for irrigation[12]
- *Declining genetic diversity* of crops and their wild relatives[13]
- *Diminishing marginal effectiveness of fertilizers*[14]
- *Pesticide problems*[15]
- *Increased ultraviolet-B radiation*[16]
- *Toxic air pollution*[17]
- *Climate change and sea-level rise*[18]
- *Biodiversity loss and a general decline in the free ecosystem services* supplied to agriculture by natural ecosystems[19]

Losses of Farmland

Earth's 5.7 billion people now occupy or use some 90 percent of the land surface that is not desert (receiving less than 10 inches—250 mm—of rain per year) or under permanent ice cover. About 17 percent of that land is planted in crops; about 2 percent is urban or otherwise built on, and the rest is used as pasture, or covered by forests that are exploited to one degree or another. The remaining land covered by "natural habitat," especially that under tropical moist forests, plays important roles in supporting human societies such as storing carbon and controlling the hydrologic cycle.

The remaining natural habitat is almost all "marginal"—poorly suited to intensive agriculture—as is indicated by the small fraction of the increase in food production since 1950 that is attributable to an expansion of cropland. In 1950, 593 million hectares were planted in grains; by 1990, that had increased by 21 percent to 720 million hectares, but production had increased by about 175 percent due to more than a doubling of average yield, from about 1 ton per hectare per year to about 2.5 tons per hectare per year.[20]

In fact, the global land area dedicated to cereals production has declined by roughly 5 to 6 percent since 1981.[21] Although reliable statistics on their relative importance are difficult to obtain, it is clear that there are four primary reasons for the decline: (1) switching land to the production of other foods; (2) conversion of farmland to nonfarm uses; (3) set-asides of marginal land in the United States, Europe, and the former USSR; and (4) outright abandonment of severely degraded land.

In response to a general diversification of diets, farmers in a number of important agricultural regions have shifted cereal cropland to the cultivation of oilseeds and pulses, such as soybeans. In addition, in northern China, extensive rice-producing areas were recently converted to potatoes.[22] These changes do not represent any disappearance of farmland—they just complicate the accounting of land use.[23]

Prime farmland is disappearing rapidly, however, increasingly sacrificed to meet the growing demands of urbanization. Population growth, urban migration, and industrialization are driving the expansion of cities over the rich agricultural land near which they typically were founded. This loss of farmland has occurred in disparate places. In the United States, for example, the American Farmland Trust recently named twelve regions as highly threatened by population growth and urbanization, including the Central Valley and coastal regions of California, south Florida, the mid-Atlantic coast/Chesapeake Bay area, and the Chicago–Milwaukee–Madison metro area. Although the twelve most threatened regions collectively represent only 5 percent of U.S. farmland, they account for 17 percent of total U.S. agricultural sales, 67 percent of domestic fruit production, 55 percent of vegetable production, and 24 percent of dairy products. Yields are nearly six times greater in these twelve regions than in less threatened areas.[24]

In East Asia, more than 1,900 square miles are lost to urbanization annually.[25] In China, for example, total farmland has declined by almost 20 percent since the late 1950s, swallowed up by factories, highways, housing, and other development.[26] China has 21 percent of the world's population but only 7 percent of the world's arable land, and the accelerating loss of farmland is prompting renewed government fears of food shortages.[27]

The cropland-living space competition is not limited to city margins. NASA scientist Marc Imhoff and his colleagues have analyzed that competition in rural Bangladesh, where people live on bulwarks raised above the water level of the countryside, which is flooded annually by rainfall and runoff.[28] During the dry season, soil for the bulwarks is dug from the centers of the paddies. The area of the bulwarks increases at the expense of paddy area as the population grows, and the water in the excavated paddies deepens to the point where the highest-yielding rice strains can no longer be planted, leading to a decline in rice production.[29] Ongoing research on deepwater rice cultivars may ameliorate the latter problem, but the need for more living space is sure to continue. Imhoff proposes that such losses of paddy area are widespread in South and East Asia. They have gone undetected largely because of the "noise" in statistics created by fluctuations in harvests related to weather and economics, and because they are partly offset by increased yields from better agricultural technologies and intensified cultivation.

A potential for further conflict over land use arises from the need to move toward a sustainable global energy economy. Today, about 75 percent of the world's energy is supplied by fossil fuel combustion.[30] The associated release of greenhouse gases and consequent threat of global warming has spurred research into other options, of which one of the most discussed is energy from processed biomass (basically plant matter). The burning of biomass fuels involves no net CO_2 emissions when done sustainably.

The trouble is that for biomass fuels to supply a substantial part of a sustainable global energy budget, vast areas of land—on the order of 10 percent of the world's croplands, managed forest, and permanent pasture—would have to be dedicated to biomass energy crop production.[31] It is highly questionable whether the benefits of converting to biomass fuels would outweigh the associated negative impacts on food security. Moreover, the intensive production of energy crops causes many of the same environmental impacts generated by the production of food crops.[32]

Nonetheless, it appears certain that the competition between food production and other uses for fertile land will intensify. At the same time, the amount and quality of productive land is deteriorating at an alarming rate. In some regions, including the United States and Europe, erodable land has been set aside largely with short-term economic goals in mind (a desire to stimulate grain prices), as opposed to recognition of environmental constraints on sustaining agricultural productivity.[33] In the United States, however, the set-asides were also for conservation purposes and were intended to be permanent.

Even with removal of fragile lands from production, human-induced land degradation could hardly be more apparent. Land degradation involves changes in soil, vegetation, topography, and climate that result in lowered potential for the land to supply benefits to society. Degradation is usually measured in terms of lost productivity. Since World War II, a full third of the world's vegetated land surface has become degraded to some degree, and global rates of degradation appear to be accelerating. Worldwide, an area of arid and semi-arid land the size of Colorado is abandoned each year due to loss of productivity.[34] In tropical regions, only half of the forest land cleared each year expands the area yielding agricultural benefits, while the other half simply (and for the short term) replaces land that has had to be abandoned.[35] Major causes of land degradation are deforestation, overgrazing, agricultural production, overharvesting of fuelwood, the invasion of undesirable plant cover, and, in some places, industrial pollution.

Fertile topsoil is arguably the most important component of agricultural productivity. Rich soils absolutely teem with microscopic life, containing billions of tiny organisms in each pinch. For example, a gram of fertile soil can yield over 2.5 billion bacteria; 400,000 fungi; 50,000 algae; and 30,000 protozoa.[36] These organisms play critical roles in recycling and mobilizing nutrients and transferring them from soil to crops. Soil organisms are utterly indispensable to successful farming, and yet modern industrial agriculture is probably their greatest threat.[37] Topsoil is now being eroded away at a rate of roughly 25 billion tons per year, or possibly as much as 7 percent per decade.[38] On our recent trip to China, we learned that by 1990 some 14 percent of that nation's

174

farmland had suffered severe soil erosion, up from 12 percent in 1978.[39] The loss of this precious element of natural "capital" that humanity inherited is essentially irreversible—it takes thousands of years to generate sufficient topsoil to form productive land.[40]

When topsoil is lost, one is forced to grow crops on subsoils. One of the best-documented early examples of this transition was in Greece, where Plato recorded the devastation of Attica's soils following deforestation: "what now remains compared with what then existed is like the skeleton of a sick man, all the fat and soft earth having wasted away, and only the bare framework of the land being left."[41] The Greeks were forced to switch from topsoil crops like grains to subsoil crops like olives and grapes.[42]

Fortunately, "severe" and "extreme" forms of soil degradation—such as that in the hills of Attica, from which there is virtually no chance of recovery on a time-scale of relevance to society—are presently limited to just 4 percent of the total degraded area. If efforts to halt degradation are not made quickly, however, this figure could quadruple by 2020. Humanity has lost roughly 8 percent of the potential value of productive land to degradation already; continuation of present trends would put that figure at about 20 percent by 2020.

The most famous example of land degradation occurred in the Sahel, the vast region bordering Africa's Sahara on the south. The people of the Sahel are nomads who migrate with their herds (pastoralists), peasant farmers who depend on rainfall rather than irrigation, and "agropastoralists," peasants who also tend livestock. They live in a climate that is arid or semi-arid, in which the scant rainfall occurs very irregularly. There are often long sequences of drought years, broken by years of unusually heavy rainfall.

The 1950s and early 1960s were wet years in the Sahel.[43] In the late 1960s and early 1970s, a string of low rainfall years was followed by a short break and then more drought. The results were desertification, severe food shortages, and world attention captured by pictures of skeletal people, dead livestock, and sandy landscapes punctuated with the skeletons of bushes. The event generated a scientific debate about its origins—was it part of normal climatic variation or was the climate changing? What part had human activities played in reducing food supplies?[44] While there is still some debate about the role of climate change, the basic sequence of desertification appears to have its roots in human activities, the expansion of farming populations, and the period of relatively wet weather.[45] Governments tried to use the Sahel as a safety valve on population growth, encouraging farmers to spread into the then newly lush southern fringes of the Sahara. Pressure from northward-moving farmers, and expansion of cash crops (like peanuts) onto fallow lands traditionally used for grazing the herds of nomads in the dry season, pushed Tuareg pastoralists northward. The fabled "blue men of the desert" left the lands where long-term climate favored grazing over growing rainfed crops, and were themselves made more vulnerable to droughts that could decimate their herds.[46]

In short, both farmers and Tuaregs moved onto land that could not sustain their activities. Governments sought to support those activities by drilling large numbers of deep wells to supply water year-round. As a result, during the dry years the pastoralists' livestock did not die of thirst; rather they died of hunger as the sparse vegetation was destroyed by trampling, overgrazing, and firewood gathering. Since the wells were insufficient for irrigation, peasants were forced to abandon their farms as crop yields dropped and their fields were invaded by the surviving livestock of pastoralists once again moving south. Winds eroded away the soil, which had lost its protective plant cover, and the marginal lands were desertified, making a continuation of the Sahara and giving the impression that the desert was marching southward.[47] Basically, "drought followed the plow,"[48] "as cultivation was extended into less productive land and as population increased, so the region's ability to meet its people's food needs was jeopardized."[49]

The pattern of drought and desertification resulting from attempts to farm marginal lands has also been seen in the northeast of Brazil, which "continues to be plagued by the problems generated by an increasing population trying to sustain itself on a dwindling resource base (i.e., increasingly marginal areas)."[50] The basic story is the same in Ethiopia, where people have moved in and started farming marginal areas with highly variable rainfall, and in the Maghreb (Tunisia, Algeria, and Morocco).[51] It happens also in richer nations: South Africa, the Soviet Union, and Australia.[52] In the United States, an area roughly the size of the thirteen original states was estimated to be undergoing severe desertification in the late 1970s.[53]

Perhaps the ultimate in land degradation is found in the unfortunate nation of Haiti, once largely covered with dense forests. They now cover only about 2 percent of the land, the rest having been cleared to make way for the activities of a population whose density has reached over 660 people per square mile (255 per km²). That is about the population density of many countries of western Europe, but there the similarity ends. Haiti is not a flat land of rich industrial farms with a benign climate where three quarters of the people live in cities. More than two thirds of Haiti's farmland slopes more than 20 degrees, and more than two thirds of the people are trying to scratch out a living on slopes that were largely denuded of soil by tropical rains after deforestation. The amount of arable land has declined by 40 percent since 1950, and per-capita grain production has been cut in half.[54] Haiti has been largely converted into a tropical simulation of a desert.

Ecologist John Terborgh described the Haitian scene as viewed across the border from the highlands of the neighboring Dominican Republic:

> On the Haitian side, there was no vegetation. The border sharply divided the scene, a boundary between green and gray. The landscape on the other side was naked and bleak. Even more to my astonishment, when I scanned with binoculars, I could see that the bare rock in Haiti was dotted with houses, not just a few, but many, as far as I could see.[55]

Small wonder over a million Haitians, a fifth of the population, have become ecological refugees. The Dominican Republic has two thirds the population density of Haiti, and two thirds of its people are in cities. The land accordingly is under considerably less pressure. The average Haitian woman is now having six children, and the Dominican TFR is only 3.3.

Inequity has not helped the Haitian situation. Besides having a long history of repressive government, Haiti's land is badly distributed. Half of the land is owned by 4 percent of the farmers, hardly a recipe for agricultural efficiency. The United States sent an armada to Haiti in 1994 to remove a dictatorship and help the people. If it were interested in converting Haiti into a sustainable society, it should have included cadres of land reformers, medical/ family planning teams (with contraceptives), ecologists, soil scientists (and saplings), agriculturalists, hydrologists, and teachers. Making Haiti sustainable can never be accomplished by the United States Marines.

The Haitians themselves understand many of their most basic needs. American anthropologist Catherine Maternowska interviewed Ginette and Celeste, Haitians who had fled their deteriorating farm to live in Port-au-Prince's Cité Soleil, the densest slum in the western hemisphere. There, 300,000 people live in a couple of square miles in shacks made partly of cardboard where thigh-deep sewage flows in the rainy season. Ginette and Celeste said, "What will save Haiti is water and roads. Then our crops will grow again and we could sell produce in every corner of our country. . . . Schools for our beautiful children, clinics for the ill, water pipes that descend into the rocks underneath the earth." Asked about trees, they slapped their knees laughing: "Does Haiti need trees? Heey! Yes! We need trees and engineers to help us plant them, and roads to bring them to us, and water to help them grow. . . ." Celeste has had eight children, three of whom have died; Ginette, now twenty-nine, has five.[56]

The details of land degradation differ from place to place, and Haiti is an extreme case. But basically, population and other pressures (such as neglect of agricultural sectors of the economies of less developed nations) in many areas have led to breakdowns of traditional cropping and grazing patterns, reduced fallow times, loss of soil nutrients, trampling of soil by cattle, deforestation, and massive land degradation.[57] That degradation, in turn, lowers soil moisture and exacerbates the impacts of droughts caused by natural fluctuations in weather. Fernand Braudel put it very succinctly: "The desert lies in wait for arable land and never lets go."[58]

Land degradation in its most common form is much more insidious than in places like Haiti and the Sahel. It is often masked by intensification of production, which can exacerbate degradation in a positive feedback loop. For instance, a recent report on

India summarizes some of the environmental impacts of that nation's green revolution:

> Yield increases in the [past two decades] have been supported by ever rising input subsidies, which have already exceeded a fiscally sustainable level. At the same time, pressures on the agricultural resource base from ecological degradation and expanding population are more pronounced. Overdrafts on groundwater, deterioration of surface irrigation systems, soil degradation through salinization, erosion, loss of nutrients and organic matter have proceeded virtually unchecked for the past two decades. In the coming decades, climate change could disrupt hydrological systems in India in unpredictable and potentially calamitous ways.[59]

The signs of degradation are now taking on an unmistakable form even in some of the world's most productive agricultural systems. Over 90 percent of the world's rice is grown and consumed in Asia, home of more than 3 billion people. But Asian rice production has only been increasing at a rate of 1 percent annually, while the region's population is growing at 1.7 percent.[60] The needed boost in production of 60 percent by 2020 (2 percent per year) must be achieved through higher yields since virtually all of Asia's cultivable land has already been turned to human use.

There are, however, disturbing indications that a "cap" has been reached on rice yields. The most recent varieties produced on International Rice Research Institute (IRRI) experimental plots do not exceed the yields of IR-8, a strain developed almost three decades ago. Economist Prabhu Pingali, head of IRRI's irrigated rice program, believes that the problem traces to deteriorating soil conditions in paddies that have now been subjected to decades of intensive cultivation. Deterioration has been caused by the reduced time that paddy soils are allowed to dry between crops (because of multiple cropping), possibly exacerbated by the replacement of organic with inorganic fertilizers.[61] The latter cannot supply key micronutrients (nutrients needed in tiny quantities), which are gradually mined from the soil by continuous cultivation and removal of crops. Nor do synthetic fertilizers restore soil carbon as do rotation systems in which nitrogen-fixing organisms are grown and the residues plowed back into the soil.

Commenting on the rice yield cap, Mahabub Hossain, head of the Social Sciences Division of IRRI, wrote in 1994:

> The race to avoid a collision between population growth and rice production in Asia goes on, amid worrying signs that gains of the recent past may be lost over the next few decades. . . . while populations of major rice-consuming nations continue to swell, growth in rice production has slowed dramatically in the 10 countries that account for 85 percent of global output. If these trends continue, demand for rice in many parts of Asia will outstrip supply within a few years.[62]

Land degradation is not a new phenomenon in the world; it clearly contributed to the fall of many civilizations in the past. The difference today is that it seems virtually universal. Development specialist Rathindra Nath Roy, head of the Catalyst Group in Madras, India, summarized the situation very well: "It should be remembered that what separates contemporary civilizations from those which have collapsed in the past, due to soil and water crises, is that today we understand our predicament and know that our development path is not sustainable. It remains to be seen whether we are going to apply this knowledge effectively."[63]

Freshwater Resources

The abundant water needed by thirsty high-yield crop strains often must be supplied by irrigation. About 33 percent of the crops harvested today come from the 17 percent of cropland that is irrigated.[64] But the rate at which land is being brought under irrigation around the world has slowed dramatically in the last decade, in part because low commodity prices relative to the costs of energy and other inputs have discouraged investment in agriculture. In addition, the marginal cost of installing irrigation systems is rising, since the best sites for water development were the first to be exploited. Another factor in areas such as the arid western United States is competition with urban users for scarce supplies. Allocation among regions of the Colorado River's water, for instance, was made in an unusually wet period early in the century.[65]

The rate at which irrigated land has lost fertility or gone out of production has been rising, primarily a consequence of waterlogging and salinization caused by irrigation without proper drainage.[66] Recent estimates indicate that more than 10 percent of the world's irrigated land is suffering reduced yields due to salt buildup. In the U.S., where the problem is better documented, salt accumulation is lowering yields on 25 to 30 percent of irrigated farmland.[67]

Meanwhile, the consequences of overdrawing aquifers are becoming increasingly evident as more and more irrigation water is pumped out at rates far beyond those at which the aquifers can be recharged. In the United States alone, the overdraft of aquifers, mostly for irrigation, is estimated to be between 6.5 and 8 trillion gallons annually. A fifth of the irrigated land in the nation is supplied by overdrafts on groundwater. As water tables drop to the point where pumping is no longer economically worthwhile, irrigation from those aquifers ceases and food production declines.

Pumping at rates above recharge is not necessarily bad, *but it is necessarily temporary.* Groundwater was the basis for a rapid expansion of irrigated grain production in the United States in the 1960s and 1970s, but the rising energy costs of pumping from depleted aquifers have already resulted in a significant decline in irrigated area.[68] In the southern Great Plains, a major grain-producing region, increasing amounts of land are reverting to less productive and less dependable dryland farming. A similar dilemma is appearing in northern India, where green revolution success has been built in part on overdrafts of groundwater.[69] In Bangladesh, depletion of aquifers is a serious regional problem that threatens to decrease rice production.[70]

Aquifers are beset by other problems as well: in some areas, urbanization has destroyed surface ecosystems that once allowed rainwater to percolate through soil and recharge aquifers. In others, pollution by toxic chemicals from industry has made underground water supplies unsafe for most uses.[71]

Overall control of the hydrologic cycle itself is a key ecosystem service that is progressively jeopardized by deforestation, drainage of wetlands, and other activities that are destructive of biodiversity. One consequence often is the onset of floods and droughts where once there were dependable flows. In the Philippines, for example, deforestation is causing rapid siltation of the reservoirs that supply water to irrigate the rice fields of central Luzon, a pattern evident in many parts of the world. In overpopulated Rwanda, the forests of the Parc National des Volcans act as a gigantic sponge that soaks up rainfall and releases it gradually into local streams. In 1969, about 40 percent of the park—some 90 square kilometers (35 square miles)—was deforested to make room for the cultivation of pyrethrum. That daisy-like flower was to be grown as a cash crop for the natural insecticide it contains. Unfortunately, not only did the scheme fail, but a huge area of the forest, home of the mountain gorillas, was gone. Rwanda lost some 10 percent of its surface agricultural water, with several streams drying up completely.[72]

Perverse subsidies make profitable (in the short run) stunningly inefficient use of water. Although this problem occurs worldwide, some of the best examples are from the American West where water supply practices dating back to the 1902 Reclamation Act provide farmers with every incentive to waste water. For instance, the irrigators benefiting from California's billion-dollar Central Valley Project have been required to pay back only a tiny fraction (less than 4 percent as of the mid-1980s) of its cost.[73] Such subsidies promote wasteful use of water at several levels: in the efficiency of the irrigation system used, in the choice of crop, and in choosing whether to irrigate at all. In California, farmers grow thirsty crops like cotton, rice, and alfalfa that are completely unsuited to the climate.

Removal of these perverse subsidies can be a political nightmare, but recent signs of hope have appeared with the birth of water markets in California and other parts of the world that permit farmers to sell their water rights. It is important to remember, however, that water will always have the tendency to flow toward money and power,[74] an important social constraint on carrying capacity. Moreover, it is not clear how far water reallocation could efficiently and equitably free up supplies, as opposed to imposing socially undesirable limitations elsewhere.

Declining Genetic Diversity of Crops

High-yield agriculture is primarily a product of evolutionary plant genetics. This scientific discipline has refined the ancient process of selective breeding of traditional crop varieties to produce new varieties with enhanced amounts of the structures (such as nutrient-rich seeds in grains) desired by humanity, while eliminating undesirable aspects (bitter flavors, toxins, poor storage quality). The basic resource that permits the selection process to accomplish this goal is a subset of biodiversity: genetic diversity. Maintaining the genetic diversity of crop species and their relatives is vital for the continuation of high-yield agriculture. That diversity, basically a storehouse of different genes, makes it possible to create new crop strains by recombining their genes in new ways. New strains are continually needed to meet ever-changing conditions: the evolution of new varieties of pests and diseases that attack crops, changing climatic conditions, exposure to novel air pollutants, and so on.[75]

The genetic diversity of crops has been threatened in two ways. First, as farmers around the world rapidly adopted a few genetically similar green revolution crop varieties, a host of traditional ones have been displaced, causing a loss of genetic variability within the crop species being grown. Second, the destruction of natural habitat is steadily eliminating populations of wild crop relatives, another reservoir of genes that could be critical to maintaining productivity.[76] For example, the important miracle rice strain, IR-36, was developed at the International Rice Research Institute by a team under the direction of the eminent rice breeder Gurdev Khush. Two critical attributes contributing to the strain's success—resistance to blast (a fungus disease) and to grassy stunt (a virus)—were derived from a wild species of rice.[77]

Finding the plant variety with the desired qualities to breed into cultivated strains may involve literally searching on hands and knees in areas where populations of wild crop relatives still exist. Seeking a gene that would give rice resistance to grassy stunt virus, for example, involved sifting through over 47,000 varieties in gene banks! In the end, the needed gene was found in a single wild species from a valley in Kerala, India. Soon afterward, that valley and the population of wild rice relatives growing in it were flooded by a new hydroelectric project.[78]

It is not logistically possible to maintain the maximum amount of genetic variability in gene storage facilities; indeed, a great deal of that variability is a response to *existence* in complex natural communities under varying physical conditions. Moreover, the defensive chemicals of a plant that could be of use may not be discovered in "captive" individuals because the chemicals may only be produced by the plants when "provoked" by the assault of herbivores.[79] The situation with farm animals is, if anything, worse. Programs have been organized to save the genetic diversity of crops, and while these are far from adequate, no such program exists for animals.[80]

Genetic diversity is likely to be an especially crucial resource if the next few decades become, as expected, an era of unprecedentedly rapid intensification of stresses on agriculture. The challenge for plant geneticists would be severe even if a maximum amount of genetic variability were available; with that variability vanishing in many crops, their difficulties will be exacerbated. Needless to say, the problems will be greatest in the poorest nations, where populations are hungriest, and agricultural sectors are least robust and most lacking in research and development capability.

Fertilizer and Pesticide Use

While the use of synthetic fertilizers has been critical to the green revolution, it is a mixed blessing. Along with pesticides, fertilizers often cause pollution of surface and underground waters,[81] and they can damage forests and other natural ecosystems by disrupting natural nutrient cycles.[82] Synthetic nitrogen fertilizers may also contribute significantly to human-caused emissions of nitrous oxide, a potent greenhouse gas that is involved in destruction of stratospheric ozone too.[83] Of further concern, the manufacture of fertilizers (especially nitrogen fertilizer) depends on the availability of fossil fuels at appropriate prices both as raw materials and to fuel the energy-intensive nitrogen-fixing process.[84] The energy crisis of the 1970s had an especially drastic effect on agricultural production in countries such as India, which could not afford the quantities of fertilizer they needed. Fuel (and thus fertilizer) shortages in 1974 were largely responsible for about a 25 percent reduction in India's projected production of 30 million tons of wheat. In 1975, the crop was reduced by about a million tons due to a shortage of fuel for irrigation pumps.[85]

Meanwhile, in the poorest nations, traditional methods of maintaining soil fertility seem certain to continue faltering.[86] For example, Nepal's population is presently growing at 2.5 percent annually, and the density of population on agricultural land has increased two and a half times in the past two generations. The size of the average farm has dropped below one hectare, too small to support a typical farm family of six people. As a result, forests are increasingly being converted into farmland, thereby reducing the availability of firewood. That, in turn, increases the dependence of rural people on cattle dung for fuel, depriving the land of the dung's critical fertilizing role. It is the sort of downward spiral all too often found in developing nations.[87]

Pesticides (insecticides and herbicides) have been an important component of the green revolution package. High-yield crop strains are most efficiently produced in extensive monocultures, which in turn tend to be very susceptible to insect pests and crop diseases. These miracle strains have been bred for yield, often at the expense of pest and disease resistance. In addition to facilitating the spread of monocultures, deployment of pesticides has discouraged fallowing and crop rotation, both of which maintain soil fertility while also helping to suppress pest populations.

Improved pest control in itself is unlikely to contribute substantially to expanded production, however. Even developed nations have made little or no progress in reducing the fraction of crop harvests lost to pests, at least to insects, in the last half century.[88] Once farmers become dependent upon pesticides, however, the narcoticlike addiction is hard to break. The first modest pesticide applications typically induce very large increases in yield; but after a few years, much more pesticide applied much more frequently is required to counteract the proliferation of pests, because the pests quickly evolve resistance to the pesticide while their natural enemies have been decimated by the poisons.

The surviving pests from each insecticide application tend to be those that, by chance, have hereditary characteristics that make them relatively resistant to the insecticide used. Those survivors breed, and the next generation has more resistance on average than the last. After ten generations or so, the pests can use the insecticide as an aperitif.

Meanwhile, the predacious insects that normally would keep populations of plant-eating insects in check are no longer large enough to keep the herbivores in check.[89] Predatory insects are ordinarily more vulnerable to poisoning than are insects that consume plants, because predator populations tend to be smaller than prey populations and because the prey have had more evolutionary "experience" than carnivores with poisons.[90] The same is true for birds and some other organisms that help keep insect pests in check.[91] Thus the broadcast use of pesticides not only doesn't control the original pests, it "promotes" previously innocuous species to pest status by decimating the predators that once controlled the new pests' populations. Small wonder that in California in the late 1970s, twenty-four of the twenty-five top agricultural pests were creations of the pesticide industry![92]

The effects of this "pesticide treadmill" can be seen in the experience of most Latin American countries. By the 1960s, 40 percent of the pesticides exported from the United States went to Central America alone. Soon, the production of cotton and other cash crops was requiring such high pesticide inputs that large areas went bankrupt and production collapsed. In the Matamoros-Reynosa area of northeast Mexico, for example, the area of land under cotton fell from 1.75 million hectares to under 3,000 hectares. At the same time, thousands of cases of pesticide poisoning were occurring each year; in Guatemala, El Salvador, Honduras, and Nicaragua alone, 14,138 poisonings were reported between 1972 and 1975.[93] Many thousands more poisonings very likely occurred but were not reported. Most directions and warnings for pesticide use are written in English, whereas the majority of Central American farmworkers tested at that time could not read Spanish, much less English.

One of the worst effects of the abuse of pesticides is the disruption of natural pest control—an ecosystem service crucial to agriculture. A classic case occurred in the Cañete Valley of Peru just after World War II. DDT and its relatives were used in the cotton fields of that valley in 1949, and at first had great success in controlling cotton pests—by 1954, yields were up 50 percent. The farmers decided that if more pesticides were used, more cotton would grow. A respected entomologist from the University of California, Ray Smith, described what happened. Insecticides "were applied like a blanket over the entire valley. Trees were cut down to make it easier for the airplanes to treat the fields. The birds that nested in the trees disappeared. Other beneficial animal forms, such as insect parasites and predators, disappeared. As the years went by, the number of treatments was increased; also each year the treatments started earlier because of the earlier attacks of the pests."[94]

Pests became resistant to the insecticide and, freed of restraint by their natural enemies, came roaring back. Other plant-eating insects that had never been pests, also relieved of predator attacks, were "promoted" to pest status. By 1955–1956, six brand-new pests had appeared that were not found in nearby cotton-growing valleys where pesticides had not been used.[95] Cotton yields dropped to only two-thirds of their pre-DDT level. A high price was inflicted for the disruption of the pest-control service of that ecosystem, but the lesson seems hard to learn.

Agriculture continues to be plagued by the increasingly widespread resistance of pests to pesticides and by unwanted side-effects of pesticide overuse. Unfortunately, the pesticide industry is a powerful, well-financed, and well-organized force in the United States, backing legislation designed to weaken public health standards for pesticide residues in food. The industry collectively contributes millions of dollars annually to members of Congress to promote its interests. Threatened by legislation to restrict the use of some seventy cancer-causing pesticides in 1994, it doubled its contributions and successfully coopted 224 members of the House and twenty-two members of the Senate as sponsors of bills to repeal the health standards.[96] The industry doesn't need to go to such efforts in the developing world, where human and environmental health regulations are unenforced or nonexistent—25 percent of all pesticides exported from the U.S. are heavily restricted or banned in developed nations.[97]

In Asia, socioeconomic changes are driving a dramatic increase in herbicide use. Economic growth and urbanization have increased the costs of farm labor, effectively cutting the supply available for the extremely labor-intensive weeding of rice paddies.[98] Weeds can reduce rice yields 45 to 95 percent if they are not controlled.[99] Even with careful weed management, yields in Asia are normally reduced about 10 to 15 percent, amounting to an annual loss of some 30 million tons of milled rice. That is enough rice, if properly distributed, to solve the continent's current hunger problems.[100]

Lacking cheap labor, farmers are substituting herbicides for weeding; at the same time, they are also shifting away from the practice of hand-transplanting rice seedlings into paddies. Traditionally, soil was softened by flooding the paddy, then plowed. Seedlings started elsewhere about a month earlier were then laboriously planted in rows by stooping workers (usually women). In technologically advanced Japan and South Korea, however, machine transplanting now prevails. In the least labor-intensive system, very young ("infant," 8 to 10 days old) seedlings germinated in nursery boxes are machine transplanted. This is now done in some 99 percent of Japanese fields,[101] and about half the fields in South Korea. The remainder are machine transplanted with 35- to 45-day-old seedlings.[102]

The substitution of herbicides for hand-weeding of Asian rice is also being accelerated by a switch to direct-seeded varieties. Attempting to overcome the yield cap, plant breeders are increasingly shifting to strains designed for seeds to be broadcast over a field, rather than germinated elsewhere and transplanted into the paddy. Under optimal conditions, these strains can produce higher yields because, among other things, they do not suffer the shock of transplantation. But direct-seeded rice suffers two to three times as much from competition for sunlight and nutrients by weeds as does transplanted rice, since both rice plants and weeds germinate simultaneously. In the traditional system, rice is transplanted into paddies where weeds have been controlled and thus gets a head start. As a result, losses to weeds are generally about 10 percent higher for direct-seeded than for transplanted rice.[103]

The trend toward increased herbicide use in Asian rice cultivation clearly will continue. Sadly, the human health effects from the use of insecticides in Asian rice agriculture are severe enough that some claim the health and other economic costs outweigh the economic benefits from the gains in yield.[104] Less is known about the potential direct health impacts of herbicides than of insecticides, but they are far from trivial.[105] Given the record of sloppy use of pesticides in developing (and developed!) nations, there is every reason to believe that a significant price in human illness and death will be paid for switching to a wide-scale use of herbicides.

Additional environmental impacts can be anticipated, perhaps most obviously in aquaculture. Agricultural chemicals are already a serious threat to pond culture of fishes in Asia, by direct poisoning, by changing the physical characteristics of the water (such as the levels of dissolved oxygen and acidity), and by altering the food webs upon which the fishes depend.[106] Fortunately, herbicides are less toxic to fishes than are insecticides, and most adult fishes are not at risk of acute poisoning. On the other hand, larval fishes and small organisms upon which fishes feed may be greatly affected by normal applications of herbicides. Indeed, all aquatic organisms are threatened by the spills of various sizes that inevitably accompany large-scale use of agricultural chemicals.[107]

Beyond the environmental impacts, perhaps the greatest hazard of a massive switch to the use of herbicides is the high probability that weeds will evolve resistance to herbicides, posing a threat to yields in the future—just as insects' resistance to DDT and other insecticides has played a major role in the resurgence of malaria over much of the globe, and antibiotic resistance has restored the once-suppressed danger of death from bacterial infections. If, as many people expect, the task of feeding the Asian population becomes increasingly difficult in the future, having to deal simultaneously with a faltering weed-control situation could contribute to a disaster. At the moment, the situation in Asian rice agriculture could be viewed as similar to that in Western agriculture after World War II, when synthetic organic insecticides and antibiotics first went into use. The reward is essentially immediate, but the penalty shows up after several decades of dependence on the technology.

Since human beings rarely (if ever!) act *purely* altruistically, it is possible that the green revolution strategy was about the best that could have been developed once the population explosion had been triggered by a *largely* altruistic campaign to conquer disease in poor nations. On the other hand, some observers claim that traditional land races of rice could have produced the necessary increases in food production without destroying the integrity of peasant communities or creating a vast agricultural enterprise that threatens to be unsustainable in the long run.[108] We have our doubts.

Nonagricultural Sources of Damage

As opportunities for further increases in food production narrow in future decades, new threats to maintaining the last half century's impressive gains are appearing. Besides the constraints already being imposed by environmental degradation of agricultural resources, environmental insults from other sources have adverse effects on agriculture. Some environmental problems have now reached global proportions, and their effects on agriculture could be critical in determining humanity's success in feeding itself in the twenty-first century.

Many of the substances that societies emit into the atmosphere have deleterious effects on agriculture. Locally and regionally, air pollutants such as sulphur dioxide, peroxyacetyl nitrate (PAN), and nitrogen oxides (NO_x, an important precursor to ground-level ozone) can reduce productivity substantially because they are directly toxic to crops.[109] For instance, ground-level ozone in the lower atmosphere is estimated to have caused a 5 to 10 percent loss of U.S. crops during the 1980s.[110] Worldwide, three regions in the northern midlatitudes—North America, Europe, and eastern China—account for most (60 percent) of the world's food-crop production and exports. These same regions also produce most of the world's NO_x emissions and are thus prone to ground-level ozone during the summer growing season. Between 10 and 35 percent of the world's cereal crops are exposed to ozone concentrations above the threshold of damage. By 2025, this fraction is predicted to approach 30 to 75 percent, implying significant agricultural losses to air pollution.[111]

Acid deposition, resulting from the injection of oxides of nitrogen and sulphur into the atmosphere, can also be directly damaging to crops and freshwater fisheries. Globally, the nitrogen fertilizer effect on crops may currently compensate for crop losses from direct damage. As the human population continues to grow, though, so most likely will emissions of these pollutants, especially in developing nations that lack the resources to deploy sophisticated pollution-control technologies but are nonetheless committed to industrialization.

Depletion of the stratospheric ozone shield is a potentially serious threat to future increases in food production. The thinning of the ozone layer in the upper atmosphere allows increased amounts of dangerous ultraviolet-B (UV-B) radiation to reach the surface. Some two thirds of the two hundred plant species (most of them crops) that have been tested are negatively affected by increased UV-B flux.[112] Legumes such as soybeans, which supply essential protein to humanity, are among the most sensitive. Fortunately, soybeans are genetically variable in their sensitivity to UV-B, and it may be possible to develop more resistant strains if that genetic diversity is adequately preserved. Some recent work suggests that UV-B damage in plants will become a critical factor only if ozone depletion proceeds further than presently predicted.[113] One must note, however, that the production of shielding compounds (such as anthocyanins) in UV-B resistant crops requires energy and thus probably will lower yields. Resistance to UV-B also may give the crops undesirable qualities from the viewpoint of consumers.

Recently, there have been ominous reports of a 6 to 12 percent reduction in the productivity of phytoplankton—tiny organisms whose photosynthetic activities make them the base of the food web—in the Antarctic Ocean. Presumably this was because of the dramatic decline in stratospheric ozone over the region.[114] Ozone depletion therefore represents one more threat to already faltering oceanic fisheries as well as to agriculture.

Thanks to the Montreal Ozone Protocol, it appears that a severe thinning of the ozone shield may be averted, although the peak level of depletion will not occur until the end of the century. Recent patterns of depletion reinforce the need to limit strictly the flow of ozone-destroying chemicals into the atmosphere, since the ozone over northern midlatitudes has been depleted twice as rapidly as was predicted.[115] The Antarctic spring of 1992 saw the greatest expansion of the ozone hole yet.[116] In 1993, the hole was even deeper, but did not cover more territory; and in 1994, for unknown reasons, the hole expanded further and remained deeper than expected.[117]

Climate Change and Sea-Level Rise

Rapid climate change almost certainly represents an even greater threat to food production than ozone depletion. It is possible that climatic zones will shift as much as fifty times faster than they ever have in the ten thousand years since the dawn of agriculture.[118] The shifts will not constitute a mere redealing of the climatic cards with some areas losing (becoming less productive) and others winning. Rather, if the flow of greenhouse gases into the atmosphere continues relatively unabated for the foreseeable future, then agricultural systems will be faced with the stresses of continual adaptation to rapidly changing conditions.[119] As in the case of stratospheric ozone depletion, successful adaptation will require ample genetic variability and ample scientific talent, as well as flexible management. But even these will almost certainly not be sufficient to prevent serious drops in harvests in some places and at some times, beyond the drops that would occur in response to normal fluctuations in weather.[120]

Climate models suggest that some of the most serious disruptions of agriculture will result from drying of the central parts of northern continents, regions that now constitute humanity's principal breadbaskets.[121] Potentially compensating yield increases in areas that could become more climatically suited to agriculture may not be realized because of inadequate soils. For example, the Canadian shield, to which Iowa's present climate may eventually "migrate," has thin, nutrient-poor, acidic soils.

Finally, it is likely that global warming will cause substantial rises in sea level, although the timing and extent of the rise is largely unpredictable at present. A major impact on food supplies could result from even an 8-inch (20 cm) rise, however. (Climatologists predict a global average rise of 8 inches and 26 inches by 2030 and 2100, respectively, with significant regional variations.)[122] Coastal farmland would be flooded or threatened with more frequent storm-surge inundation in many low-lying and heavily populated areas (the Nile delta, Bangladesh, etc.). Salinization of coastal aquifers would increase, reducing sources of irrigation water. Even more critical, coastal wetlands and estuaries would be rapidly altered, and many would be unable to "migrate" inland because of human-imposed barriers. Their disruption would further damage oceanic fisheries that often depend on such areas as nurseries or food sources.[123]

While the fertilizing effects of the greenhouse gas carbon dioxide might be beneficial, this is by no means certain,[124] and the benefits probably would be insufficient to compensate for the overall negative impacts of climate change. Even though agriculture would undoubtedly benefit from climate change in some areas, it is difficult not to conclude that global warming poses the most serious known environmental threat to food production. Humanity may prove extremely lucky, with all the uncertainties about the warming being settled in its favor. But the uncertainties cut two ways, and an equal chance also exists that they could all be settled in the worst-case situations from the human perspective. It is well to keep in mind that: "Projected population growth rates and the ensuing food demands, *even in the current global climate,* will make it difficult to provide for human sustenance and food security in the 21st century."[125]

Controlling the emissions of greenhouse gases will prove a much more difficult task than limiting the release of ozone-destroying chlorofluorocarbons. The flow of greenhouse gases is very tightly linked to human population size through the burning of fossil fuels, deforestation, and agriculture itself. For instance, one of the most potent greenhouse gases is methane, and among its major sources are rice paddies and the guts of cattle. The increase of another important greenhouse gas, nitrous oxide, may be partly due to the use of nitrogen fertilizers and to land-use changes.[126]

Yet many of the efforts required to reduce greenhouse gas emissions constitute "no-regrets measures"; that is, they would be beneficial even if global warming were not a threat. Increasing energy-use efficiency, developing and deploying sustainable energy technologies, reducing air pollution, and halting deforestation, for example, are all badly needed in their own right.

Loss of Biodiversity and Faltering Ecosystem Services

Biotic diversity is the most irreplaceable component of our resource capital, and the least understood and appreciated. Plants, animals, and microorganisms are organized, along with the physical elements of the environment with which they interact, in ecosystems. These organisms thus help to provide indispensable, free *ecosystem services,* which support civilization. Many of these services are essential to agriculture, including:

- Maintenance of the gaseous composition of the atmosphere
- Moderation of climate
- Control of the hydrologic cycle (which supplies fresh water)
- Detoxification and disposal of wastes
- Cycling of nutrients and replenishment of soil
- Control of the great majority of pests and diseases that could attack crops
- Pollination of crops and wild plants
- Maintenance of a vast "genetic library" containing many millions of kinds of organisms, from which humanity has "withdrawn" all manner of benefits, including all the crop and livestock species on which agricultural systems were built. And if preserved, the genetic library potentially could provide enormous further benefits in the future.[127]

Yet biodiversity resources are being lost at an accelerating rate that may cause the disappearance by 2025 of one quarter of all the species now existing on Earth.[128] Every species that disappears is a marvel gone forever—often without humanity ever knowing what potential direct economic value it might have possessed, much less its role in providing ecosystem services. Every genetically distinct population that is exterminated reduces precious living capital and potentially weakens nature's ability to support humanity. Even if the evolutionary process that creates diversity continued at rates comparable to those in the geologic past, it would take tens of millions of years for today's level of diversity, once seriously depleted, to be restored.

The enormous potential for developing new, improved crops and domestic animals from the world's vast storehouse of biodiversity is being compromised by the global extinction episode now underway. Only a score or so of plant species are really important as crops today; at most a few hundred supply humanity with significant quantities of food. Yet there are at least a quarter million species of higher plants, many with substantial untapped potential as crops. About 75,000 are known to have edible parts, and 7,500 or so have been used by human societies as food.[129] Furthermore, selection often can create cultivated strains that are edible even if their wild ancestors are not. The current wholesale destruction of populations and species of wild plants, however, is very quickly foreclosing the potential for developing new food sources.

Atmospheric changes that damage agricultural ecosystems also damage natural ecosystems, impairing the essential services those systems supply to human societies, and especially to agriculture. Increased UV-B radiation may disorient pollinators and interact with climate change, acid deposition, and other forms of air pollution to weaken trees and make them more susceptible to attacks by insects and diseases, or even kill them outright. Such synergistic impacts can lead to the gradual death of forests, which is underway in parts of Europe and may also have begun in eastern North America.[130] Among the consequences of forest removal are local disruption of the hydrologic cycle, increased soil erosion, a reduction in free pest-control services (with the loss of some species of birds and other predators), local (and possibly regional and global) climate change, and a widespread loss of biodiversity.

Land degradation on Australia's wheatlands resulting from the deliberate removal of natural vegetation illustrates the importance of biodiversity and ecosystem services to agricultural productivity. Over the past century, nearly all native vegetation was cleared to maximize the area planted in wheat; only small fragments on ir-regular terrain were left in place, making up just 7 percent of the land area. This decimated local biodiversity, since the food sources and habitats of many native animals were destroyed along with the vegetation. It did much more than that, though. Removal of the deeply rooted native trees and shrubs greatly disrupted the hydro-logical cycle, basically cutting off the natural pump that kept the water table safely below the land surface. As the water table rose, it brought naturally occurring salts into the topsoil, where they have been concentrated by evaporation, poisoning the wheat. Sa-linization and waterlogging have forced farmers to take many fields out of production. They paid a heavy price for the thought-less destruction of biodiversity. Salinization and waterlogging, al-though each affects only about 3 percent of the cleared land in Western Australia, are estimated to cost farmers there some $150 million annually.[131]

Ironically, the destruction of biodiversity continues at a high rate in many areas of Australia. In late 1994, we joined in a cam-paign to prevent the Australian government from issuing export permits that would have allowed the destruction by woodchipping of hundreds of critical areas of old-growth eucalyptus forests. Al-most all the benefits would go to the few owners of highly mech-anized chipping mills (which have greatly reduced the number of jobs in the industry) and to the Japanese. The latter get the chips at a little more than $10 per ton, while they carefully protect the forests of their home islands. The vast majority of Australians and the rest of humanity would be the losers as critical natural capital is destroyed.

We and many others talked to Environment Minister John Faulkner, who recommended against issuing the permits. The short-sighted decision to go ahead was made by the resources minister, David Beddall, and Prime Minister Paul Keating. They are now paying a political price for continuing the Australian tra-dition of demolishing its natural resources as quickly and for as little national benefit as possible.[132]

Social Constraints

We have stressed the constraints on food production in the bio-physical environment. As the woodchip issue illustrates, however, constraints built into the social, political, and economic environment are at least as daunting. Those constraints are also evident in an urban-oriented industrial development policy in many developing nations that has ignored the environmental needs of agriculture and rural citizens. These policies often deliberately depress food prices, which leads to a lack of incentive to improve farming technologies and a widening urban-rural income gap, and in turn causes distortions in the food distribution system. Related social conditions inhibiting expansion of agricultural production include an inequitable distribution of land, lack of access to inputs and farm credit, widespread unemployment (and thus lack of economic demand for food), inequities in the world food market, and political neglect of the agricultural sectors of many poor economies.[133] These issues have been dealt with extensively by others, most recently in an excellent study by the World Institute for Development Economics Research (WIDER).[134]

International trading policies clearly have pervasive influences on local patterns of food production and environmental deterioration.[135] United States agricultural policies in particular can profoundly affect human and environmental well-being in poor countries because of the powerful role the U.S. plays as the world's leading exporter of grains and other important commodities. Yet sorting out the environmental and equity impacts of trading relationships is extremely complex, involving profound historical effects and a daunting set of two-edged swords.

Establishment of the global grain market created the theoretical possibility of a sustainable increase in global carrying capacity through the relaxation of local resource constraints and specialization by nations in accordance with their "comparative advantages." In many respects, however, the world trading system seems designed more to perpetuate the power relationships that fueled the development of today's industrialized world by exploiting the resources of then-colonial nations. Rather than sustainably enhancing the planet's capacity to support humanity, the global trading system has too often facilitated the unsustainable export of local carrying capacity from poor to rich parts of the world. For example, India's natural resources have been sequentially plundered by the British Empire and then by the institutions remaining in its wake.[136] In the past two decades, rich nations have erected a growing set of barriers to trade in order to protect domestic jobs, at least in the short term. These barriers are imposed most heavily on poor countries with the least bargaining power.[137]

The influence of international trade on patterns of agricultural production is fraught with issues of equity. Despite the overall trend toward protectionism, a basic popular misunderstanding persists that free trade is an unalloyed good.[138] This certainly is not the view of professional economists familiar with the issue.[139] The ideal conditions required to realize many of the benefits of free trade (e.g., internalization of externalities) are not even approached now, and are unlikely to be approached in the future.

Trade has not raised most poor nations out of poverty; indeed, it has caught many of them in a "specialization trap," exporting commodities.[140] In order to earn more foreign exchange, commodities production has been increased, lowering prices in the face of low demand elasticity and often reducing equity in land distribution. For instance, there is normally little domestically generated political or social pressure to expropriate the holdings of small subsistence farmers. But when opportunities appear to make fortunes from large-scale production of crops for export, subsistence farmers are immediately at risk.[141] Prevailing terms of trade can thus exacerbate inequities within poor nations. Cash crop production, which increasingly finances food imports, has often resulted in the further impoverishment of both the rural poor and the environmental resource base that sustains production.[142]

Changes in the patterns of land use in Latin America illustrate this effect. In Honduras, one of the worst cases, commercial cotton cultivation (for export), beginning in the late 1940s, displaced many poor farmers from the most productive agricultural land. Cotton production required many hands, however, and growers relied heavily on seasonal farm labor. In the 1960s, World Bank lending policies shifted economic incentives to cattle ranching.

The management of vast pastures required very little labor compared to that needed in cotton production. Landless laborers were displaced to the highlands, areas poorly suited to agricultural production, while much of the best agricultural land (48 percent of the valley land) was converted to pasture. A subsequent decline in the price of beef on the world market then induced a shift to cantaloupe and shrimp. The rate of growth of production in both of these nontraditional export crops exceeded 20 percent per year during the 1980s, at considerable environmental and social cost. The area in shrimp farms increased a hundredfold, forcing small farmers and local users out. The resource base supporting each commodity in this series of boom and bust cycles has substantially deteriorated.[143]

The international debt crisis has intensified cash crop production, the exploitation of natural resources, environmental deterioration, and rural impoverishment. While there is considerable controversy over the cause of the debt crisis, its consequences cannot honestly be disputed.[144] Expressed in financial terms (which externalizes and thereby greatly underestimates true costs to human and environmental well-being), developing nations remitted in debt service alone $1.345 trillion dollars to creditor nations between 1982 and 1990. Yet, at the start of the 1990s, debtor nations were collectively 61 percent deeper in debt than in 1982. During the 1980s, sub-Saharan Africa's debt increased by 113 percent; that of the most impoverished developing nations collectively increased 110 percent.[145] Debt service now accounts for up to 87 percent of the export earnings of poor nations.[146] Whereas both parties to the loan agreements initially expected mutual benefit, the results have been disastrous all around.

All these factors interact. External debt and other economic pressures spur government support of cash crop production for export at the expense of subsistence agriculture. A frequent result is consolidation of farms, displacement and impoverishment of farmworkers, and increased unemployment, hunger, and environmental disruption as displaced people move to forested uplands or other marginal land in the struggle to survive. Poverty promotes population growth rates and further environmental deterioration; the latter diminishes the productivity of agricultural systems, thereby inducing further impoverishment.[147] A positive feedback system, in which deleterious trends reinforce each other, is set in motion.

Trade is certainly not all bad—far from it. In fact, many of the problems of poor nations that might at first glance be assigned to adverse terms of trade (e.g., the debt crisis) often trace more to incompetent development policies. All too often, poor countries have opted for national airlines over farm-to-market roads. All too often, they have squeezed the agricultural sector by taxing its exports while protecting local industry, increasing the costs of farmers who have to purchase local products. With more sensible policies, the terms of trade would improve. One partial solution to these problems would be to work toward trading arrangements that encouraged more value to be added to agricultural (and forestry) products exported by poor countries. This would take some of the pressure off the land, since less would need to be grown to yield the same revenues. In the next chapter, we explore other ways of overcoming some of these paralyzing social constraints and working sustainably within biophysical limits to expand the availability of food to all of humanity.

Chapter Seven

EXPANDING FOOD AVAILABILITY

Humanity faces a formidable challenge in trying to provide adequate diets for the 8 to 10 billion or more people destined to inhabit the planet in coming decades. Yet there is substantial room for improvement in both biophysical and socioeconomic aspects of food production, storage, and distribution. On the biophysical side, while many of today's farm practices are unsustainable, that does not imply that production levels per se are unsustainable. Rather, there are various opportunities for expanding the harvest, in part through further development and deployment of green revolution and other biotechnologies. The greatest potential gains could be achieved through restoration of productivity to degraded lands and the development of "alternative" agriculture, including improvement of traditional crops and farming techniques.

A substantial increase in food supplies could be gained by reducing the portion of crops devoured by pests. Losses in the field could probably be lessened by spreading the use of integrated pest management (IPM). At the very least, that change would lighten the collateral environmental damage from agricultural chemicals. But the greatest gains, especially in the short term, could be made through limiting post-harvest losses by providing better storage and transport facilities and protection against pests of stored food.

The increasingly dismal situation in oceanic fisheries cries out for more effective regulation of harvests, with moratoria on fishing to allow valued stocks to regenerate. Aquaculture holds considerable promise, but is very sensitive to environmental degradation, such as pesticide runoff. The net *contribution of aquaculture will also depend on controlling its environmental side-effects such as the destruction of coastal mangroves, which can decrease fisheries production elsewhere.*

Economic, social, and political transformations will also be critical to solving the food problem. The most fundamental change required to capitalize on any of the foregoing biophysical opportunities or to seize even greater ones is to increase socioeconomic equity at all levels in the global agricultural system: between the sexes, between rural and urban regions and economic sectors, and between nations. Much more investment must be directed toward the world's poorest farmers to halt the downward spiral of population growth, poverty, and environmental destruction. The same is required to alleviate the plight of small-scale fishers, trapped between stock depletion by industrial fishers and poisoning of fishing grounds by coastal pollution.

Above all, rich and poor nations alike must make a large investment in pulling the destitute out of poverty. Today much hunger is caused by maldistribution, and maldistribution traces primarily to a shortage of buying power among the poor—they cannot generate the required "demand" for food. Narrowing the vast gulf between rich and poor will be essential for feeding all of humanity and establishing a sustainable society. Investment in equity is ultimately in everyone's best interest; we may not be united now, but we will be in the end if we fail.

Clearly, humanity cannot afford to continue rapidly degrading Earth's productive potential—its natural capital in the form of land and soil, freshwater supplies, atmospheric quality, biodiversity, free ecosystem services, and fisheries. Even without such degradation, global nutritional security will be exceedingly difficult to attain. Nutritional security comprises several important components:[1]

- *Food security,* "to ensure that all people at all times have both physical and economic access to the basic food they need"[2]
- *Drinking water security,* to reduce the incidence of water-borne disease
- *Minimum income security,* involving minimal social security programs to assure access to adequate food
- *Nutrition intervention,* to protect vulnerable groups (e.g., the very young and very old, pregnant and nursing mothers)

- *Nutritional education,* to enhance the ability to choose foods wisely.
- *Population stabilization,* to make all this possible

In view of the expected growth of the global population, a substantial portion of which is poorly nourished, achieving nutritional security will require, at a minimum, a doubling of food production over coming decades. Meeting the needs of the poor and satisfying as well the likely demand of rich and middle-income groups for meat and other luxury products would actually require a tripling of output by 2050.

This translates into a tripling of yields on present farmland, since virtually all potential farmland has already been converted to human use. Indeed, while urban expansion gobbles up vast tracts of prime farmland, people in rural areas are pressing into steeper terrain at both extremes of the moisture gradient, precipitating tremendous environmental destruction. Ecologically sound ways must be found to amplify the amount of food derived from today's farmland.

Ultimately, the plow will win its race with the stork only with tremendous help from the system that divides up the pie; no more seconds and thirds for some while others go hungry. The socioeconomic inequities prevailing in agricultural systems worldwide at every level of organization are not only stifling production, they are also an important root cause of the degradation of the natural underpinnings of productivity. The socioeconomic constraints on further expansion of food supplies are at least as daunting as the biophysical constraints, and failure to address either kind is guaranteed to perpetuate and intensify human suffering.

Increasing Yields

The impressive increases in grain yields obtained in the past few decades are principally due to the widespread deployment of "green revolution" technology. Between 1950 and 1990, grain production roughly tripled.[3] Green revolution technology depends on specially bred strains of major cereal crops that produce much higher yields than traditional strains if heavily fertilized, watered, and protected against the pests to which they are notably susceptible.

The prospects for further expansion of food production through current green revolution technology may be signaled by recent trends in the production of rice, the world's most important food grain crop. Japan serves as a model of production possibilities if cost were not a factor, employing the latest technology and enjoying firm government support for agriculture. In fact, the government subsidizes rice at five or more times the world market price, enabling farmers to use inputs at otherwise uneconomical levels. Since 1970, however, rice yields in Japan have risen an average of only 0.9 percent annually. Lester Brown and John Young of Worldwatch Institute concluded in 1990: "Japanese farmers have run out of agronomic options to achieve major additional gains in productivity."[4] Japan's failure to increase yields despite every advantage suggests that substantially boosting output of major grain crops will be difficult where green revolution technology is already well established: in virtually all industrialized nations and increasingly in developing nations.

Only serious efforts to strengthen the agricultural sectors of developing nations, with special attention to the needs of the poorest farmers, could bring a further large-scale spread of green revolution technologies in the remaining suitable farming areas. Even if this were done, it seems doubtful that anything like past rates of growth in food production, based on the green revolution, could be generated for very long. How likely is it, as Brown and Young point out, that farmers outside Japan can outpace the performance of the Japanese rice farmers over the past two decades?

When we visited the International Rice Research Institute (IRRI) just south of Manila in the Philippines in 1992, we got a sense of the urgency of succeeding in that seemingly impossible task. The facilities are impressive: broad expanses of experimental fields, a large cold-storage unit for maintaining a seed bank of some seventy thousand rice strains, greenhouses in which wild relatives of the cultivated rice species are being grown—living sources of genes for improving cultivated strains.

The crew of dedicated scientists is even more impressive for both their competence and concern. They are teasing apart the intricacies of virtually every aspect of rice ecosystems to make them more efficient, safe, and sustainable. Researchers are examining the effects of agricultural chemicals on farmers' health; investigating ways to reduce emissions of the greenhouse gas methane from paddies; studying the effectiveness of fallowing and intercropping for maintaining soil fertility; and trying to develop a biological control for sheath blight, a nasty fungus disease of rice. (They are hoping to find a bacteria that will inhibit the growth of the fungus to help poor farmers who cannot afford chemical fungicides.)

The list of projects goes on and on, but the most urgent ones were focused on developing new rice strains that could break through the present "yield caps." "Cap" is actually something of an understatement. According to IRRI economist P. L. Pingali, rice yields have *dropped* even where the crop is produced under careful scientific management. Moreover, varieties of "miracle rice" released recently are not performing as well as earlier ones.[5]

A research team of geneticists at IRRI under the direction of the eminent plant breeder Gurdev Kush has been working hard on this problem. In 1994, a breakthrough was announced. A new "super" strain has been developed with the potential of increasing yields as much as 25 percent. Like other high-yielding varieties, the super rice is a dwarfed plant, with short stalks that are less susceptible to "lodging" (when wind or rain causes the stalks to fall over from the seed head's weight). The new strain, unlike current ones, has fewer stalks, but each bears seed heads, and each seed head contains about twice as many rice grains as today's HYVs. Thus, significantly more of each plant's biomass is edible. Finally, the plants are more compact so they can be grown more densely.

Much work still must be done in transferring genes for disease resistance into the new plants and tailoring strains to the varied local conditions of Asian rice culture. When the super rice is available to farmers, beginning in about five years, it is hoped that it will boost output about 10 percent under field conditions.[6] It's a promising start, but it would only make up a small part of the nearly 100 percent increase needed over the next half century or so, and it's not clear what can be done for an encore.[7] Agriculture policy analyst John Walsh recently assessed the overall situation in rice agriculture, concluding that the introduction of improved rice varieties cannot compensate for "increased pressure from pests, depletion of soil nutrients, and changes in soil chemistry caused by intensive cropping."[8]

Fertilizers are, and are likely to remain, key elements in efforts to increase farm output. The new super rice will require fertilization; without generous applications of fertilizers, green revolution crop varieties generally produce yields no better than those of traditional ones. Between 1950 and 1990, worldwide chemical (synthetic) fertilizer use increased tenfold, which underscores the importance of fertilizers in generating the great expansion in grain production during that period. Unfortunately, though, the chances of repeating that performance are small.

In developed nations, fertilizer use has long since passed the point of diminishing returns.[9] Indeed, fertilizer use has essentially plateaued in rich countries and in many areas of poor nations where green revolution technologies have taken hold. Grains cannot utilize more fertilizer, and the quantities consumed now fluctuate according to how much land is withheld from production and changes in fertilizer prices and the complex patterns of agricultural subsidies.[10] Worldwide there was a decline in total use between 1989 and 1993 in response to those sorts of factors. In Norway, the government recently mandated a 50 percent cut in both fertilizer and pesticide use as part of a sweeping effort to make agricultural practices more sustainable. The result was actually an increase in yields![11]

Nonetheless, there is still considerable room for fertilizer-induced yield increases in many parts of the developing world. The trouble is that most farmers who have the means to acquire fertilizers are already using them, although often not very efficiently.[12] A substantial further expansion of the green revolution, therefore, would require a complex of political and economic changes, globally and nationally, to make green revolution technologies available to subsistence farmers who still lack access to them.

Biotechnology is often heralded as the next green revolution, but its potential for expanding food supplies actually appears relatively limited. Its value is likely to lie in giving crops better protection against diseases and pests, enabling crop plants to use some nutrients more efficiently, or providing a capacity to thrive under adverse conditions such as low rainfall or salty soils.[13] Within the next two decades, genetic engineering may enhance the nutritional quality of diets by increasing the diversity of foods available, by making some products more nutritious, or by developing qualities that make food safer to consume and easier to ship and store. Genetic engineering also could play an important role in maintaining the genetic diversity of crops, since it allows for the simultaneous introduction of a given useful trait into all varieties. Thus, locally adapted varieties could be genetically enhanced while remaining in production.

One biotechnological improvement that could make a tremendous difference would be to give major grain crops the ability to fix nitrogen, thereby greatly reducing their need for nitrogen fertilizer. Unfortunately, for biological reasons this is very difficult at best, and two decades of effort have brought little progress. Even comparatively easy tasks have not been accomplished quickly. As summed up by Alex McCalla, director of the Agriculture and Natural Resources Department at the World Bank, "while biotechnology holds promise of significant genetic improvements, that promise is becoming reality much more slowly than earlier forecasts suggested."[14]

An aspect of biotechnology not often discussed in developed nations is its potential to bring about swift and devastating economic dislocation. For example, bacteria are now being engineered to produce vanilla flavoring, so it can be "manufactured" rather than grown. Large-scale deployment of that technology would likely do away with the jobs of many farmers in Madagascar and other poor countries who raise vanilla beans as a cash crop.[15] The long-term social and environmental costs and benefits of such a substitution are exceedingly difficult to evaluate, but should not be completely ignored. There are also serious, unresolved ecological and evolutionary questions about the long-term effects of genetic engineering on the sustainability of the agricultural enterprise and on the environment.[16]

The bottom line is that another spectacular increase in food production—multiplying yields of staple crops two- to fivefold—is probably beyond the reach of biotechnology alone. It can be a help, but not likely a cure, for humanity's food problems.

Restoring Productivity to Degraded Lands

Restoring the productivity of the world's degraded lands is absolutely crucial.[17] Preservation of land quality in the first place is always far cheaper and easier; but today, restoration is the one sensible way in which substantially more land could be brought into cultivation. Supplying an expanding human population not only with food, but also with many other agricultural, forestry, industrial, and medicinal products, hinges upon increasing terrestrial productivity in general.[18] If biodiversity and ecosystem services are to be preserved, a sustainable increase of productivity on human-controlled land will be necessary to ease the pressure on remaining natural habitats.[19] Furthermore, human-induced reductions in land productivity can profoundly influence important global biogeochemical cycles with uncertain, but probably negative, consequences for civilization.[20] Finally, the amount of productive land is frequently a limiting factor on economic output, and its degradation thus threatens to undermine the development of many poor nations.[21]

What are the prospects for restoring productivity? Fortunately, an enormous potential for natural recovery exists in most land types. This is apparent in the long-term historical success in many tropical regions of shifting cultivation, which effectively exploits natural recovery mechanisms. Although many mechanisms of regeneration are impaired by cutting, burning, and weeding, recovery from cultivation impacts is relatively swift (twenty to forty years).[22] This is because plot size is typically small (roughly one hectare), so dispersal (the primary mechanism by which forest species reestablish themselves) from surrounding source areas can readily occur.[23] In addition, shifting cultivation burns usually do not destroy the layer of humus that is critical to both soil fertility and resistance to erosion, and soil is only briefly, if at all, bare of vegetation.[24] Finally, the impact of cultivation is relatively brief and benign, usually allowing the seeds of recovery to be preserved within the plot, in forms such as slash piles, stumps, and roots that can initiate and propagate succession.[25]

These favorable conditions for recovery unfortunately do not characterize a lot of the world's degraded land. In the Amazon rainforest, millions of hectares have been converted to pasture and then abandoned after only four to eight years of use as productivity rapidly declined. Natural recovery processes are seriously impeded by a number of factors. For one, a single pasture may be hundreds or even thousands of hectares in extent. This makes re-establishment of the forest problematic, since 90 percent of the tree species depend for seed dispersal on forest animals, very few of which ever venture into the open.[26] But seed and seedling predators, particularly some ant species, are much more abundant in open pasture than in closed forest and will remove experimentally placed seeds within minutes of placement.[27] Microclimatic conditions (air and soil temperatures and humidity) are harsh in devegetated areas, which further limits seedling survival.[28] Depending upon the intensity of use of the pasture before abandonment, recovery of the forest can be expected to take from one hundred up to five hundred years or more.[29] These estimates assume no further human impact. In many situations worldwide, recovery of productivity on abandoned land is prevented by burning or by sporadic human exploitation of regrowth as it occurs.[30]

Even without continued human disruption, however, regrowth of forest may not occur at all. Grasses often invade areas cleared of other types of vegetation, for instance, establishing fire cycles and other conditions that favor their persistence. A state of arrested succession and lowered soil fertility often results, which can strongly hinder efforts to establish crops or other vegetation.[31] An agricultural area of about 3.5 million hectares in eastern Amazonia that was abandoned early in this century had little vegetation aside from scrub and brush fifty years later.[32] In India, trees have failed to become established in abandoned, desertified areas adjacent to sacred forest groves despite ample seed sources.[33]

So we cannot count on natural processes to do the job of restoration on their own within the necessary time frame. Human intervention may be essential to ensure a desirable path and rate of recovery.[34] The potential for accelerating recovery is difficult to assess, however, since the majority of degraded areas with known histories have not yet recovered. Moreover, the recovery process is nonlinear (with respect to time), and intervention can only accelerate some phases of it. Where land is suited to direct human use and has not been stripped of topsoil, substantial recovery may be achieved in as few as three to five years with intensive management.[35] More typically, though, recovery of moderately damaged but basically productive land takes twenty years.[36] On the other hand, recovery of self-sustaining, mature ecosystems in relatively fragile areas that are unsuited to continuous direct human impact (such as modern agriculture) may take a hundred years or more. If the natural recovery potential of degraded land is not taken advantage of quickly enough, irreversible deterioration may be the result.

In most cases today, social constraints are limiting the prospects for recovery much more than biophysical ones. On one hand, reclamation of degraded lands would seem very economical, even in the very short run. For instance, the United Nations Environment Programme estimates the total losses of direct annual income due to desertification of all drylands to be $42.3 billion. UNEP's estimates of the direct annual cost of all preventive and rehabilitational measures range between $10 billion and $22.4 billion.[37] Comparing these estimates should hardly be necessary, though; given the utter dependence of human well-being on productive land, allowing it to continue being degraded for short-term gain is folly of the highest order.

On the other hand, the social barriers to investment in restoration often are complex and formidable. This is nicely illustrated by the circumstances in one of the world's most severely degraded regions, the Sahel. The nomads of the Sahel do not view their cattle as beef on the hoof, but as wealth—capital to be hoarded against need in hard times. Cattle are central to the traditional social life of the group, given as bride payments and generally serving as a living record of social negotiations and kinship links.[38]

Pastoral nomads therefore usually do not look kindly on stock reduction schemes, especially those promoted by governments dominated by traditional enemies of nomadic culture who have tried with great determination and substantial success to destroy it. We have seen Tuareg nomads living a marginal settled existence in the outskirts of Timbuktu, rather than pursuing what was once a very successful way of life.[39] That way of life may be doomed to disappear everywhere—representing a tragic loss of cultural diversity as well as a likely reduction in regional carrying capacity, since in many areas it may not be possible to establish an alternative sustainable food production system.

Ecologist Peter Warshall has suggested that, although tree-planting programs in the Sahel have thus far been disasters, a change in incentives could greatly enhance the success rate and simultaneously reduce costs.[40] For instance, if local animist Muslim peoples were offered a Haj (trip to Mecca) for every so many trees that survived twenty years, Warshall believes reforestation efforts would be successful. Indeed, overgrazing in some parts of Africa could be reduced simply by stabilizing monetary systems. Where the value of money is uncertain, even non-nomadic people tend to store value in cattle, adding to herds as a form of savings, without much regard for range carrying capacity.[41]

Another solution would be a shift from cattle grazing toward game ranching. Antelopes, unlike cattle, are adapted to semi-arid lands. They do not need to trek daily to water holes and so cause less trampling and soil compaction. They feed on a wide variety of plants, browsing (eating the leaves of bushes and trees) as well as grazing (eating grasses); cattle, in contrast, feed on a few selected grasses, which they quickly deplete. Antelope dung comes in the form of small, dry pellets, which retain their nitrogen and efficiently fertilize the soil. Cows, in contrast, produce large, flat, wet droppings, which heat up and quickly lose much of their nitrogen (in the form of ammonia) to the atmosphere. Cow pats also can dry to form a "fecal pavement" blocking the development of sprouting grasses. Moreover, antelopes are naturally resistant to many diseases that seriously affect cattle. An experimental game ranch in Kenya has been a great economic success while simultaneously restoring the range.[42]

Here, as with the Sahelian nomads, tradition creates a formidable barrier to success. The problem of persuading African peoples to break with tradition and take up game ranching will be enormous—we imagine roughly like persuading Americans to give up commuting in automobiles. Indeed, cattle have represented wealth to Africans much longer than cars have represented independence to Americans.

Even in the face of great socioeconomic barriers, though, determined local leaders can have considerable impact. Recall the region of Tammin, Western Australia, where removal of native vegetation has severely disrupted the hydrological cycle and caused salinization of wheatlands.[43] There Joselyn ("Jos") Chatfield, a farmer, has gone against all tradition—not just Australian, but worldwide—and organized her community to plant native vegetation on land that had been converted to wheat. In Australia, as everywhere else, rates of natural ecosystem destruction still outpace those of ecosystem restoration by orders of magnitude. The loss of native vegetation in Western Australia had caused local water tables to rise, bringing salts in the soil to the surface. Restoring some native vegetation reversed the process and flushed salts well below the root zone of the wheat.[44] The small loss of crop area to plantings of native trees and shrubs has been more than compensated for by the restoration of previously salinized fields to wheat production.

It is not too late to recover much of Earth's lost productive potential. As efforts are made to do so, however, it is critical to prevent degradation on the roughly two thirds of the vegetated land surface that have not yet been so affected. Both so-called "alternative" and traditional forms of agriculture may offer feasible approaches for doing that.

Alternative and Traditional Agriculture

There is a great need to make industrial agricultural systems more ecologically sound, lest the gains of the last half century be eroded away, literally and figuratively. That need is just starting to be met. Farmers in the United States and in parts of Europe are increasingly moving to alternative forms of agriculture and away from the green revolution's chemicalized monocultures.

The term *alternative agriculture* actually applies to a variety of farming systems. In general, these systems attempt to limit the use of synthetic pesticides and inorganic fertilizers by taking advantage of the natural pest-control and nutrient-cycling services of ecosystems. The emphasis is on sustainable management of water, soil, and other resources, fitting their use closely to local conditions.[45]

Various "conservation tillage" practices are used to reduce soil erosion and husband moisture. An example is ridge tillage, whereby seeds are planted on the slightly flattened tops of ridges. The space between the ridges is protected by residues from previous cover planting—often legumes (members of the pea family),[46] which can add nitrogen to the soil through the action of bacteria living in nodules on their roots. Ridge tillage, like some other forms of conservation tillage, suppresses weeds and protects the soil, making the large doses of herbicides used in no-till systems unnecessary.

Perhaps even more promising is the development of new, more sustainable farming systems in poor nations, based on ecological principles and often on traditional systems, many of which maintained productivity for centuries.[47] These have the potential for raising overall production in virtually all developing nations, but they should prove especially helpful in the tropical regions where the green revolution has scarcely made a dent.[48]

The International Institute for Tropical Agriculture (IITA) in Nigeria has been experimenting with methods of raising the yields of agriculture in the moist tropics. Its basic approach has been to find ways of closely imitating the traditional African bush fallow system.[49] Shifting cultivators in Africa have traditionally planted crops with the help of a digging stick and little or no soil preparation. A mixture of crops, often a combination of shrubs, root crops, and cereals, is normally planted with little or no fertilization or weeding. The crop mixture provides ground cover, protecting the soil from erosion caused by tropical downpours and from overheating by direct sunlight. Overall output from the mix is generally higher than would be achieved with a single crop.[50]

IITA is developing techniques for no-till agriculture, in which fragile soil is protected by using herbicides rather than plowing to control weeds, and by mulching to enhance soil fertility and prevent erosion by rain. Other work is focused on "alley-cropping," in which crops are grown between rows of fast-growing leguminous shrubs. The shrubs are pruned frequently so they do not shade the crops too much, and the clippings are used as a mulch. Studies on the use of legumes as "living mulches"—ground covers among which crops can be planted—are also being carried out.

There are difficulties with these and other techniques being developed for use by both small-scale and large-scale farmers in Africa. For instance, it is much more difficult to control tropical weeds with herbicides than with plowing, and herbicides are expensive and have unwanted environmental side effects. The shrubs used in alley cropping are sensitive to herbicides, making a no-till system hard to use and requiring careful spacing of the shrub rows if soil is to be tilled between them. Some potential ground covers turn out to decompose too quickly to give the soil lasting protection.

Nonetheless, these and other advances, especially the development of improved crops designed for African conditions, show considerable promise. For example, IITA has developed hybrid plantains that are highly resistant to black sigatoka, a devastating fungal disease of plantain and banana. The new hybrid has more than twice the yield of existing varieties and is expected to improve benefits to farmers growing it tenfold over growing the older varieties that require expensive and environmentally disruptive fungicides.[51]

Another example of encouraging work on improving traditional farming systems is that of agroecologist Martha Rosemeyer. Her work is on enhancing the sustainability and yield of black beans, a staple of the Central American diet. Rosemeyer was based in southern Costa Rica, working on the farm of Darryl Cole, an American who had moved to the area in the 1960s.[52] We visited her experimental plots several times in the early 1990s.

In the traditional system, beans are grown with a thick mulch and generally require little care during the growing season. The Costa Rican agricultural establishment in recent years has encouraged farmers to grow their beans by "modern" methods, using synthetic fertilizers and no mulch. Rosemeyer discovered that the most productive regime includes both mulching and just a small amount of fertilizer; this also minimizes soil erosion and weed growth. Recent expanded experiments with other scientific collaborators involve planting small trees, some of them leguminous, adjacent to the bean fields, also a traditional farming practice in the region. The leguminous trees help supply nitrogen; all of them draw nutrients from deep in the soil and shed their leaves into the mulch. The scientists have confirmed local farmers' lore on the comparative benefits of different tree species and are developing a more productive system using a combination of modern inputs and traditional practices.[53]

There is a great tendency for Western agriculturalists and scientists (regrettably, probably including ourselves) to underrate the knowledge and skills of "less sophisticated" peoples. But people who live close to nature generally know it well and understand such things as the food and medicinal values of plants far better than do city dwellers, especially in rich nations. Peasant farmers often are excellent local ecosystem managers.[54]

The Kouranko farmers of Guinea's Kissidougou prefecture are a case in point. They carefully manage trees by selectively conserving certain species to control water levels in swamps, to shade coffee, to provide fruit and oils (oil palms), and to coppice for fuelwood. They restrict the hunting of animals that disperse seeds of valued trees, raise and transplant seedlings of desired species, and manage fire to maximize the agricultural value of the trees. The same farmers integrate their farming practices with broader aspects of the landscape—for instance, carefully managing watercourses so as to trap silt and enrich the swamps in which they grow rice of carefully selected varieties.[55]

Indigenous knowledge can be very impressive.[56] Farmers in the Northern Province of Zambia recognized twenty-seven groups and seventy-one subgroups of soils, and were familiar with the relationships of each soil type to different crops.[57] Farmers in the Philippines rejected a government recommendation on fertilizer application because in their experience the recommended procedure would not increase yield, a conclusion that was confirmed by further research.[58]

Research increasingly involves the participation of local farmers at the seventeen institutions of the Consultative Group on International Agriculture Research (CGIAR), which include IRRI, IITA, and Mexico's CIMMYT (Centro Internacional de Mejoramiento de Maíz y Trigo). Another CGIAR institution, the International Potato Center (CIP) is conducting research involving Peruvian potato farmers; the International Center for Living Aquatic Resources Management (ICLARM) is collaborating in research with Filipino and Malawian cultivators; women farmers in India are doing research on pigeon peas and pearl millet jointly with the International Centre for Research in the Semi-Arid Tropics (ICRISAT).[59] At IRRI, farmers' knowledge is being incorporated into the Institute's research programs on a variety of topics ranging from elimination of weeds and dealing with soils of poor physical qualities to refinement of farm implements.

Unhappily, the programs of farmer participation in research at the CGIAR centers are threatened by current budget constraints—just when they are both critically needed and beginning to pay off.[60] The total budget of those institutions in 1988 was only $243 million, well under one thousandth of the U.S. military budget or about equal to the defense expenditures of Bangladesh. A few additional tens of millions of dollars annually allocated to the institutions in CGIAR could pay huge benefits. Consider that the relatively small amount, slightly over $1 billion, spent for IRRI and CIMMYT since their establishment can reasonably be claimed to have saved civilization by generating the green revolution. The billion or more dollars that were spent on famine relief operations by the U.S. military in Somalia, by contrast, could not possibly help to solve the basic food problem.

A fundamental problem is that the latter approach—and massive "quick fixes" in general—is just the sort of program that governments like. Rescue operations are relatively easy to deploy, benefit first the most powerful segments of society,[61] and often work well in the short term (if only then). Programs such as participation of traditional farmers in research, by contrast, are designed to deal with complexity, diversity, and long-term results. They must avoid becoming institutionalized and standardized; if they lose their flexibility, they lose their value.

The foregoing are just a few examples that indicate the potential for creating more productive and sustainable agroecosystems. Such changes are especially valuable in areas where great yield increases in staple crops have thus far not occurred, which often are areas of fragile soils or undependable rainfall. There are many others, illustrating how much "scientific" agriculture has to learn from traditional methods of cultivation.[62] Agropastoral systems, which combine animal and crop production, agroforestry systems in which trees and shrubs are grown in combination with perennial or annual crops and/or animals, and mixed-species plantations that yield a variety of tree crops are examples of others that can be made both sustainable and productive.[63]

Another promising source of substantial gains in food supplies in the long term will be through improvement of previously neglected traditional crops. This is especially the case for the humid tropics, where the prime green revolution crops are either unsuited to the climate or are unfamiliar and unaccepted as food.[64] But the impact of colonialism on later agricultural development often led to the suppression and neglect of nutritious traditional crops, which potentially could make important contributions to food supplies in many less developed regions.

Grain amaranths, members of the cockscomb family, once were widely cultivated in Latin America. One amaranth had religious significance for the Aztecs; consequently, after the Spanish conquest, Catholic missionaries suppressed amaranth cultivation and replaced it with maize. Some amaranths have protein-rich leaves, and their seeds frequently contain high-quality protein. Many of the eight hundred species of amaranths are tropical, and their potential for feeding humanity has been recognized but has barely begun to be explored. Ironically, they are increasingly in production in Asia, particularly India, rather than in their native haunts in the western hemisphere.

Many other examples can be cited of scarcely exploited plant groups of great agricultural potential, such as quinoa in the spinach family.[65] Crash programs to take advantage of them should be an important part of humanity's food strategy. One important way to help ensure the conservation of the full spectrum of these traditional crops, and domestic animals as well, would be to configure subsidies to retain many diverse types within agricultural systems.[66]

Careful consideration should be given as well to regional opportunities for substituting more productive green revolution crops for less productive ones, as China has done in switching from rice to potatoes and corn in northern and higher altitude areas.

The key point is that the system adopted in each area must be suited to the local ecological, social, and economic context. A different kind of green revolution is needed, one that enhances overall production in traditional agricultural systems, rather than a wholesale replacement of indigenous methods with alien systems.[67] Production gains must be won locally and regionally in small increments, because what works in one area may not work in another. There are no panaceas. This means enormously more attention must be paid to agriculture in general, and in particular to taking advantage of the human capital represented by experienced farmers. Technology-driven agriculture must be supplemented by, and eventually often replaced by, systems based on the power of resource-poor farmers to adapt, experiment, and innovate.[68]

Foreseeable gains in production from alternative agriculture and improved traditional crops and practices, augmented by some further spread of green revolution technologies and advances from biotechnology, are still not likely to increase crop production at the rate required to satisfy expected needs. Neither are they likely to be able to compensate fully for the threats to future production that may arise from the depletion of essential resources underpinning agriculture or from global climate change. With bad luck, they might prove incapable even of maintaining today's level of production over the long run, let alone provide a sustainable harvest two, three, or more times larger.

Civilization cannot safely count on increased food production alone to provide a decent diet for everyone. Great attention must also be given to protecting the ecological resource base of agriculture, reducing crop losses to pests and spoilage, and to other ways of assuring availability of sufficient food for all people.

Protection of Ecosystem Services

Ecosystem services are probably the least appreciated component of agricultural productivity, yet they are critical to sustainability. Maintaining soil quality and fertility, protecting watersheds, preventing rapid climate change, reducing UV-B radiation, and curbing low-altitude pollutants that are toxic to plants are all crucial to increasing—or even maintaining—current production levels. Similarly, control of water pollution and protection of coastal wetlands would raise the sustainable harvests of fishes.

Ecosystem services often can be either disrupted or enhanced by relatively small changes. For example, traditional coffee plantations shaded by native trees mimic a natural forest and provide reasonable habitat for Latin American songbirds that migrate in summer to the forests of North America.[69] In those forests, the birds are essential to the pest-control function of the ecosystem, and they play a key role in maintaining the health and productivity of the forest trees.[70] As small a step as moving away from shade-grown coffee to sun-grown coffee in Central America could have dramatically deleterious effects on the forests of the United States and Canada—and on local farms and fisheries partly dependent on services from those forests. The shift also would have a great impact on the sustainability of coffee-growing, since the more productive unshaded fields require heavy inputs of fertilizers and pesticides. They also lose much more soil to erosion with no canopy trees to break the force of torrential rains.

One of the most important ecosystem services, of course, is the provision of food from the sea. Unless marine scientists are all wrong (possible, but unlikely), humanity will not soon be able to harvest more food sustainably from the sea than it does today. Indeed, the problem will be to maintain the total catch in the vicinity of 100 million metric tons while, ideally, increasing yields of valuable fishes such as salmon and tuna. The solution to this problem is not difficult. It consists of two fundamental elements. The first is to implement careful monitoring and strong regulation of all important fish stocks so overharvested ones can recover and once again produce yields that are basically "interest," not capital. In most cases, overexploited fisheries will recover if the pressure is removed, as have Atlantic salmon and mackerel, North Sea herring, and Pacific halibut. Occasionally, though, the depletion of stocks may change the composition of the fish community in ways that make recovery much more difficult.[71]

The other element necessary to assure humanity a continuing (but probably not growing) supply of marine fish is to control ocean pollution and protect coastal areas. Unfortunately, both efforts run into a familiar series of severe economic and social problems. For instance, some 200 million people in the world make their living in the fishing industry. When cod stocks plummeted in the North Atlantic, fifty thousand Canadians were put out of work; the shutdown of Georges Bank idled thousands of others, although many had already given up as the stocks declined. Such numbers generate strong support for keeping the industry going in the short run, even if that creates long-run disaster. The parallel to loggers' pleas to cut the last of the old-growth forests in the U.S. Northwest in order to preserve jobs for a few years longer is painfully apparent.

Nonetheless, there are some hopeful signs. New Zealand's oceanic fisheries are now controlled under an "individual transferable quota" (ITQ) system. Quota owners are given the right to harvest a specified amount of fish by paying a resource royalty to the government. The owners may harvest the quota themselves or trade or sell their quota. There have been many wrinkles to iron out in the system, but it does seem to have led to some reduction in the New Zealand fleet.[72] Similar market-based systems are being adopted elsewhere.

At the same time, more and more people are moving to the margins of the oceans, where about 70 percent of human beings now live.[73] This population growth not only boosts demand for seafood, building the pressure to overfish, it also intensifies the pollution and coastal wetland destruction that threaten the very existence of the fisheries. Here is a stark example of the negative impacts of human numbers on a critical natural resource. Strict controls on coastal development and pollution emissions would help, but they will be difficult to enforce, especially in developing countries. Restrictions would also help protect coastal aquaculture operations.

In some areas, fish yields can actually be increased by empowering poor fishers—some with substantial collateral benefits. In Calcutta in the 1950s and early 1960s, a group of people began to lease part of the garbage-strewn Mudialy wetlands from the Calcutta Port Trust, plant trees, shape ponds, and exploit local fish populations.[74] After some problems, the Mudialy Fishermen's Cooperative Society was organized in 1961. At first slowly, and then in the 1980s rapidly, they transformed the wetlands, revising the hydrologic system of the entire area. Sewage water flowing in is treated naturally in a series of ponds until it is nearly drinkable, and the nutrients in it are channeled into fish production. Tens of thousands of planted trees have now matured and can be harvested. Fish yields and earnings have almost trebled.

As the society's accountant, Astam Kumar Bag, put it: "When I joined the society in 1961 our living conditions were dreadful. I lived in a one-room mud hut and couldn't afford to eat much. Since 1986, our standard of living has risen dramatically." Asked if he now considered himself rich, he replied, "Yes, more or less, I suppose." His children live with him in a three-room house with a garden and have three good meals a day with fish for every lunch.

The story of the Mudialy Society is only a small part of the story of the transformation of the East Calcutta marshes. They have been changed from the sort of dismal swampy area one might expect adjacent to one of the world's poorest cities, to, in the words of sanitary engineer Dr. Dhrubajyoti Ghosh, "the largest and finest traditional sewage and waste disposal system in the world." The system now also supplies fish, wood, and vegetables to Calcutta. It is a monument to what can be done if human capital is given a chance. If organizations such as the Mudialy Fishermen's Cooperative Society could become the rule rather than the exception in poor countries, they would be making a gigantic contribution to alleviating poverty and improving the distribution of food.

Reducing Food Losses from Pests

No one knows exactly how much of the world's agricultural production is lost to pests before and after harvest, given enormous variation from place to place and year to year. But a range of 25 to 50 percent seems a reasonable estimate.[75] Assuming very conservatively that the average loss were no more than 30 percent, with great effort that loss might be halved. In that case, the human food supply could be increased about 20 percent—enough to feed roughly a billion more people.[76] Such an increase would buy roughly a decade to make progress in population control and find sustainable ways of expanding food production.

Two distinct strategies exist for significantly expanding the share of food produced that appears on people's dinner tables. One is to reduce the enormous fraction that now is lost to spoilage and pests after being harvested; the other is to improve protection of crops against pests and diseases before harvest.

Improvements in crop storage and transport facilities, especially in less developed nations, could result in substantially reduced losses after harvest to pests and spoilage, which now claim an estimated 40 percent of the global harvest (up to 50 percent in some regions).[77] This solution is probably the easiest, cheapest, and quickest way to make major increases in food supplies.[78] Current losses could be substantially lowered by controlling rodents, insects, fungi, bacteria, and other organisms that attack food in storage and cause spoilage.

Exactly how great the potential gain from improved facilities might be is difficult to estimate for several reasons. First, the amount of food lost after harvest is not precisely known, and the potential gain will vary greatly from crop to crop. In addition, important biological, economic, and social uncertainties are associated with improving food storage and distribution systems to control the losses.

This strategy seems especially promising for expansion of food supplies in developing nations where food is short, vulnerability to losses is apt to be greatest, and facilities are at best inadequate and often nonexistent. Modern pest-proof facilities for drying, storage, and distribution would also diminish vulnerability to local food shortages and famine, and lower the incidence of contaminants (such as fungal toxins) that can seriously threaten human health.

In farm fields, the broadcast use of pesticides remains a principal means of controlling insects and other pests, which ordinarily cause losses of 5 to 25 percent. But spraying fields repeatedly by ground-based sprayers or crop-dusting airplanes not only carries with it too many direct threats to human and ecosystem health, it is too expensive for many farmers in much of the developing world. Worse, in the long run, it is ineffective.[79] Establishing integrated pest management (IPM) systems therefore should be promoted wherever the technical ability to do so can be mobilized.

Integrated pest management is a blend of old-fashioned biological and cultivation techniques (developed by farmers over centuries to control pests and weeds) and modern plant breeding and chemical technologies. IPM systems are being adopted quite widely today, as the bloom increasingly fades from the purely chemical approach.[80] In IPM, predators and diseases of pests are used as "biological controls," while crop strains are developed with multiple resistance to the attacks of diseases, insects, mites, and nematodes. Crop rotation and interplanting are devices that inhibit the attacks of pests and diseases, since both tend to specialize on one crop. Alternating the crops grown in a field from season to season, or fallowing—letting a field lie idle during a crop cycle— limits the ability of pests to build large populations and reduces the possibilities of disease transmission. Planting two or three crops together in a field also can help suppress weeds and pests, while protecting soil from erosion as well. Planting and harvesting times can be carefully scheduled to minimize pest and weed impacts. Finally, pest populations are carefully monitored so that chemical controls (pesticides) are used only when critically needed—as scalpels rather than bludgeons.

A recent study showed that in Latin America major obstacles to implementing IPM include political and economic instability; lack of legislation regulating the sale, purchase, and use of pesticides (and failure to enforce existing regulations); heavy subsidization of pesticide use by the chemical industry and aid programs; and socioeconomic inequity, lack of education, and conservative attitudes among farmers.[81] Similar obstacles obviously would prevail in other parts of the less developed world.

Here once again, we find socioeconomic factors, including equity, to be extremely important. There are no severe biophysical constraints; ecologists know how to suppress pest populations more sustainably, but the socioeconomic situation generally militates against deployment of IPM. In Mexico, only 11 percent of small farmers use pesticides, while 99 percent of large farmers do.[82] There, as in many other developing nations, large farmers get trapped on the "pesticide treadmill"—needing to apply more and more poisons to try to control pests that are becoming increasingly resistant. Pesticide manufacturers resemble drug dealers, hooking clients on a product that must be used in escalating doses to achieve the desired effect.

As the broadcast spray system breaks down, experiment stations (whose fields are similar to those of large farmers) develop IPM systems that suit local conditions, and the large farmers are financially able to hire the advisors necessary to implement them. Most commonly, the system is applied to cash crops. As is all too often the case, the poor farmers are left with serious pest problems and low yields on the food crops they grow.[83]

That does not mean that poor farmers are incapable of adopting IPM systems; indeed, they may be more likely to than rich ones. In Indonesia, the government has been supporting IPM training programs for local farmers. "In agroecosystem analysis, the trainers guide the farmers on how to identify, draw and count the average number of pests and predators, to relate the results to an 'economic threshold' of each pest, and to reach conclusions on whether there is a need to use pesticides or natural enemies."[84] The farmers learn to deal with local rice pests such as the white rice stem borer and infections of rice with diseases such as bacterial red stripe. In the village of Ciasem Baru, farmers began collecting unknown insects on their own, uncovered previously unknown features of the life histories of pests, and shared their knowledge. Here, as elsewhere, the potential for more efficient use of the land rests with the empowered family farmer, who can pay close attention to the microconditions experienced by his or her crops.

IPM systems also clearly could make a significant contribution to reducing food losses after harvest. Some IPM strategies for controlling pests of stored food have been developed, but much more research is needed in this area.[85] Again, the greatest need is in developing nations, most of which are in warm climates where spoilage, insects, and rodents are serious year-round problems.

It should be noted that some pest problems are not amenable to IPM in the usual sense. In Africa, for instance, crops are lost to waterfowl, granivorous birds, rodents, vervet monkeys, baboons, elephants, warthogs, hippos, and ibex (in Ethiopia), among others.[86] There is a great need in Africa for experimentation with various techniques to keep large animals out of fields without harming them. Possibilities include solar-powered electrified fences, various kinds of scarecrows, and hired guards. In parts of West Africa, biological control of rodents is practiced using snakes, and a simple field guide to helpful (and harmless to humans) snakes might encourage the practice. No good IPM method seems to exist for stopping locust plagues; spraying breeding grounds with short-lived malathion is currently the best answer.

Control of weeds is also a major problem for poor farmers. Systems like ridge tillage tend to be machine intensive, and so are not likely to be adopted by those farmers who are often the main suppliers of local foods as opposed to cash crops. Yet it is the production of local food crops that has the greatest potential for increase, and we suspect that, given some support, family farmers would find simple, safe, efficient, and labor intensive (rather than machine intensive) ways to accomplish it.

Diverting Grain from Livestock to People

The strategy of cutting food losses is promising for a significant short-term gain in food supplies. Even greater gains could be made as a short-term, one-time increase in food supplies by using less grain to feed livestock (which today consume more than a third of the world's grain harvest) and growing more to feed people instead. This would be socially and economically much more difficult than improving storage facilities, because it would entail reducing or removing a component of standard of living perceived as important. It also entails shifting from feed grains to food grains preferred by people.

In recent decades, citizens of some rich countries have reduced both the levels of grain-feeding and their meat consumption for health reasons, and as a result the fraction of the world's grain produced for feed has dropped since 1980. But many developing nations are achieving levels of prosperity that allow growing portions of their populations to consume more animal products.[87] With even moderate per-capita income growth, there typically is a marked shift in consumption, and meat and dairy food consumption is rising in the developing world. Still, many more people could be fed directly on a ton of grain than can be nourished by the meat produced by chickens, pigs, or cattle fed a ton of grain.[88] A significant gain could be realized just by emphasizing more efficient meat products (more poultry, less beef), less grain fattening of livestock, and diversification of vegetable foods in diets.

The Action on the Social Side

Two things should be crystal clear at this point. The first is that no technological "silver bullet" exists that can provide the doubling or tripling of food production needed in the next half century without devastating consequences for the natural underpinnings of the agricultural enterprise.

The second is that social, political, and economic constraints are at least as important as biophysical limits and environmental degradation in limiting the amount of food that can be produced. In many developing nations, these social constraints are rooted in population growth, poverty, and inequality: in shrinking farm sizes as holdings are divided among many sons; in huge haciendas owned by the few and pitiful plots owned by the many; in discrimination against women, who are often the primary contributors to food production; in lack of access by poor farmers to inputs, farm credit, and markets; in long-standing political neglect of agricultural sectors of many developing nations; in unfair food distribution within nations at all levels from within the family to between regions; and in growing inequities in the terms of trade (the ratio of international export to import prices). There are no silver bullets here either.

Furthermore, the social and biophysical problems interact—rapid population growth, poverty, and environmental stress—each often worsening the others and reducing the ability of agricultural systems to provide food.

The argument has sometimes been made that poverty causes population growth, not the reverse, and that curing poverty not only will solve the population problem but is the *only* cure that will succeed. It's a magic formula: just cure poverty and all will be well. Experience in developing nations over several decades has shown, though, that the relationship between poverty and higher birth rates is much more complex. Poverty, associated with illiteracy, lack of health care, and high infant mortality rates, clearly helps keep birth rates high. But those same high birth rates also impede the alleviation of poverty and accelerate environmental deterioration.

It has also often been asserted that the food problem is only one of maldistribution. This is another magic formula. Certainly everyone does not have similar access to the world's food supply—perhaps a billion people are seriously underfed while another billion are, if anything, overfed. Although the studies by Robert Chen and his colleagues (described in Chapter 1) have shown that better distribution would reveal no enormous surpluses, it is true that today one major cause of hunger is poverty. If there were no poverty, *and if people were willing to be near-vegetarians and share food equally,* everyone could be well fed. Unfortunately, the caveat is not trivial. We are talking about human beings; and most human beings like to eat meat and are not notorious for sharing equally. Claiming that population size has nothing to do with the food problem is like claiming it has nothing to do with smog in Los Angeles because if everyone walked or biked to work, the smog would disappear.

Nonetheless, the easiest and most sensible way to improve the nutritional status of humanity is to distribute food more equitably. Reducing the gap between rich and poor people would help provide social stability and also encourage a fair international trade system. Both will be necessary to attain universal nutritional security. An effective global war on poverty must therefore be a component of humanity's efforts to establish a sustainable society.

Worldwide, the trend in living standards has been upward, but concealed in that global trend is an enormous amount of human suffering. In 1978, Robert McNamara, then president of the World Bank, coined the term *absolute poverty* to describe that suffering. He defined it as follows: "a condition of life so limited by malnutrition, illiteracy, disease, squalid surroundings, high infant mortality, and low life expectancy as to be beneath any reasonable definition of human decency."[89]

Such a situation is almost unimaginable to the average American; with very few exceptions, even the poor in the United States are well off by comparison. Not only is the depth of such poverty nearly incomprehensible, the number of people in the world who exist on the very brink of survival is just as mind-boggling. The number of absolute poor in recent years has been variously estimated at slightly under a billion to 1.4 billion—as much as a quarter of the human population.[90] More than a billion people have also been estimated by the UN Food and Agriculture Organization (FAO) to be undernourished, more than half of them too malnourished to carry out normal activities.[91] Some 1.2 billion people in poor countries lack access to clean, safe drinking water;[92] more than 2 billion have no decent sanitation facilities.

For the most part, these are all the same people—McNamara's "absolute poor." The great majority of these destitute, hungry people live in sub-Saharan Africa and South Asia, although substantial numbers are also to be found in Latin America, other parts of Asia, and North Africa. Inexcusably, several million also live in the rich countries.[93]

In the post–World War II world, enormous strides seemed to be made in increasing wealth throughout the world. Global average income per person doubled between 1950 and the late 1980s. Of course, people in the rich industrialized nations had the greatest increases, almost tripling their average per-capita incomes, while those in middle-income societies such as Mexico or Turkey more than doubled theirs. Some developing nations, such as Syria or

Egypt, saw small increases in average incomes. But the real tragedy is that many poor countries, including much of sub-Saharan Africa and the Indian subcontinent have seen no improvement at all in their per-capita incomes.[94] Moreover, especially in many poor countries, small or moderate rises in average income were based on significant increases for the wealthy and perhaps an emerging middle class, but no improvement for the poor majority. Too often, as the saying went in the U.S. Great Depression years, the rich got richer and the poor got children.

This is the essence of the world's widening rich-poor economic gap. While the world population more than doubled and global economic output quintupled, the ratio of the share of global income going to the richest 20 percent of the world's people to that going to the poorest 20 percent skyrocketed from an estimated 30 to 1 in 1960 to 150 to 1 in 1990.[95] By 1990, the richest 20 percent absorbed 83 percent of the total global income, while the poorest 20 percent had only 1.4 percent.[96] In 1994, Sandra Postel of Worldwatch Institute properly labeled the income gap a "chasm of inequity."[97]

Poverty and the Environment

The connection between poverty and environmental deterioration was poignantly described by Elizabeth Dowdeswell, UN Under-Secretary General and Executive Director of the United Nations Environment Programme (UNEP), in her World Environment Day message in 1993:

> Stand with your back to the water, in an increasing number of countries, and you will see the impact of poverty on the environment.
>
> In the foreground is a city whose most basic services have broken down. Raw sewage is being pumped into the bay. Garbage collection is nonexistent, and the carriers of infectious diseases are scurrying from one garbage pile to another. Lift your eyes and you will see denuded hillsides, cleared by people so desperate for fuelwood that they have destroyed the forests that hold the soil. If it is raining, you will see a river of mud—

once rich topsoil—being washed through the town and into the sea.

Turn around to face the sea or a road and you will see the effect of this destruction: boats or vehicles with cargoes of refugees preparing for a journey to anywhere that will take them.

This picture, of people driven by poverty to destroy their own means of sustenance, is repeated in many countries. Around the world, more than one billion people live in poverty, and a part of being poor means destroying today what could have sustained you tomorrow. Not by choice, but as a necessity for immediate survival.

In Somalia, farmers are eating their seed grain. In the Sahel, women are walking 15 kilometers for a bundle of fuelwood. In West Asia, overirrigated land is becoming salinized. In Latin America and throughout the humid tropics, genetically rich forests are being felled far beyond the margin of increase. Everywhere marginal land is being overexploited, leaving its occupants more vulnerable to the vagaries of climate. . . .

The poor use fewer resources, create less waste and do less harm to the global environment than the rich. Wasteful overconsumption remains the single most powerful threat to the world's environment. There is a threshold of poverty, however, below which the poor, too, become disproportionately destructive. There comes a point when present survival means destroying resources which could have nurtured the poor for years. The most vivid image of this is the farmer eating his seed grain. Other examples, less dramatic, are being repeated around the world.

Wherever extreme poverty is a problem, there you will find the poor sacrificing long-term benefits—what we call *sustainability*—to the short-term need to survive.

Thus, extreme poverty is not only a result of environmental degradation, it is also—to some extent—a *cause*. . . .[98]

Poverty interacting with population growth is a major cause of environmental degradation in rural areas. Economic pressures often force poor rural families off the better-quality land in many less developed countries and into marginal areas or off the land altogether. Richer farmers go into cash crops and, with the profits earned, buy out their poorer neighbors' land. They have an advantage to begin with in being able to afford the improved seeds, fertilizers, pesticides, and other inputs and technical advice that quickly put them ahead. Displaced, landless farmworkers are forced either to join the legions of job-seekers in cities or to clear and farm a piece of "unused" land, claiming squatters' rights (a common practice in Latin America). The land is usually of poor quality; in the humid tropics especially, it may provide a living for only a few years before erosion or simple exhaustion of the soil ends productivity. Wasteland and more poverty are the results.

The deforestation and degradation processes have created poverty in other ways as well—principally by displacing (or in the worst cases, essentially exterminating) the native populations that dwelled and subsisted in the forest. The Yanomami and other tribes in the Amazon basin, the hill people in Southeast Asia, and still other tribes in equatorial Africa are being squeezed out of living space.[99] While their ways of life might not seem rich to a Westerner, they were more than adequate in their own eyes. Yet the cultures and lifestyles of these defenseless people are being shattered in a frenzy of selling off natural resources in the name of "development."[100]

The key to interrupting these vicious cycles of population growth, poverty, and environmental and cultural destruction is ensuring greater socioeconomic equity at all levels: between the sexes, between households, between rural and urban areas, and internationally.[101] Without this, the numbers and suffering of the poor will continue to grow, and eventually the remaining rich will be brought down as well, as the planet's life-support systems are progressively dismantled and destroyed.

Equity Within Nations and Food Production

Prominent among the powerless and dispossessed are women, who contribute a disproportionate share of the farm labor in large regions of the world. At the extreme, African women are claimed to produce 90 to 95 percent of the continent's domestic food.[102] In South Asia, where women supply most of the labor in rice agriculture, the workload of women farmers (including household chores) exceeds that of men by 1.5 to 2.5 times (in terms of time spent). Children also contribute substantially to farm and household labor.

Agricultural productivity may be influenced in several ways by gender-based inequity. First, the amount of time women have available to farm after tending to immediate household needs may limit productivity. In both South Asia and sub-Saharan Africa, the domestic energy of roughly 90 percent of households is supplied by fuelwood (and to a lesser extent, cow dung). Limited studies have indicated that rural women in those regions now expend 10 to 25 percent of their daily energy on biomass fuel collection. During the dry season, an additional 10 to 25 percent may be expended on water collection—up to 17.5 hours per week in a village in Senegal. The workload of women in both regions is increasing as a consequence of diminishing and receding fuel and water resources, particularly in arid regions.[103]

Another important gender-based inequity resides in women's relative lack of access to farm credit, agricultural extension services, material inputs to farm productivity, and hired or child labor, especially during seasons of labor scarcity. Farmers in poor nations in general lack access to credit and other forms of support, which is available to fewer than 15 percent of farmers in Latin America, Africa, and Asia.[104] Female farmers are even more disadvantaged; the FAO estimates that women's share in agricultural credit is 10 percent or less.[105] The huge unmet need by women farmers for these services and inputs no doubt has a direct effect in constraining productivity.

Such an effect manifested itself dramatically in a survey done in Burkina Faso, where men and women farm different plots of land. Taking into account plot quality, crop, and year, plots cultivated by women had yields 30 percent lower, on average, than plots farmed by men in the same household. Overall, female-controlled plots had significantly lower yields than male-controlled plots, despite a curious observation that the women's plots scored better, on average, on measures of land quality. The disparity in yields appeared to result from a large difference in access by men and women to farm inputs, particularly to manure and children's labor.[106]

A third, nearly universal inequity is lack of decision-making power by women over the management of natural resources (agricultural resources, forests, water, etc.) that are essential for meeting the household's needs. It has often been observed that sustainable use of local resources is more likely to occur if the right to manage them is granted to the women and men whose subsistence depends upon their long-term condition.[107] Many development specialists believe that the quality of natural resources might best be sustained if the state returned decision-making power to the poorest of the poor. At present, the shadow prices (prices reflecting the real costs and benefits to society) of natural resources often greatly exceed their market prices, implying that their exploitation is massively subsidized by the most disadvantaged members of society (e.g., the forest dweller or the tenant farmer).[108]

The history of changes in land distribution and the need for land reform were discussed in Chapter 5. Our conclusion from contrasting the incentives faced by small and large farmers was that a fairly even distribution of land among family farmers was the likeliest arrangement to promote the maximum food production in a region (other things being equal), while minimizing the deterioration of natural capital. Farmers who own their land normally wish to pass it on to the next generation in as good or better condition than they received it. This desire encourages, among other things, a level of careful soil conservation that often is not achieved in the high-discount world of industrial farming.

Recent events in southern Mexico testify to the difficulties of accomplishing land reform. The armed uprising in Chiapas, rumored outbreaks of guerrilla violence in Veracruz and Guerrero,[109] and the assassination of a presidential candidate and another high official of the governing party call into question the picture of a prosperous, stable, modernized Mexico in which issues of poverty in general and land reform in particular are no longer important.[110]

"Development" efforts so far would not be too badly caricatured as a long succession of errors by development experts. Much more of the process should have been entrusted to local people. They can make errors too—but not on the scale of, say, trying to power the further development of China with coal. For a series of case histories of local empowerment and how that can help alleviate poverty, we recommend a fine book called *The Wealth of Communities*.[111] Our only caveat would be to recognize that local empowerment is a scattered and relatively impotent force today. Should it prove to be the wave of the future, it would greatly improve humanity's prospects—but it would not be a panacea. Too many of our problems now are truly global and will require global as well as local solutions.

One of the outstanding, pervasive inequities between groups is that between urban and rural populations. It traces back clearly in many developing nations to colonial policies, but one could find much deeper roots in the history of agriculture as a way of life. To understand how inequity at this level influences food production, one must first realize that the economic incentives provided to farmers often do not reflect local need for agricultural products, but are shaped instead largely by powerful external forces. Domestically, these include economic demand for imported goods and political considerations; international forces include debt, terms of trade, and the dynamics of world commodities pricing, among others.

Rich nation models of industrialization, after which most developing nations fashioned their economic and political objectives, required a vast expansion of economic resources available to the state.[112] The only economic sector large enough to finance the expansion was agriculture, which soon came to be managed to maximize the short-term revenue-earning potential of the state. Funds squeezed from the agricultural sector were and still are invested largely in other sectors, leaving desperate and growing needs unattended in the designated engine of development.[113] This bias against agriculture has resulted in lower rates of return on investment in other economic sectors. In a survey of about 1,650 public-sector investment projects, for example, the rate of return averaged 11.5 percent in nations with a strong bias against agriculture and 18 percent where the bias was moderate or low. Rates of return on private-sector projects in these two classes of nations were 13 and 16 percent, respectively.[114]

The neglect of the rural sector in many countries is shameful. On average, less than a quarter of the education, health, water, and sanitation services are received in rural areas of developing countries, although two thirds of the people live there.[115] Even more skewed is the fraction allocated to local governments of total national spending and of social expenditures, less than 10 and 6 percent respectively.[116]

In developing nations, services and subsidies such as cheap food are concentrated in the cities in part because access and communications are easier, but also because the often insecure governments of developing nations cannot afford to enrage the urban populace, which is likely to be the source of support for any coup d'état. And, even though cities in the developing world, from Nairobi to São Paulo to Karachi, for decades have been swamped by rural immigrants, squatting in self-built tarpaper and packing-box shacks in shantytowns without sanitation or transport services, the majority of the world's absolute poor remain in the countryside.[117]

The implementation and continuation of policies that are plainly detrimental not only to agriculture but to the entire economy are motivated by a set of imperatives common to most developing regions. These include: (1) rapid and massive expansion of public services; (2) promotion of the industrial sector; and (3) responsiveness to powerful and potentially volatile urban interest groups.[118] Thus, earnings from agriculture have been expended on health, education, infrastructure, and employment, which is not entirely inappropriate although it often has been accomplished through the creation of unwieldy government bureaucracies. Less appropriately, agricultural earnings have also been used for capital investment (particularly for imports) in urban manufacturing and other industrial enterprises. Scarce agricultural earnings have all too often also been pumped into military institutions and other "public goods" that benefit only a small fraction of the population.

Yet it seems that political survival, more than any other factor, has driven adverse governmental intervention in agriculture. Farmers are heavily taxed, directly and indirectly. Food prices are fixed at artificially low levels to subsidize the cost of urban labor and living.[119] This has induced and exacerbated migration to the cities, thereby increasing the threat of political unrest among the urban poor, further necessitating artificially cheap food.[120] Basically, urban political leaders, industrialists, and workers all have an interest in keeping food prices as low as possible. As M. Lofchie nicely summarized it, "Thus, suppression of the agricultural sector is a policy that unites the total ensemble of urban interests."[121]

Meanwhile, the prospects for transforming agriculture into a sustainable enterprise, or for achieving needed increases in yields and overall production, remain dim in the absence of credit, farm inputs, adequate storage facilities, locally based agricultural research, farm-to-market roads, and other rural infrastructure. This neglect of the agricultural sector, and badly implemented policies when the sector is not so neglected, directly lowers nations' carrying capacities by limiting their ability to feed themselves.[122]

International Equity and Food Supply

Until the 1980s, the fraction of absolute poor in the world population seemed to be falling, although the numbers continued to rise because of population growth. The economic problems of developing nations were compounded in the 1970s and 1980s when many of them borrowed funds from banks in rich nations to help finance development, counting on returns from the export of commodities to repay the loans. But the bottom dropped out of most commodity markets while interest rates soared, and many developing nations found themselves saddled with debts they simply could not repay. Even with refinancing, some debt forgiveness, and other adjustments, many debt-ridden nations have suffered setbacks in their development.[123]

The gap in wealth between the richest and poorest nations widened appreciably during the 1980s as a result. The stringent economic policies required for refinancing by private international banks, the World Bank, and the International Monetary Fund (IMF) undermined much of the progress that had been made toward closing the rich-poor gaps *within* many of those nations. Those policies, "structural adjustment programs," were intended to extract developing nations from foreign debt crisis through economic "reform," including measures to curb inflation, increase the private sector's role in the economy, increase exports, and reduce government deficits. But a basic goal, of course, was to repay creditor nations.

Many economists believe that structural adjustment will have long-term benefits, but it also has imposed some severe short-term costs. More than other segments of developing societies, the poor suffered as budgets for government health, education, and antipoverty programs were cut. Whereas the single most important element in helping the plow outrace the stork over the next half century would be to strengthen the agricultural sectors of developing nations, structural adjustment policies have often done just the opposite.[124]

UNICEF, the United Nations Children's Fund, blamed the debt burden and economic restructuring policies of the 1980s for a half million of the 14 million deaths of children the agency estimates occur each year from the interacting effects of hunger, poverty, and disease. Unlike the usual bland UN agency report, UNICEF's 1989 annual report minced no words: "It is essential to strip away the niceties of economic parlance and say that what has happened is simply an outrage against a large segment of humanity. . . . Allowing world economic problems to be taken out on the growing minds and bodies of young children is the antithesis of all civilized behavior. Nothing can justify it. And it shames and diminishes us all."[125]

Costa Rica, a country noted for its political stability and comparatively high standard of living, is now considered "the latest casualty in the structural adjustment arena."[126] Under the usual guidance of the International Monetary Fund and the World Bank, and with funding from the U.S. Agency for International Development, it underwent its first structural adjustment in 1980, and in 1994 was evaluated as being in poorer condition than it was then. As described in a recent Office of Technology Assessment report:

Costa Rica's trade deficit has increased, fiscal deficit and inflation continue at high levels, and the increased orientation toward an export economy has exacerbated rural poverty, environmental degradation, and economic instability. Further, gross domestic product has increased while real wages have decreased, indicating inequitable distribution of benefits. Infant mortality rates have begun to rise as has the incidence of disease, which many link to the declining government budgets for health, nutrition, and sanitation under the structural adjustment program. . . . Import duties have been decreased and tax concessions and other subsidies drain resources from the government. The impact on small producers is large: prohibitive costs prevent their entrance into the nontraditional export production and price supports, credits and protection from imports have been slashed for traditional crops [domestically consumed]. Costa Rica now imports nearly one-half of its food supply, as opposed to being nearly self-sufficient as it was in the early 1980s.[127]

It is clearly up to the creditor nations to implement as soon as possible some sort of debt relief/foreign aid package that doesn't simply intensify the devastation of poor nations. In a global society, everyone's futures are interconnected. The flow of wealth from poor to rich nations occurring today was once described by Willy Brandt, the former chancellor of West Germany, as "a blood transfusion from the sick to the healthy," and it carries similar risks.

International trade in food has been a major part of the dismal trade picture and has similarly worked to the disadvantage of less developed nations. In many cases, the trouble has begun with food aid. Humane people naturally wish to send food to the starving. But food aid is a stopgap emergency measure that may end up killing more people in the long run than it helps in the short run. Donated food usually is distributed mostly in the cities and often fails to reach the hungry in the countryside. By depressing prices, the influx of food can worsen the urban–rural inequities that already depress the agricultural sectors of less developed economies. Food aid without comprehensive assistance that strengthens agriculture and rural development is short-sighted humanitarianism at best.

But food aid has many times been the camel's nose in the tent of food import dependence. Indeed, the U.S. Department of Agriculture in the 1960s and 1970s was only too happy to provide food donations under Title II of the Food for Peace Act; recipients almost always later became paying customers for American subsidized exports of surplus grain.[128] The availability of cheap imported grain in developing countries has continued to depress prices for subsistence crops, injuring poor farmers and the agricultural economy. To pay for the imports, luxury foods, coffee, tea, and other commodities could be exported to rich countries to earn foreign exchange—until the bottom dropped out of the commodities markets.

It is long past time for rich countries to revise their trading policies that hurt the poor. Unfortunately instead, developed nations are increasingly finding substitutes for the commodities exported by the less developed ones. Examples include synthetic rubber and bioengineered vanilla. Or they find ways to keep prices too low to benefit exporters, as the proliferation of coffee plantations throughout the tropics has done. Or they rapidly change patterns of demand, clobbering overspecialized commodity-producing less developed countries. The dietary avoidance of saturated fats in "tropical oils" has devastated the economies of South Seas copra (dried coconut meat) producers. In innumerable subtle ways, the economic and trade policies on the part of developed nations exert a profound influence on human, agricultural, and environmental well-being in developing nations.

Developing nations that now avail themselves of a comparative advantage, raising and exporting commodities such as tropical fruits and coffee rather than growing food crops for home consumption, could soon find themselves in very tough circumstances. More than a hundred developing countries are now net food importers. If the prices of foods that developing countries normally import to feed their populations remain stable (as they more or less have been for the past decade) or rise, while exported commodity prices keep falling, feeding their populations will become increasingly difficult. Worse yet, global climate change might reduce harvests in the grain-exporting nations, causing prices to rise precipitously, as they did in 1974–1975 when the first serious worldwide grain supply shortages after World War II occurred.[129]

Can We Do It?

An extremely important element in assuring adequate nourishment for everyone would be to overhaul agricultural economics, both internationally and within nations, including the abolition of many counterproductive agricultural subsidies. To do so would be both complex and politically daunting, especially since more free trade would be the easiest way to accomplish that. But it is necessary nonetheless.

In the wake of the patent failure of the communist experiment, governmental intervention in domestic food markets is likely to decline. This could prove of great benefit to developing nations where policies of keeping food prices low in the cities have been so inimical to agriculture. Of course, all governments have a responsibility to ensure that their people are fed and to avoid political instability caused by hunger. Such instability could threaten international peace, an especially grim prospect as nuclear weapons technologies continue to spread. The importance of agriculture and food distribution systems in this regard has been made very clear in the wake of the collapse of the Soviet Union.

Nutritional security is essential for environmental security; hungry people are in no position to consider the long-term health of Earth's life-support systems. Nutritional security for all peoples requires negotiations and agreements among governments at least as much as does political security. While these arrangements should utilize market mechanisms where possible, that does not mean that totally unregulated "free trade" in food is ideal. Efficient as markets can be, government interventions (including international agreements) are required when significant costs and benefits are not privatized.

Mechanisms are needed that place an appropriate value on preserving ecosystem services and that allocate the costs of doing so among the people who can both afford them and benefit most from them. Somehow, developing nations must get fair compensation for preserving tropical forests, just as a citizen of the United States is compensated when his or her land is taken for a nature reserve.

Properly valuing the resources and ecosystem services essential to agriculture will require the development and deployment of a set of indicators of economic and environmental sustainability. Conventional economic accounting does not incorporate the degradation or depletion of natural resources. A reevaluation of Indonesia's economic growth taking resource depletion into account by economist Robert Repetto, now at the World Resources Institute, was quite illuminating. He showed that when depletion of natural resources (such as forests, soils, or minerals) was considered in measuring GNP, economic growth was considerably reduced.[130] The average increase in gross domestic product (GDP; essentially the same as GNP) between 1971 and 1984 was estimated by conventional calculations at just over 7 percent per year; but when resource depletion was accounted for, it was only 4 percent. Given that population growth was over 2 percent per year during that period, not much progress was made on a per-capita basis. Indonesia's government reportedly has taken the study seriously and is carrying out further work on natural resource and environmental accounting; whether policy changes will ensue is yet to be seen.[131]

The results of this kind of reassessment show clearly that consuming renewable resources on a one-time basis in the process of "development" may be a key to fame of a sort, but not to fortune. Such assessments capture only a small proportion of critical environmental externalities.[132] But they offer a glimpse of the reasons that natural scientists see a far more urgent need to curb population growth than do most social scientists. Rapid population growth helps drive the destructive forces, as new land must be cleared and water sources tapped to increase food production and accommodate more people. When the population growth rate is subtracted from the corrected rate of economic growth, little real progress can be seen. Rather, the mounting losses of forests, wetlands, soil, and groundwater—and mounting stresses on the atmosphere, oceans, and land surface—can be seen as a rising mortgage against the future.

The new environmental accounting system is starting to catch on. It has so far been applied mainly to developing nations, which are often under economic pressure to sell off natural resources (frequently at unrealistically low prices to boot) to earn foreign exchange, thereby stripping themselves of resources for future generations. But the problem also applies to rich countries, as witness the battles over wetlands preservation and the tiny remaining fractions of old-growth forests in the United States, or the scramble to exploit Siberian forests and minerals after the Soviet Union fell apart.

Such indicators are needed to provide feedback between the economic system and the environmental system that supports it. The closing of the international rich-poor gap must be tackled head-on, however. An outline of how such closure might be possible from a biophysical perspective has been developed by Professor John P. Holdren, of the Energy and Resources Group of the University of California, Berkeley.[133] In his scenario, poor nations would develop fast enough to increase their per-capita energy use by 2 percent per year between 1990 and 2025, doubling it from 1 to 2 kilowatts (kW, 1,000 watts). Simultaneously, rich nations would strive to *reduce* their per-capita use by 2 percent annually through increased efficiency, dropping their average use per person from 7.5 to 3.8 kW (while maintaining or increasing benefits).

Rich and poor nations would converge on an average per-person energy use of 3 kW during the remainder of the century. Since energy use is a reasonable surrogate for estimating availability of the various physical ingredients of human well-being, that represents an elimination of *international* inequity. Meanwhile, the world population peak size of 10 billion people would be reached around 2100; then a slow decline would begin.[134] When the peak was reached, total energy use would be 10 billion × 3 kW, or 30 terawatts (TW, a trillion watts). Holdren's scenario is summarized in the following table:

THE HOLDREN SCENARIO

		Population $\times$ *Energy/Person* = *Total Energy Use*		
		(billions)	(kilowatts = kW)	(terawatts = TW)
1990	RICH	1.2	7.5	9.0
	POOR	4.1	1.0	4.1
		5.3		13.1
2025	RICH	1.4	3.8	5.3
	POOR	6.8	2.0	13.6
		8.2		18.9
2100		10	3	30

The scenario assumes that population size can be limited to 10 billion; but with sufficient effort and luck, growth might even be stopped at somewhat less than that. The scenario also assumes that a high standard of living can be achieved with a per-capita rate of energy use of only one fourth to one third of that now seen in the U.S. This assumption seems reasonable based on technologies already in hand. It might involve, for example, redevelopment of the United States into a society built around people rather than automobiles, so that virtually everyone would eventually be able to walk or bicycle to work. Such a change might seem disastrous to those who think GNP must grow no matter what, but it would be highly beneficial for the quality of life.[135]

Even with far greater efficiency than today, of course, Holdren's scenario yields a total energy use more than twice that of 1990, a situation that would still produce catastrophic environmental impacts, unless the mix of energy technologies were substantially altered from today's. Fortunately, that the mix must be changed is already widely recognized by policy analysts, if not by decision-makers.[136] The main thrusts behind this recognition are the clear limits to readily accessible supplies of petroleum and natural gas, and increasing public opposition to unacceptable environmental risks and trade-offs (such as oil spills in fragile coastal or polar areas or the sacrifice of prime farmland to stripmine coal).

The Holdren scenario is a broad overview of how the gap between rich and poor nations might be closed. We concede that, as a basic plan for increasing equity (at least at the international level), the scenario is *very* optimistic. After all, most of humanity has little grasp of the basic dimensions of the human predicament, and the rich show not the slightest sign of willingness to change their behavior to help either the poor or their own descendants. Yet there is reason to believe that such a mix of ignorance and callousness can be cured.

Will Equity Be Enough?

While mounting evidence points to the importance of increasing equity and social justice in resolving the human dilemma, such changes clearly will not be enough by themselves to bring about a sustainable world. Indeed, even heroic efforts both to increase food production and bring about an early halt and reversal of the population explosion might not suffice. Only time will tell—so much depends on successes in limiting population growth and slowing the rate of global change, especially land degradation and loss of biodiversity. Luck with the climate may also be a major factor. While there are divergent views about the means and possibilities of expanding the global food harvest, there is near unanimity among knowledgeable scientists that *there is not the slightest room for complacency*.[137]

In an ideal world, not only would enough food be produced and properly distributed, a substantial carryover stock of grain and other foods should always be maintained as a buffer against unusual climatic events (such as the drought of 1988 and the Mississippi valley floods of 1993, each of which decimated the U.S. corn crop) and unanticipated crop disease epidemics (such as the southern corn-leaf blight, a fungus that wiped out 17 percent of the U.S. corn crop in 1970).[138]

Such events can reduce the world harvest to below the level of consumption, eating into the reserves built up in years of surplus production. So far, fortunately, years of shortage have followed bumper crops, but often reserves have dropped perilously close to "pipeline" supplies—roughly enough to meet demand for six weeks or so until the next harvest is in.[139] At that point, prices begin to rise on the world market, putting the squeeze on food-importing poor countries. Two or three consecutive years of production below consumption are not impossible. The result in that event could be a considerable rise in the death rate among the vulnerable poor.[140]

That is why we must tackle immediately the daunting tasks of lessening the global maldistribution of food, establishing a more adequate system of reserves, and boosting production by all sustainable means. The specter of a world in which widespread chronic hunger and food surpluses can exist side by side, combined with insufficient buffering against production drops, must be ended.

Chapter Eight

TOWARD GLOBAL SECURITY

In human beings, as for all other sexually reproducing animals, issues related to reproducing and eating pervade virtually every aspect of existence. This means that keeping the plow ahead of the stork cannot be viewed simply as a face-off between agricultural and contraceptive technologies. People's relationships with each other—social, economic, and political—strongly influence their patterns of reproduction. Those relationships, along with the size of the population, in turn shape the all-important biophysical, social, economic, and political environments in which the agricultural system must function.

Consequently, succeeding in the tasks of having fewer babies and producing more food will require a seemingly wrenching series of changes that go far beyond reproductive biology and agriculture. These are complicated problems, and they are not amenable to simple solutions. People of different cultures will inevitably respond to them in different ways, and a great deal of patience with others will be necessary if a transition to a sustainable society is to be achieved.

Some circumstances will work in our favor. One is that many of the needed actions kill two or more birds with one stone. Teaching women to read helps counteract sexist repression, makes women more receptive to new ideas in agriculture and elsewhere, and gives them opportunities to make positive contributions to their societies and their families' well-being, as well as reducing their family sizes. It's also a moral thing to do. Giving local peoples more control over their resources expands Earth's carrying capacity, reduces poverty, and makes rich people less likely to be killed by ri-

246

oters. It's also a moral thing to do. Saving biodiversity increases the possibilities for expanding agricultural production, preserves valuable renewable resources, protects against floods and droughts, enriches peoples' lives with beauty, and enhances their chances of staying well. It's also a moral thing to do.

Despite a superficial similarity to past human dilemmas, the situation since the 1960s has been utterly unprecedented. For the first time, a global civilization, with a population several times what most would consider to be optimum, is rapidly reducing Earth's long-term capacity to support people. This generation basically has one shot at doing the right thing—at turning the situation around and making a transition to a sustainable society. If we're lucky, the window of opportunity may remain open for a decade or two more, but all too soon it will slam shut. As the twentieth century winds down, the signs are not good.

Doing the right thing may seem unlikely at the moment, but there are signs both in history and in current events that it can be done. It will take more than science and economics to solve humanity's basic predicament—it will take an ethical revolution and strong commitment. Whether we'll manage to pull it off remains to be seen.

> Noah had ample warning from a respected authority to build his Ark, and he used his time to good advantage. Skeptics laughed, ridiculed, and drowned—but Noah, the original prophet of doom, survived. We too have been warned that a flood of problems now threatens the persistence of industrial society, but this time the ark cannot be built out of wood and caulking. We must ensure our survival by redesigning the political, economic, and social institutions of industrial society. If a new institutional ark cannot be made watertight in time, industrial society will sink, dragging under prophets of doom as well as skeptics and critics.[1]

Are such dramatic transformations really necessary? Noah saved only two of each species, but we'd like to save all of humanity. Might it be too late already to create a stable world in which the population-food equation is humanely balanced? The newspapers don't make cheery reading on the state of the world these days. Local wars seem as common as ever; genocide is alive and well, even in Europe. The threat of nuclear weapons becoming available to tin-pot dictators and terrorists is increasing. Hunger is endemic in many regions, and famines repeatedly plague Africa. Ozone depletion, the threat of global warming, plunging sperm counts, and arguments over endangered species create a cacophony of confusion over the environment. AIDS is now a global epidemic, while Ebola virus, pneumonic plague, and other infectious diseases seem to be waiting in the wings to gain that status. The medical community is very worried about their ability to contain them.[2] Crime, often connected with drugs, plagues many areas of the world.

On the other hand, maybe we're just being alarmist. This may merely be business as usual. Violence and political repression are the stuff of which history is made. The minions of Hitler, Stalin, Pol Pot, Chairman Mao, Queen Victoria, and Genghis Khan, as well as the frontier forces of the U.S. Cavalry, Zulu Impi, Arab and European slavers, Spanish conquistadors and inquisitors, British imperialists, Roman legions, and many less well-remembered groups, slaughtered and displaced people ad lib.

The Bad Old Days

Civilizations have risen and fallen repeatedly, accompanied by famines and plagues. Within our own, things have rarely been rosy. The West has had its share of starvation and plague. Stories of the Irish potato famine, largely generated by British policies, make tough reading even from a distance of a hundred fifty years—cabins full of dead bodies, women and children foraging in turnip fields "like a flock of famished crows, devouring the raw turnips, mothers half naked, shivering in the snow and sleet, uttering exclamations of despair while their children were screaming with hunger."[3]

Making a living during the Industrial Revolution would have been considered a vision of hell by modern workers in Europe or North America.[4] Working conditions in mines and mills were nearly unbearable and extremely dangerous. In the 1890s, a million American workers (about one in sixty Americans) were killed or injured on the job each year.[5] About one in every three hundred railroad employees was killed annually, and ten times as many were injured. Air pollution was horrendous and sanitation minimal, food was often poisonous, street crime made it extremely dangerous to walk the streets of many nineteenth-century cities, and police were often totally corrupt. Drug addiction was rampant and legal in the United States until after the turn of the twentieth century. Opium "slaves" were everywhere; opium (commonly as laudanum) and heroin (as cough medicine) were sold in drugstores, especially to Civil War veterans who had first been given opium to ease the agony of their wounds. Life for most people in the Middle Ages was even shorter, more miserable, and terrifying—if the Black Death didn't get you, the foraging armed men of a warring noble were likely to.

The situation was not dramatically different in other parts of the world. Despotisms of varying dimensions controlled Asia from the Middle East to Japan. Nor were the "savages" so noble.[6] Many New Guinea tribes were forced to live on mountaintops to protect themselves from their neighbors. Ritual warfare was constant, and slaying women and children descending from hilltops to obtain water was commonplace. When German imperialists hanged a New Guinea native caught stealing to make an example of him, another man was hustled forth from the native spectators with a request that he also be hanged because everyone had enjoyed the first one so much. In Africa, tribal rulers willingly sold their own people into slavery. King Mtesa of Buganda had thirty brothers burned alive to celebrate his coronation. He habitually slaughtered innocent peasants who came to his capital; as many as two thousand a day were sometimes speared or roasted to death.[7] In the Aztec empire, human sacrifice and cannibalism were major features of everyday life.[8] Conflict with Amerindian groups forced the Inuit to live in the relatively inhospitable tundra, where men murdered each other with some regularity in order to steal wives.[9] Intense warfare helped to control the number of people living in the Hawaiian Islands after the population reached the islands' carrying capacity around 1600.[10] Indeed, there is no compelling evidence that a world largely colonized by Bugandans, Incas, Inuit, or Chinese would be one iota better off than the one we have that was largely colonized by Europeans.

But the received wisdom of society is not to worry about it. People are a miserable lot; always have been, probably always will be. They can sometimes control their destructive behavior and get their affairs in order. Things get worse, but then they get better. War is followed by peace. Famine is followed by plenty. Stock markets eventually rebound from crashes; booms follow depressions. Yes, many parts of Africa seem to be in bad shape, and hunger stalks much of the sub-Saharan region, but Africans are clever and persistent farmers. Given half a chance, they could feed themselves.

Surely we should do our best to improve our lives and those of people we care about, and otherwise just wait things out. The situation may seem bad, but not all bad. After all, the Cold War is history, South Africa seems to be making a transition to a stable, multiracial society, Middle East peace may be possible, the economies of some rich nations like the United States seem in good shape, infant mortality is declining in most places, and Science (or some other religion) will undoubtedly find ways to solve all human problems. Progress occurs by fits and starts, but the trend is upward. Economic growth has generally improved the human condition, and more of it is the cure for most of humanity's problems. Furthermore, it seems unstoppable; the gross global product is increasing, as is the per-capita global product. Each new mouth to feed is accompanied by a brain to think and two hands to work. A smaller proportion of humanity probably is hungry today than in 1950.[11] Human ingenuity can make supplies of food and other resources infinite. Given the demand, the market will call forth any product. Progress is bound to continue.

Many of the key points of this received wisdom will not, of course, stand up to close scrutiny. For instance, one recent analysis documents that, in fact, economic growth has "enriched the few, impoverished the many, and endangered the planet."[12] Social scientists Richard Barnet and John Cavanagh looked at the pressures of population growth, automation, and reorganization of global employment and concluded: "A huge and increasing proportion of human beings are not needed and will never be needed to make goods or to provide services because too many people in the world are too poor to buy them."[13]

Peace anywhere, including South Africa and the Middle East, may be an ephemeral thing on a planet whose resources and environment are being stressed by growth in both population size and consumption. And, of course, people who live in festering third-world slums like "Chicago" in Abidjan or are homeless in Chicago, Illinois, don't have their economic views expounded in *Forbes* magazine or on the editorial pages of *The Wall Street Journal*, which mindlessly view the world through rose-colored glasses as long as the very rich are getting richer still.

Unfortunately, neither the cyclic nor the continuous-improvement views of civilization fits today's situation. Our global civilization faces collapse if today's trends are permitted to continue. The laws of nature will not be repealed to permit either population size or consumption per person to grow forever. But people corporately plan as if those laws did not exist—as today's situation in China demonstrates dramatically.

The New China

One of the most depressing things we have ever personally observed is the accelerating boom in consumerism in the People's Republic of China. In 1981, some 1.8 million Chinese households were registered as private businesses; by 1992, that number had gone to 15.5 million (and that may be an undercount since many businessmen still fear a shift back to suppression of capitalism).[14] One of the social side-effects of this trend has been a decrease in equity. The gap between rich and poor in China, once the smallest in the world, has widened considerably.

China already has a surplus of labor, especially in the countryside, where 860 million people live. The number of unemployed in rural areas is about 100 to 120 million, and a "floating population" of 70 to 150 million people from the countryside is roving among towns and cities searching for employment.[15] These migrants are allowed to work in the cities, but their household registration as peasants remains in effect, and they have no access to local welfare as they work in factories, building freeways, street peddling, and so forth. They constitute a new and problematic "third category" (along with peasants and urban dwellers) in Chinese society. They are the source of most urban crimes, thefts, prostitution, and the creation of slums.[16] In addition, a large portion of government workers are redundant.[17] China's Ministry of Labor expects that, despite the rapid economic growth, by 2000 there will be 200 million unemployed in rural areas and 68 million in urban areas.[18]

While sensible people in the West struggle to correct the development mistakes of the past, China seems hell-bent on repeating them. We watched bulldozers moving stacks of coal at a power plant, symbolic of China's increasing dependence on what may be the worst possible energy source from an environmental and public health standpoint.[19] Where it is clear that nations like the United States should be struggling to reduce dependence on automobiles and redesign cities so that commuting could be done largely by mass transit, bicycle, and on foot, the Chinese are now buying 350,000 cars a year and plan to be adding 3 million annually to the fleet by 2010.[20] The nation officially reversed its support of public transportation in May of 1994, with the announcement that a capital-intensive industry producing family cars would become an "economic pillar of its emerging socialist market economy."[21] Auto manufacturers from Mercedes and Porsche to Volkswagen, Peugeot, Toyota, Nissan, and General Motors are all seeking a piece of the Chinese action.[22] At the moment, however, cars are out of reach of most private citizens (only about 1 percent of sales are to individuals); most purchasers are government institutions and state and private enterprises. For the time being, most powered vehicles in private hands will roll on two wheels.

The eagerness of the Chinese to acquire (if not own) automobiles is spectacular—almost as spectacular as the traffic jams that already make moving around Beijing a nightmare. At an outdoor new-car sale, we mixed with shoppers in icy cold weather. We asked a car salesman why China was building cars when traffic was at a near standstill and there was no provision for parking anywhere. He explained to us that cars, roads, garages, and parking spaces all must be built simultaneously. And that's what they are doing. Some 150,000 laborers were brought into Beijing to complete a ring-road freeway in six months, working around the clock.

We visited the plant of Beijing Jeep Corporation, a joint venture of Beijing Autoworks and Chrysler Corporation, where a staff of about sixty-five hundred people is busily producing about eighty thousand new cars annually for the Chinese fleet. The production was divided between 2020S (the Beijing Jeep), the most common "light cross-country vehicle" made in China, and the fancier Jeep Cherokee, many of which were destined for export. Our guide was Liu Zhiqing, the engineer in charge of environmental protection. Outgoing, fifty-eight years old, and clearly proud of Beijing Jeep's efforts to keep poisonous effluents out of the local environment, Liu was the father of a son and daughter. Although "too old" to learn to drive (driving schools are booked far in advance and show preference for younger students), he suspected that his own children might eventually have cars, but was more doubtful about virtually everybody driving. He volunteered that he had been personally very critical of Chairman Mao's "big mistake"—not getting started on population control soon enough.

Cars are only part of the picture. In the high-priced embassy district of Beijing, in the midst of the many shining new skyscrapers, the Kulun Shopping Center has a top-of-the-line Harley-Davidson in the window and a wide selection of Italian fashions for the elite Chinese shoppers living nearby. In less wealthy areas of the city, there are crowded stores full of microwaves, refrigerators, perfumes, cosmetics, fancy candies, and other luxury goods—as well as closed-circuit TVs perpetually touting the advantages of various products. Everywhere, men could be seen using cellular phones. Beepers were much in evidence in restaurants.

The situation is not without its humor. On the street, we were handed an advertisement for a new Churrasqueria,[23] "Rodizio El Gaucho," which said in part:

El Gaucho means "cowboy" from Pampas in Argentina. Bold and uninhibited Gauchos gallop across grassland in day time and in the evening they set campfire with big pieces of beef on it. With delicious barbecue and good wine, they sing with guitar, dance samba, enjoy themselves until late at night. Now we are introducing this special flavor from South America to Beijing. We will select different parts of good calf and put them on charcoal fire. Waiters will hold a big barbecue fork and pare barbecue which is on the fork to your plate. If you are "meat eater", you will think it is real barbecue after eating more than ten different parts of calf, tongue, tenderloin, belly, buttocks, etc. If you are vegetarian, you can have cold dish, hot dish, pizza, etc. After you've finished all plates and dessert, you pay only 10 dollar/per person. We also employed cooks and waiters from South America. Maybe the waiter who serves your table is just a foreign waiter.

The growing inequity of the new China, exemplified by the ad for a dinner in the Churrasqueria, which would be far beyond the financial reach of perhaps a billion of its citizens, was also much in evidence. A photographer hired by our group made almost as much in one day as an average Chinese did in a year. Around the fancy hotels and stores, one had to run a gauntlet of aggressive beggars, mostly children, an element completely absent when we first visited China in 1980.

The signs were not limited to the city; some areas of the countryside are moving upscale rapidly as well, leaving others in desperate poverty. We visited Liu Min Ying village in Da Xing County, outside of Bejing, a town (previously a commune) that clearly represents an ecofarming development model for the rest of China. It was described to us as "typical of the best." Our visit was hosted by Zhang Guanghui, the vice mayor of the town, a pleasant fiftyish woman who was clearly fully liberated. Over a snack of delicious mandarin orange segments and bananas (delicacies in midwinter), peanuts, and roasted pumpkin seeds, she told us she did not agree with the women who thought things were moving backward in the realm of women's rights; she believed they were getting better. She described the town's biogas system for digesting animal manures and pulling off the methane to use as cooking fuel. That, and some use of solar energy, had ended a long-standing energy shortage in the village.[24] The village was able to reduce the application of chemical fertilizer while tripling the use of organic fertilizer between 1982 and 1987; meanwhile, per-capita income doubled and forest cover increased by 60 percent.

But Liu Min Ying village's hopeful model is only that. Peasants in the relatively rich east (near Shanghai) had about 2.7 times more income per capita than those in the west (Ganzu) in 1978. By 1991, that gap had widened to 4.5 times.[25] China seems in clear danger of fractioning into several "nations." One, primarily coastal, might (at least temporarily) support Hong King–like islands of great affluence surrounded by great poverty, while others in the hinterlands remained mired in utter misery.

Costs of Inequity

China is not alone under that threat. Increasing inequity, hunger, disease, environmental problems, and resource shortages are likely to be harbingers of social collapse. In many cases, they may cause or exacerbate armed conflicts. The 1967 Arab-Israeli war was fought in part over water, and supplies are even tighter today.[26] In studies commissioned by the American Academy of Arts and Sciences and the University of Toronto, thirty experts concluded (*before* the Rwandan disaster and the revolt in Chiapas) that "scarcities of renewable resources are already contributing to violent conflicts in many parts of the developing world. These conflicts may foreshadow a surge of similar violence in coming decades, particularly in poor countries where shortages of water, forests and, especially, fertile land, coupled with rapidly expanding populations, already cause great hardship."[27]

More conflict is likely to make it very difficult to promote the kinds of cooperative behavior and development of sound national and international policies that will be essential for dealing with the population-resource-environment predicament. A principle formulated by demographer Nathan Keyfitz is even more certain to hold true: "If we have one point of empirically backed knowledge, it is that bad policies are widespread and persistent."[28]

Unhappily, racism, sexism, and economic inequality seem as typically human as trying to be sexy. Just as we can culturally compensate for being sexy with contraception, so people have shown that they can culturally overcome these other human attributes—

although, as Hitler, Stalin, My Lai, Bosnia, Rwanda, metal detectors in American schools, and myriad other examples attest, the dark side of our "humanity" always seems to be lurking just beneath the surface. The point is that the human past, with its great triumphs and its dismal excesses, is what we're stuck with, and using any aspect of it to justify human failings today gains us nothing. We should look to history to help us understand today's behavior and avoid being condemned to repeat our ancestors' errors. With humanity's greatly enhanced technological capabilities and enormously larger population size, the costs of making mistakes have escalated too far beyond our capability to pay for them.

One of the trends in the world that works against improving the lot of poor people—that increases inequity—is the expanding influence of the already superpowerful multinational corporations. A new global economic order is emerging under the influence of a few hundred of these giants.[29] The biggest three hundred multinationals now control about 25 percent of the world's productive assets.[30] The annual sales of Philip Morris are greater than the gross national product of New Zealand; Ford is a bigger economic entity than Saudi Arabia and Norway combined.[31] These giants are binding together the rich—not just residents of nations like Japan, Germany, and the United States, but hundreds of millions of middle-class and wealthy people in poor nations like China, India, Indonesia, and Brazil. As the rich of the world unite in dealing with a common economy from which the poor are largely excluded (while entertainment is globalized in ways that will make the poor ever more aware of their situation), stability of any sort may be increasingly hard to attain. As Barnet and Cavanagh put it: "The world is getting smaller . . . but it is not coming together."[32] The very nature of nation-states seems bound to change as they lose more and more of their ability to regulate their own affairs. Giant corporations are increasing their ability on one hand to buy politicians and on the other to evade the jurisdiction of governments.[33]

The expansion of global corporations illustrates a sort of economic momentum that may prove to be as important as demographic momentum in shaping the future of our world. One very evident result is the failure of the new "global economy" to create jobs in sufficient numbers. By 2015, it is expected that the number of jobless men and women in developing nations will more than double to nearly 1.5 billion.[34]

Indian scientist and humanist M. S. Swaminathan described the impact of this trend on India: "Multinational industries are highly mechanized and do not create many jobs, yet India must find employment for 100 million more people by the year 2000. As Gandhi said, what we need is production by the masses, not mass production."[35] A village blacksmith cannot compete with a modern industrial plant turning out tools on an assembly line. What is needed in places like India is an economy largely based on cottage industries.

The reasons for rising unemployment in the world were summarized in a different way by German poet Hans Magnus Enzensberger: "more and more people are being permanently excluded from the economic system because it no longer pays to exploit them."[36] The basic problem facing the new global economy will be market saturation. Population growth and downward pressure on wages mean increasing numbers of poor people, which is not a long-term recipe for expansion of markets. Sooner or later, a system driven by spreading frenzied consumerism is bound to grind to a halt.

Under these circumstances, it seems very unlikely that the human predicament can be solved without a coordinated effort on the part of all peoples, rich and poor—and the global corporations whose future is certainly threatened by both instability and unsustainability. At the moment, the prospects for such an effort seem to be fading rather than brightening. Some of the other obvious interrelated factors in the overdeveloped world that hinder finding solutions include increasing social and ethnic fragmentation, the rise of political correctness and fundamentalism, the decline of traditional education, and the near absence of education on how the world really works. In addition, we face a natural human tendency to discount events far from us in future time or in geographic or cultural distance.

Social Traps

In numerous situations (social traps), immediate local incentives are inconsistent with the long-term best interest of both the individual and society, and with the maintenance of Earth's carrying capacity.[37] The substantial hidden subsidies promoting the use of fossil fuels are a good example. The use of those fuels not only has grave effects on the biophysical environment, but, by promoting dependence on the automobile, on the social environment as well.

In the face of such perverse incentives, individuals and firms find it almost impossible to take the long view.[38] Consequently, guarding against such social traps should be an important function of governments. Governments have learned to take the long view in areas such as national defense and social security. Now decision-makers must learn to recognize social traps that affect sustainability.

One of the most pervasive causes of social traps is the natural human tendency to discount costs that appear remote, either in time or space. The most straightforward reason for discounting over time is to adjust for the "time value" of money: a thousand dollars delivered today is worth more than a thousand dollars to be delivered in ten years because of benefits that can be derived from investing the money over the decade. Discounting is done routinely in the context of cost/benefit analysis and has enormous influence on fiscal policy in every arena.

Although discounting is valid in principle, two problems make discounting over a substantial time horizon (several decades or more) a gamble with the welfare of future generations. Estimating future costs and benefits is difficult, and selection of an appropriate discount rate is subject to both subjectivity and uncertainty. Economists cannot easily assign monetary value to many of today's environmental amenities (such as biodiversity, soil, and groundwater) and risks (global warming, ozone depletion, antibiotic-resistant strains of bacteria), much less those of the future. When future costs are uncertain, a risk-averse government would encourage discounting less than if risks could be predicted with certainty. When analysts cannot agree on the uncertainties, however, too often they make no adjustment at all in the discount rate. The result is an underestimate of potential future costs, such that projects that imperil future generations appear more favorable than they should. These uncertainties are compounded over the period for which the calculation is made; the longer the time horizon, the greater the gamble. And when essential resources are involved, that gamble is with future carrying capacity.

The problem with discounting is not simply that decision-makers often fail to apply it appropriately. Beyond that, the very process of discounting (especially at rates as high as 10 percent) encourages the public to underestimate the importance of future costs and defer their payment. Consider the problem of determining whether society would profit by taking measures now to deter the onset of global warming. Suppose that inaction will result in a known and certain cost of $100 billion to be incurred in 100 years. Discounted at 10 percent (on an annual basis), the present value of that cost is reduced to a mere $7.2 million. In a cost/benefit framework, investment in any deterrent whose net immediate cost exceeded $7.2 million would be irrational. But that discounted cost is so small that society may simply ignore a potentially serious future problem.

Of course, people might assume that posterity will be richer than we are, easily able to pay the $100 billion. While in the recent past, successive generations have indeed enjoyed ever greater average wealth, this trend is already showing signs of faltering. It may well not continue until the time comes to pay for those deferred costs. Furthermore, pressing economic problems often cause developing nations to apply higher discount rates to the future costs of depleting essential resources or of accepting toxic wastes and environmentally damaging industries rejected by rich countries. Although some of these problems have been the subject of a great deal of attention by academic economists recently, standard practices have thus far changed little.[39] And standard discounting in a context of long-term resource management constitutes a recipe for a growing burden of ecological debt, resulting in a lower future carrying capacity and possibly the breakdown of civilization.

Another form of discounting, also important and inherent in policy judgments relevant to creating a sustainable world, is discounting over distance. The significance of events (including the magnitude of benefits and costs) occurring at a distance is discounted. The distance may be measured in strictly geographic terms, or it may be remote in a social, economic, or political sense.

Discounting over distance is reflected in several dimensions of human behavior and judgment. Consider how societies value domestic environmental health relative to that abroad. Japan is using timber stripped from virgin forests in several nations (including the U.S.) for low-quality products such as concrete forms, while carefully protecting its own forests. Twenty-five percent of all pesticides exported from the United States are heavily restricted or banned from use in the U.S. and other industrialized nations. The German government made little effort to control industrial emissions until the effects of acid precipitation were manifest in its own forests and soils (to the tune of $1.4 billion per year). By then some eighteen thousand Swedish lakes had acidified to the point that fish stocks were severely reduced, in part due to German emissions.

While in some instances, discounting by distance is clearly in the best interest of the discounters, misjudgment of the distance relevant to a situation may exact a penalty. In the United States, overestimation of distance contributes to the extraction and sale, at below market values, of natural resources (such as timber) from regions that are geographically and socioeconomically remote from policy centers in Washington, D.C. (such as Alaska and Colorado), which clearly results in a net cost to the nation. It seems logical to the Japanese to take advantage of the low prices that some nations charge for their forest products, because discounting by distance allows the Japanese to ignore the consequences. But those consequences are borne by the whole world, although disproportionately by victims other than Japan. The direct costs include the loss of potential medicinal or industrial products, the destruction of potentially important crop relatives, local land degradation, and exacerbation of global warming. These, along with numerous indirect social, economic, and ecological costs, are likely to outweigh by far any profits made by the buyers or sellers of cheap wood and woodchips.

Overestimates of the relevant distance of discounting have led to profound environmental problems with direct implications for sustainability. For example, until recently, the upper atmosphere was considered so remote as to make it appear safe to emit airborne pollutants that did not cause local or regional smog problems. It came as a surprise that, though seemingly distant, the connections between the gaseous composition of the stratosphere and our day-to-day lives are actually very tight. Similarly, the ability of humanity to disrupt global biogeochemical cycles through local and regional habitat conversion has only become apparent in recent decades.

Currently, the many indications that human society has exceeded Earth's carrying capacity and is beginning to pay a price for it are barely noticed. The negative impact of human activity on the planet usually manifests itself first to those whose lives are tightly dependent upon the health of fragile local ecosystems. Yet by the time many current environmental problems directly affect decision-makers, whose lives are buffered by distance and wealth, it will be far too late to correct them. Discounting over distance fosters the illusion that wealthy nations and individuals can afford to ignore the increasingly desperate plight of the poor.

Ecologist Thomas Lovejoy's program of taking policy-makers and celebrities to tropical forests has helped make apparent their intimate connections to parts of the biosphere that are often misperceived as remote. Senators and movie stars who have been to the Amazon with Tom have become powerful allies in the struggle to preserve biodiversity. Such programs should become parts of government's normal functioning; like national defense, the benefits are too widely dispersed for them to be automatically undertaken by private enterprises.

Discounting over both time and distance encourages behavior today that reduces Earth's carrying capacity for future generations. Governments are the best agencies we have today for helping us to avoid those social traps—but they must learn to recognize them. With luck, global corporations might also learn, although at the moment we fail to see any way in which they can be persuaded to act in response to any imperatives except those provided by markets.

Despite these handicaps, concerned human beings have no choice but to press ahead trying to deal with the human predicament.

Beating the Stork

The basic elements of a strategy for helping the plow beat the stork have been given in foregoing chapters. In our view, they make it clear that there is no single magic cure for the human predicament. We can't just follow slogans like "Think globally and act locally" or pretend that recycling is the answer to all environmental problems. Only by taking many different approaches to the predicament—by addressing problems at various levels in many different ways—can the chances of creating a sustainable society be greatly improved.

It is also important that everyone understand clearly what the primary goal is: to organize a world in which all people can explore their full potential as human beings for many, many generations into the future. The short-term, desperate means of getting humanity on track toward that goal is to attempt simultaneously and quickly to reduce human fertility and substantially increase food availability while carefully protecting Earth's life-support systems. That is no small order, and it clearly involves many policy measures carefully tailored to fit the needs and attitudes of different nations, regions, or ethnic groups. Trade-offs and compromises will often be necessary between the primary goals themselves and other goals that people value. Obviously also, solutions that will work in one place may be totally ineffective in a different economic or cultural context. But overall, steps to increase the equity of human societies are clearly needed at all levels. At the most elementary level, poor people need more money to buy food. Equity is needed now as never before because humanity has not previously exceeded the biophysical carrying capacity of Earth.

The needed actions, as you now must realize, amount to a nearly total transformation of human activities. We can't attain a sustainable society where all can be adequately fed simply by passing out condoms and pouring on fertilizer. We propose some key actions here quite briefly; even so, they dismay even us, since they resemble a catalog of moral imperatives. We place them in several categories, the first of which is education. Education is not only essential in itself, but also is needed to provide the necessary groundwork for putting the other actions high on the human agenda.

EDUCATION

Much more of humanity's resources should be devoted to education in both rich and poor countries. If nothing else, this is the one time-tested way to narrow the rich-poor gap; educated people tend to find opportunities or make their own. Education builds human capital, promotes productive work, and enables people both to limit their reproduction and to obtain food.

By education we include not just education in schools and universities, but adult education through public media, conferences and workshops, computer networks, job sabbaticals,[40] and any other means available. That education should have greatly expanded coverage of equity issues—why, for example, the notion that skin color indicates innate intelligence is nonsense. Education must deal directly with reproductive issues, including the need to limit families to an average of two or fewer children and the use of contraceptives to achieve that end. It should discourage women from childbearing before the age of twenty, which is justifiable on health grounds alone. And education should strive to inform everyone about the fundamental principles of how human beings and their economic activities relate to the ecosystems that support civilization.

The actions that education should help move toward the top of the global agenda include:

REDUCING CONFLICT WITHIN AND BETWEEN NATIONS

This point is self-evident, and in some respects pie in the sky, but we think it should be mentioned. Much of the starvation and impoverishment in recent decades can be traced to warfare. The level of cooperation required to solve the human predicament clearly requires a global society in which armed conflict is largely suppressed. And a global society already stressing its life-support mechanisms can hardly afford the damage that modern warfare inflicts on them. Increased political stability thus is an essential key to moving toward sustainability. This is, unfortunately, something of a chicken-and-egg problem, since many of the inequities that must be reduced themselves tend to generate conflict.

INCREASING INTERNATIONAL EQUITY

All nations should cooperate to close the rich-poor gap, with overdeveloped nations reducing their material consumption in order to make room for needed physical development in less developed countries. Speed is of the essence, since the scale of the transition is vast, and the lead time required for such tasks as reorganizing cities, redesigning transport systems, and deploying new energy technologies is on the order of a half century.[41] Speed may also be of the essence, because if the activities of global corporations get entirely beyond the control of governments, a transition to sustainability may well prove impossible.

Development assistance should become a major item on the agendas of all rich nations. The stream of wealth now flows from poor to rich; that must be reversed if civilization is ever to be sustainable. To make aid appropriate, minimize waste, and see that help gets to the proper segments of poor societies is an enormous challenge for both donor and recipient countries. But for both it is an investment in survival.

The international trade system needs to be continuously revised with a view to increasing equity (especially providing employment), reducing environmental impacts, and enhancing food security.[42] The problems here are especially complicated and will require even more careful and broader negotiations than have already gone into NAFTA and GATT.

INCREASING EQUITY WITHIN NATIONS

Efforts should be redoubled to suppress racism, sexism, and religious intolerance.

Unequal treatment of the sexes and gross economic inequity should be attacked by expanding opportunities for employment, especially for women, but also for men.

Providing universal education and literacy should be given high priority everywhere.

REDUCING FERTILITY

All industrial nations should move immediately to make the safest possible contraceptive techniques accessible to all sexually active people. Informational campaigns should be started and, if necessary, economic incentives provided to move population sizes in those nations to rates of *shrinkage* in the vicinity of 0.3 to 0.4 percent annually—rates already achieved in Hungary and Estonia. TFR goals in all developed countries should be 1.5 or less. This is hardly a formula for rapid depopulation. If Hungary continued to shrink at 0.3 percent, it would take fifty years for its population to decline from 10.3 million to 8.9 million. A similar 14 percent shrinkage in the United States—back to 224 million people—would remove roughly as much pressure from Earth's ecosystems as a reduction of 400 million people in poor countries.

All developing nations should generate reasonable timetables for reaching a replacement TFR and then proceeding to 1.5. The steps necessary to achieve those targets should involve a minimum of coercion and social disruption.

Development aid to poor nations should include, where requested, help with family planning services.

Family planning services should be combined or coordinated with comprehensive health care services, especially for women and children.

In rural areas, special effort must be put into supplying basic sanitation and health services, elementary education, clean water, and cooking fuel (or solar cookers) to rural families, all of which may lessen the demand for children.[43]

Constraints on research and testing of fertility control technologies should be relaxed, so that risks can be more appropriately judged against the risks of pregnancy and childbirth *and* the risks to everyone from increasing overpopulation.

INCREASING FOOD PRODUCTION

Economic assistance to less developed countries should focus on the agricultural sector, supplying poor farmers (and especially female cultivators) with credit, extension services, farm-to-market roads, and other needed improvements in infrastructure.

Development that takes prime agricultural land out of production should be restricted everywhere.

Efforts to increase crop production should continue to focus on raising yields on land already under cultivation, rather than bringing more marginal land under the plow.

Much more should be invested in agricultural research aimed at creating ecologically sound, highly productive agriculture region by region, with special attention to improving the yields of poor farmers growing subsistence crops and taking advantage of traditional systems (such as shifting cultivation) where appropriate. Involvement of farmers in that research should be increased everywhere.

Major investments should be made in restoring degraded lands to higher levels of productivity.

Integrated pest management should be deployed everywhere, coupled with government policies that discourage the broadcast use of synthetic organic pesticides.

Local control of resources essential to agriculture, such as watershed forests and woodlands and freshwater sources, should be promoted as much as possible. In general, "globalization from below"—the sorts of citizens' movements already described designed to provide employment and preserve resources and cultures—should be encouraged as an alternative to the superconsumerism promoted in the globalized economy.[44]

Ways must be found to strengthen and effectively implement Law of the Sea provisions to convert all harvesting of marine fisheries to a sustainable yield basis and to arbitrate international disputes over shares of the catch.

PROTECTING THE NATURAL INFRASTRUCTURE OF FOOD PRODUCTION

Rich nations should move to develop sustainably along the lines suggested in the Holdren scenario on energy consumption to lessen their assault on vital life-support systems.

As far as feasible, the world economic system should be designed so that environmental costs and benefits accrue to the same people. In some poor countries, this may involve promoting private rather than communal ownership of resources, since the latter often leads to unsustainable exploitation.[45]

Effective programs must be established to preserve populations and species of plants, animals, and microorganisms, which are our most precious natural capital—the one resource critical to agriculture and forestry that is not replaceable even on a time-scale of hundreds of thousands of years.

More comprehensive programs are needed that specifically strive to preserve the vital genetic diversity of crops, crop relatives, and domestic animals.

In the sustainable development plans of both rich and poor nations, serious efforts should be made to reduce greenhouse gas emissions to help stabilize the climate. Emission reduction targets should be strengthened and incentives provided to recruit the enthusiastic cooperation of businesses.

The implementation of the Montreal Protocol should be carefully monitored to guarantee reestablishment of the integrity of the ozone shield.

Indicators of sustainability should be developed and adopted as soon as possible. In particular, national resource accounting systems should be adopted to capture as fully as possible the depletion of natural capital.

INCREASING FOOD AVAILABILITY

Rich nations should help poor nations to undertake "grassroots" development programs that can generate many jobs and thus increase the ability of people to buy food while limiting impacts on critical ecosystem services.[46]

Efforts should be made to shift diets in overdeveloped nations toward the vegetarian end of the spectrum and away from intensive grain feeding of animals. Similarly, a great increase in meat consumption (and grain feeding) should be discouraged in developing countries. The marketing in those countries of junk foods, replete with empty calories, should also be discouraged, as they are adding to the already high burdens of malnutrition.[47]

Improvement of food drying, storage, transport, and marketing systems should also receive substantial investment, concentrating on regions where food losses are great and hunger is prevalent.

Agricultural subsidy systems should be carefully reexamined and revised with an eye to reducing hunger and increasing food security (including providing more famine insurance in the face of global change).

At a minimum, "barefoot doctor" health care systems should be universally established, along with greatly enhanced medical research and public health systems to battle resurgent and novel diseases of humanity. Epidemics can disrupt both food production and food distribution.

Where Should We Start?

This list is long and ambitious, but far from exhaustive. It gives some idea of the scale of the transformation we are recommending. We think that the keys to starting the transformation are held by a relatively few countries. Some, such as the United States, Japan, Germany, China, and India, have the potential because of their sheer size and economic power. Some others are small, but could wield the power of example: Australia, Canada, Costa Rica, and Sweden come to mind immediately. Even tiny and obscure Seychelles was able to make a positive difference in international negotiations on global warming.

For everyone, the best place to start is at home. For us, that means the United States, which has one of the world's strongest economies, some of the richest endowments of key natural resources (especially agricultural land), a stable democratic government, and the best university-level educational system in the world. It also has a potentially healthy tension between a capitalist economic system and constraints based on the understanding that the market unaided cannot supply sustainability and other social goods that society needs but that central planning simply does not work. It is not absurd to assume that the U.S. potentially holds the key to the human future.

We see no choice but for Americans who recognize the predicament to become heavily involved in politics, to take voting very seriously, and to *pay attention to fundamental issues* rather than the crime, petty politics, accidents, and nonsense that pass for news in most of the media. Rising carbon dioxide in the atmosphere is infinitely more important than rising prices on the stock market. The current decline of biodiversity is a truly cosmic issue, an "event" that will mark the planet for millions of years after the breakup of the former Soviet Union is totally forgotten. The changing evolutionary and ecological relationships between *Homo sapiens* and the viruses, bacteria, and fungi that feed upon it will almost certainly affect many more human lives than the changing relationships between Israel and the PLO or between Catholics and Protestants in Northern Ireland. Rapid population growth, which threatens the future of everyone in our already overpopulated nation, should be much bigger news than which politician is trying to get elected by promising tax cuts or a war on drugs.

Since education is so critical, and since there is too little time to effect the needed changes through schooling, changing the electronic media, especially television, should be a major political target. Congressmen and women should be urged to reimpose regulation on the broadcasting business with the aim of returning network TV news to the control of people in the news business. Today, television networks tend to be subsidiaries of vast conglomerates, controlled by accountants who care only about ratings and the bottom line. The deterioration of TV news into "entertainment," featuring almost continuous coverage of celebrity murderers, the sexual activities of "stars," and other trivia is a direct result. The bean-counters and corporate owners won't make the money available, for example, for investigative reporters to spend a few months getting to the bottom of critical stories such as who benefits from, and who pays for, the subsidized desertification of much of the western United States.

The claim is made that entertaining trivia is what the public wants, that there is little demand for coverage of important but nonsexy issues. In part that is true. But it is what the public has been *trained* to want. Someone fed a diet of fast-food hamburgers for his or her entire life "wants" fast-food hamburgers and is not likely to patronize a new Chinese or French restaurant or a sushi bar. Hamburger eaters generate no demand for Peking duck. People must know what they are missing and how they could benefit from having it supplied. Well-informed people want news that deals in depth with the real issues that are shaping their lives and those of their descendants. The popularity of CNN and the MacNeil/Lehrer News Hour, with all their warts, is one sign of this. Greater demand for a flow of significant information *must* be generated if our democracy is to be sustained. Otherwise the "information superhighway" will be clogged by a traffic jam of electronic garbage. It is clearly government's job to make up for the market's failure to supply that need—and it is the job of voters to see that the government acts.

The Most Important Personal Steps

In the face of such a daunting menu of needs, what can a concerned individual do personally? The basic list is short:

- Have an absolute maximum of two children; preferably just one. If you want more, adopt.
- Educate yourself continuously on environmental issues. Nothing is more important to your future (including your economic future) and that of your children and grandchildren. Then pass on what you know to friends and coworkers. If possible, assemble a group and divide up areas of interest: you might concentrate on atmospheric problems, another on racial or gender equity, another on population problems, another on economic issues, and so on. That way you and your friends can greatly expand your grasp of the issues.
- Get involved in the environmental concerns of your community and join a local environmental organization. Human beings are social animals; you can get some enjoyment out of your efforts to save the world.
- Join as many national environmental organizations as you can afford—Zero Population Growth, National Audubon Society, Sierra Club, and Environmental Defense Fund are good examples. You may not always agree with all of their positions, but these NGOs are a major force generally pushing in the right directions. If you don't like certain policy positions, become active within the organization.

- Support other NGOs that are working on the equity and other issues discussed in this book that are not normally viewed as "environmental," but which we hope you are now convinced actually are. Be it the National Organization for Women or the Southern Poverty Law Center, these are groups that merit and need help.
- Change your lifestyle as much as you can to lessen your impact on Earth's life-support systems (and usually improve your own health and finances as well!). Eat less meat, walk as much as possible, be energy efficient. Take as many as possible of the steps listed in *50 Simple Things You Can Do to Save the Earth* (Earthworks Press, Berkeley, CA, 1990). By themselves they are far from sufficient, but they will all help.

A Vision of the Future

Ecologists are often accused of seeing everything that is wrong with the world and nothing that is right. We assault the dominant myths of societies by asserting that "normal" human activities are threatening the future of civilization. In the United States, this contributes to the success of the religious right, which views environmentalists as just a wing of a liberal conspiracy that since the end of World War II has largely destroyed an earlier United States of their imagination—a nation not plagued by pollution, racism, crime, "sexual deviation," and abortion. Most of these people are good at heart, longing for a world in which they can feel unthreatened. They naturally resent people who challenge their comfortable view of the world, which they acquired in schools and churches from individuals similarly innocent of the less pleasant truths of the human predicament. They are encouraged by Limbaughloid commentators and discouraged by the open cynicism of politicians, many of whom show no signs of principled behavior. But basically they and virtually all other people share the vision of the future held by every ecologist we know (and that's a lot of ecologists)—that of a sustainable society in which they and their descendants can enjoy a safe, secure, and culturally rich life.

One small but significant ray of hope is that ecologists have been working increasingly with social scientists, trying to find solutions to the human predicament. Traditionally, the relationships between ecologists and economists have been about as cordial as those between the Protestants and Catholics in Northern Ireland. But as the human predicament grows more obvious, cooperation is starting to replace noncommunication and competition. At Stanford University, ecologists and evolutionists meet regularly with agricultural economists, sociologists, political scientists, earth scientists, engineers, law professors, and others to share views. Several of them have undertaken joint research projects. The new discipline of ecological economics is one of the fastest growing in academia,[48] and new organizations like the Beijer Institute of Ecological Economics of the Swedish Royal Academy of Sciences are promoting that growth. At a recent meeting of leading economists and ecologists in Askö, Sweden, a consensus statement was produced that illustrated the converging views of the disciplines whose names share a common root in the Greek word for *housekeeping* (*economics* deals with society's housekeeping; *ecology* with nature's housekeeping).

The Askö statement said in part:

> The economists agreed that GNP is not an ideal measure of human welfare and that it is all too often misinterpreted in the popular and business press and by politicians. They noted, however, that economists have in recent years put considerable effort into devising improved measures. The economists also shared the ecologists' concern on the importance of global-scale issues, since they agree we now have a "full" world where natural capital is increasingly becoming scarce. They agreed that there is a need for a careful reconsideration of where and how economic growth and shrinkage should be pursued.[49]

The meeting was a two-way street, with the ecologists stressing the need for members of their discipline to "take advantage of the knowledge of social scientists, in particular the importance of markets for allocating environmental resources. . . . They felt that often [ecologists] do not appreciate the underlying economic causes or other driving forces of environmental problems or the indirect effects of remedial measures proposed."[50]

Academic interchange and agreement between diverse and previously intellectually isolated disciplines may seem a small thing against the picture painted in this book. But they show that academic leaders from very different backgrounds and scholarly cultures can work together to develop a joint vision of how a sustainable society might work. In a microcosm, it is the sort of process that we believe needs to be initiated worldwide by political leaders. Academics in particular, and well-informed citizens in general, can help create an atmosphere in which political leaders can do the right thing. That happened in the struggle for civil rights, in the genesis of the environmental movement, and in the antinuclear war movement.

What do political leaders need to emphasize to help generate broad support for a transition to a sustainable society? If they can be persuaded to exercise true leadership (and thus run the obvious political risks), we suggest the following:

• We could have more cohesive and enjoyable communities if there were much less wasteful consumption per capita and if the global population's growth were halted soon, eventually reversed, and allowed to decline gradually to less than 2 billion people.[51] At Hungary's current rate of shrinkage, it would take some four hundred years for that level to be reached by the world population, assuming shrinkage could begin today. Given that considerably more growth is inevitable, it is likely to take substantially longer to reduce the population size to 2 billion.

• Resources are basically scarce; there is no fountain of gold in Washington or in any other capital city. One of the most fundamental tasks of any society is to allocate those scarce resources through markets or governmental action. Allocation clearly must be governed by considerations of fairness and equity, including equity between generations. Thus taking a long-term view is essential.

• Greater equity in resource allocation would expand Earth's carrying capacity and help us keep well within it. Closing the rich-poor gap is the only humane way of averting our otherwise mutually assured destruction.

• Countries like the United States can *enhance* the quality of their citizens' lives while redeveloping their infrastructures and lifestyles. An America redesigned around people rather than automobiles would do much less damage to the environment and make "room" for needed physical growth in poor nations. This restructuring could provide healthier, less stressful lives for Americans. The need for police, prisons, and armies of lawyers might gradually diminish as peer pressure once again took over the task of inhibiting antisocial behavior.

• It is to everyone's advantage to respect the rights and sensibilities of others—to "do unto others as you would have them do unto you." Those of us who believe in a woman's right to choose should be sensitive to the abhorrence with which many others view abortion and work with them to make it unnecessary. In general, other people's beliefs and lifestyles should be viewed as their own business as long as they are not socially disruptive or environmentally destructive.

Designing and creating a sustainable society will be far from simple and will require long and difficult public debate. That debate can be energized by nongovernmental organizations (NGOs). These have been expanding in numbers and seem to represent a new feature in social systems, working with or in opposition to governments, at all levels from village to international. They help to return power to the grass roots, to give people more control over everything from local resources to global agreements. In the population and environment areas, they often are more trusted by people than are government officials, a critical advantage in many poor nations.[52]

On the other hand, NGOs sometimes also promote the fractionation of society at a time when unified action to solve large-scale problems is needed. Some NGOs, of course, are basically lobbies for antisocial industries, exemplified by those representing the deceptively named "wise-use movement" in the U.S. It is made up of groups dedicated to the destruction of Earth's life-support systems for short-term profit.[53] But overall, we think NGOs, exemplified by environmental and civil rights organizations, have already helped change society in the right direction and hold promise for being a very positive force in the future.

In short, we believe it is possible to assure our grandchildren a world on its way to a happy and sustainable future. To do so will require a lot of good will, cooperation, hard work, and some sacrifice. But answering the challenge can be rewarding (and fun) for almost everyone. In the United States, we might see a resurgence of community spirit and sense of purpose not seen since World War II.

A Moral Dimension

Of course, our own moral values are implicit in our recommendations. One could argue (and some have argued) that the strategy of the rich nations should be to turn the wealthy world into a series of lifeboats in which their citizens can survive the environmental holocaust that threatens the good ship Earth.[54] We believe, in contrast, that every effort should be made to give decent lives to those passengers already on the ship, meanwhile doing everything possible to repair it. We would limit the flow of passengers coming up the gangplank so they will be outnumbered by those dying natural deaths on board, gradually reducing the passenger list.

In the past, we have supported the idea of "triage." We still think that, when resources are strictly limited, a system that concentrates on helping people in need and applying the resources where they will do the most good is highly moral. Nonetheless, we no longer think triage is a useful or practical way to look at international aid. Remember that since we wrote about it, hundreds of millions of people, mostly children, have perished of hunger and hunger-related disease. Few citizens of the overfed North have even been aware of it. Even now, no mechanisms are on the horizon by which Earth's resources can be allocated more rationally among nations in the face of the powerful interests maintaining the status quo. Nor are there ways that the international community can force reallocation among regions within nations. Perhaps out of frustration with the sheer size and ubiquity of the problem, public interest in helping less developed nations seems at an all-time low. We must concentrate on rebuilding that interest and trying to see that any aid that can be generated is not wasted. Above all, aid must be designed to help the poor help themselves.

Generally speaking, we believe that a world with fewer people living richer lives would be a better world than one with the maximum number of people that is biophysically sustainable, each living the life of a battery chicken. This obviously is a judgment call. Some people may feel otherwise (although we doubt that many would if they understood the choice). We would prefer a sustainable Spaceship Earth on which no one had to travel steerage.

We also disagree with the view once held by many ethnographers that all cultures are morally equal. Anthropologist Ruth Benedict put it two generations ago as: "the coexisting and equally valid patterns of life which mankind has carved for itself from the raw materials for existence."[55] According to this view, one cannot pass moral judgment on practices such as warfare, ritual murder, cannibalism (when deliberate killing is involved), infanticide, racist oppression and genocide, slavery, female (and to a lesser degree, male) circumcision and mutilation, all of which are cultural practices we (and we think many of our readers) deplore. We recognize that cultural patterns of which we approve, such as using contraception (and abortion as a much less desirable backup) to limit reproduction, are judged immoral by others. While we disagree with that judgment, we believe it is a legitimate one—in the sense that there is no reason simply to accept all cultural practices as morally valid and equivalent. Above all, people must learn that cultural practices are not just "givens." While most practices developed in response to some social need or pressure, their validity should always be open to scrutiny, discussion, and modification when conditions change. Let us not forget that *Homo sapiens* became the dominant species on the planet chiefly because of the flexibility of cultural evolution.

Finally, there is that ever-vexing issue of which "inalienable rights" to invent next. The one of greatest concern to us currently is the asserted "right" of women to have as many children as they want. We believe that rights also involve responsibilities, and that no one's rights extend beyond the point where they start impinging on the rights of others. Since all children have impacts on local, regional, and global life-support systems, the number of children a person has, like each person's patterns of consumption, are a legitimate concern of the children themselves (whose well-being may be strongly affected by the number of their siblings), their families, their communities, their nations, and civilization as a whole. Indeed, in our view, the higher an individual's level of consumption (a level presumably rather predictive of the consumptive pattern of any offspring), the more restricted his or her reproduction ought to be.

These, of course, are judgments that must be made by societies as a whole. The point here is that each individual's reproductive (and consumptive) behavior is a legitimate area of concern for the rest of humanity. "The right of couples and individuals to determine, freely and responsibly, the number and spacing of their children" is sensible only if "responsibly" means "within their ability to nurture those children well, and within the constraints of the biophysical and social carrying capacities of their society and the planet." A Beverly Hills billionaire couple with a 10,000-square-foot home and five cars would do the world a great favor by restricting themselves to at most one child. Conversely, a welfare recipient or a single high-school student should have none at all until acquiring the means to support and care for the child properly.

In other words, we don't believe there is any "right" to unrestricted reproduction *or* to unrestricted consumption. Societies should (and often do) directly or indirectly put restrictions on both; but those now in place are far from adequate to permit a transition to a sustainable society. We suspect that the need for much tougher restrictions on both will become obvious within the next few decades. But it will be late in the game by then, and many now-open doors will have closed. We can only hope that society wakes up soon and begins a general discussion of the issues, leading to more sensible and humane strategies when the time finally arrives that they can be implemented.

Equity

One of the main elements of the solution to the human predicament clearly goes along with our personal moral values: the need for greater equity in human affairs. Yet, morals aside, humanity's only chance of creating a sustainable civilization depends on global cooperation to adjust the scale of the human enterprise so that the size of the human population falls once again within Earth's carrying capacity. To see that, it is not necessary to delve into the many ways already discussed in which equity can both increase carrying capacity and help keep population size within that capacity. One need only consider the tensions between rich and poor nations generated in the course of negotiating international agreements for protecting the ozone layer and curbing greenhouse gas emissions. The current income gap between overdeveloped and less developed nations rightly makes the latter suspicious when rich, overconsuming countries start preaching resource constraints to the poor.

A sustainable world would require that nations cooperate in regulating the rates of utilization of both natural resources and the natural sinks that absorb effluents.[56] Each nation also should accept limits on the resource flows that it can command—so many gallons of water will be extracted from rivers with international basins, so many cubic meters of lumber and tons of oil imported, etc. All trade agreements should be designed to allow for differences in environmental and workers' welfare regulations without penalizing nations that have higher standards and strong controls. Similarly, each nation must pledge to control environmentally damaging emissions that have transborder impacts. All nations should agree on a global cap on carbon dioxide emissions and cooperate in reaching that goal. They should also work together to reduce other important greenhouse gas emissions. The emission shares of the poorest nations should be large enough and flexible enough to permit reasonable development until safer energy technologies can be phased in. Implicit accompaniments to those pledges, of course, will be limits on reproduction and per-capita consumption in general.

Much of this may seem to be an impossibility today, and it may be impossible in the future. But in our view, the only alternatives to making and complying with such agreements is to court disaster by permitting a continuation of the present headlong race to destroy Earth's life-support systems and exhaust civilization's natural capital. When disaster loomed as the ozone hole was discovered, the nations of the world managed to pull themselves together and negotiate a ban on the production of the chemicals that pose the biggest threat to the ozone layer.[57] Whether or not the agreements will work as intended is still unclear. Rich countries by late 1994 had not yet lived up to their agreements to help the poor financially with the transfer of ozone-friendly technologies, and it is in the poor countries that the demand for CFCs is growing most rapidly.[58] Indeed, rumors appeared in 1994 of a black market in CFCs removed from discarded air conditioners and refrigerators in developing countries and sold in rich countries where high taxes have been imposed on the recycled chemicals.

The 1992 and 1994 world conferences on environment, population, and development, as well as the environmental discussions associated with the NAFTA and GATT negotiations, show that humanity can move in a cooperative direction, if slowly. When enough people grasp the fundamental human predicament, political leaders will be faced with the most serious and difficult negotiations in history; and if the record of the ozone negotiations and those on greenhouse gas emissions are any guide, issues of international equity will be a major stumbling block. A substantial narrowing of the rich-poor gap is likely to be required, not just to help reduce birth rates and improve the human nutritional situation, but as a key element of any negotiations to reduce the scale of the human enterprise. Poor nations have little incentive to cooperate in maintaining the lifestyles of the rich while they remain mired in poverty.

Equity achieved through a scheme like the Holdren scenario is a very different matter from equity that would be reached by *leveling up* everyone in a world of 10 billion people to the average consumption of rich nations today.[59] Such leveling up would be a recipe for certain environmental collapse, as would the much higher-consuming future that was envisioned in the Brundtland Report.[60] Of course, the Holdren scenario does not require "leveling down"—reducing everyone to a similar level of poverty. Such a step might be "fair" in the context of the present generation, but it would be completely unacceptable to the wealthier segments, which would feel robbed. More important, it would prevent the savings and investment in the future (and in technology) that will be essential for maintaining an environment of reasonable quality (and thus for intergenerational equity). If nothing else, the failure of communism clearly demonstrated that leveling down does not work.

Working toward international equity is an essential step in the long and complicated negotiations necessary to create a sustainable global society. It is possible biophysically, and it could in theory be accomplished while increasing the health and satisfaction of citizens of both rich and poor nations. Needless to say, whether the sociopolitical will can be mobilized to accomplish it in time is a more difficult issue.

Sustainability Is Profitable

Many readers may ask, Where can the resources be found to redevelop the rich countries, promote the grassroots development of the poor nations, and generally increase equity worldwide? It's the wrong question, because we can profit by moving toward sustainability. The United States has 45 percent of the global environmental technology market ($200 billion per year growing at 6 percent annually). Those profits trace directly to the nation's preeminent position in environmental regulation, which supplied our market economy with the proper incentives. The U.K., which lags in that area, has no share in the market. If the accounting were done properly, with appropriate attention to social costs, creating a sustainable society would be the *only* profitable course. The breakdown of society will hardly increase GNP in the long run (although many costs incurred in the process, such as pollution-generated health care costs, do inflate the GNP). The proper question is, how can resources be rapidly reallocated so that nations, firms, and individuals incur a minimum of economic losses and a maximum of benefits during a difficult transition?

This is probably the toughest issue to be faced in trying to keep the plow ahead of the stork. It will require a long debate in most nations—a debate on social and political choices. One very obvious and often mentioned source of resources to reallocate is military budgets. Civilization now spends roughly $700 billion it cannot afford and wastes huge quantities of natural resources on war and preparations for war, while maintaining roughly 25 million people under arms.[61] The United States alone spends over $250 billion annually, despite being the sole remaining superpower. Military activities and weapons production also cause serious environmental damage, even without fighting wars. The cost of repairing environmental damage from the nuclear weapons-building program of the U.S. alone is estimated to amount to more than $300 billion.[62] Despite the end of the Cold War, the arms trade still accounts for over $25 billion a year (although it is down from $73 billion in 1984).[63] As recently as 1988, desperately poor Africa was spending almost $5 billion importing arms. Compare those numbers with roughly $8.5 billion spent annually in nonmilitary foreign assistance by the U.S., *total* U.S. nonmilitary aid since World War II of about $150 billion, and the Cairo Conference's request for annual family planning assistance from rich nations of about $6 billion.[64] It's easy to see where humanity's priorities now lie.

Of course, the United States and many other nations still have external enemies and therefore some need for military power. But rather than perpetuating a world of armed camps, it is essential to reinforce the sort of international security arrangements that are now sometimes (and too often ineffectively) provided by the United Nations and international agreements. Until the perceived need for military establishments can be greatly reduced, it will be difficult to wean humanity from its dependence on armed forces. Humanity will remain, collectively, like a poor family that keeps an elephant for a pet.

Choosing how to reallocate other resources is going to differ from nation to nation. All we'll say here is that in the United States the issue of sustainability must move to front and center in the ongoing national debate about the allocation of tax funds to the military, "entitlements," debt service, and so on.[65] The time for myopic solutions is long gone.

Can It Really Happen?

It would be naive to imagine that a drastic reallocation of society's resources, and a move away from the notion that human beings above all else must be superconsumers, would not be opposed by the relatively small portion of the population that controls the productive assets of civilization. Maldistribution of power is obvious in poor countries, but it is perhaps even more serious in rich ones, and in the world as a whole (as evidenced by the power of multinational corporations). Politicians generally look after the interests of the rich first, and the rich generally prefer the status quo. After all, if you're floating on top, it is natural to think the system that put you there is perfect. Few stop to contemplate that other things besides cream float.

So is there really any hope of getting change on the required scale? We think so, although the political battles will probably be brutal. After all, redistribution involves not just giving to, but taking from; and in a high-discount world, the benefits to those suffering the taking will be less obvious. An important reason for our optimism is that the inequities hurt the rich too, and governments and educational systems could make the well-off much more aware of their stake in reducing social costs and limiting discounting. Wealthy people are not immune to environmental or social deterioration. Increasingly around the world, the well-off are living in guarded, fortified compounds and taking courses on how to avoid kidnapping. That is hardly a high-quality life! But we suspect the situation will continue to deteriorate from the standpoint of a shrinking proportion of "haves" surrounded by a sea of "have-nots." Eventually the rich will start asking themselves what human life should be about; some already have.

The main resource that seems to be in critically short supply almost everywhere now is the political and social will to do the job. We think that if both rich and poor could see that they and their children and grandchildren would reap great benefits from a sustainable world, they might mobilize that will. We think it is possible, even though it appears extremely utopian at the moment.

Societies do not always change incrementally and in predictable ways. When the time is ripe, enormous transformations can occur very quickly. The most recent large-scale example was the sudden end to the Cold War. The unexpectedly swift and peaceful transfer of power in South Africa and the accelerating peace processes in the Middle East and Northern Ireland are others.

In the late 1960s, demographers were telling us that it would be well into the twenty-first century before the U.S. birth rate would drop to replacement level, even if the government made a substantial effort to lower it. Instead, there was a totally unanticipated drop of the TFR to below replacement level in the early 1970s. Similarly, to those of us alive in the 1940s, it seemed that a century or more might be required before blacks would be given anything like a fair deal in the United States. Lynchings were common, and nationally prominent African-Americans were rare. In the South, schools were separate and unequal, and blacks were often denied the vote. Although we still are beset by very serious racial inequities and racist attitudes, the changes in the 1950s and 1960s were truly revolutionary.

People did change their behavior and begin to treat others more fairly. Social equity has increased in the world as a whole since World War II. Yes, there have been many horrors since 1945, and huge numbers of people still suffer prejudice and disadvantage. But the average person has fairer access to the things that make life worthwhile now than he or she did in the days when colonialism, chauvinism, institutionalized racism, sexism, religious prejudice, gross economic inequality, and autocracy were the rule almost everywhere.

None of the earlier rapid transitions, of course, required behavioral changes as profound as moving to a much more equitable world in which a smaller population lived comfortably within Earth's carrying capacity. But then none of them was so essential to the well-being of *all* human beings. Ironically, the only practical solutions to the human predicament are now ones that "practical" people consider to be "too utopian," or "too liberal," or "too humanistic," or "too Christian." What is required today is, after all, what the great religions have always asked of human beings—to treat others as we ourselves want to be treated, to accept social responsibilities, and to exercise stewardship over all of creation.[66] We don't know if or when the time will be ripe for such a transformation, but we do believe that all of us should be striving to ripen the time.

NOTES

A Personal Preface

1. P. Ehrlich, 1968. *The Population Bomb.* (New York: Ballantine Books.) Early works often just carried Paul's name—but Anne was increasingly involved in developing the ideas, so "they" is more accurately used.
2. Ehrlich, 1968, p. 11.
3. Ehrlich, 1968, p. 149.
4. P. Ehrlich, 1969. "Eco-catastrophe," *Ramparts,* September.
5. P. Ehrlich and A. Ehrlich, 1970. *Population, Resources, Environment: Issues in Human Ecology.* (San Francisco: W. H. Freeman and Co.), 322–324.
6. Among the earlier ones: P. Ehrlich and J. Holdren, 1971. "Impact of population growth," *Science* 171:1212–1217; J. Holdren and P. Ehrlich, 1974. "Human population and global environment," *American Scientist* 62:282–292; P. Ehrlich, A. Ehrlich, and J. Holdren, 1977. *Ecoscience: Population, Resources, Environment.* (San Francisco: W. H. Freeman and Co.).

Introduction

1. Population Summit, 1994, *Population—The Complex Reality: A Report of the Population Summit of the World's Scientific Academies* (Golden, Colo.: North American Press).
2. G. C. Daily and P. R. Ehrlich, 1992, "Population, sustainability, and Earth's carrying capacity," *BioScience* 42:761–771.
3. M. Wackernagel, J. McIntosh, W. Rees, and R. Woollard, 1993, *How Big Is Our Ecological Footprint? A Handbook for Estimating a Community's Appropriated Carrying Capacity* (Vancouver: University of British Columbia, Task Force on Planning Healthy and Sustainable Communities).
4. See G. C. Daily and P. R. Ehrlich, 1995, "Socioeconomic equity,

sustainability, and Earth's carrying capacity," *Ecological Applications* (in press).

5. For details on those limits, see P. R. Ehrlich, A. H. Ehrlich, and J. P. Holdren, 1977, *Ecoscience: Population, Resources, Environment* (San Francisco: W. H. Freeman & Co.).

6. Daily and Ehrlich, 1992; G. Hardin, 1986, "Cultural carrying capacity: A biological approach to human problems," *BioScience* 36:599–606; N. Keyfitz, 1991, "Population and development within the ecosphere: One view of the literature," *Population Index* 57 (1):5–22; P. R. Ehrlich, 1994, "Ecological economics and the carrying capacity of Earth," in A. Jansson, M. Hammer, C. Folke, and R. Costanza (eds.), *Investing in Natural Capital* (Washington, D.C.: Island Press), 38–56.

7. World Commission on Environment and Development, 1987, *Our Common Future* (New York: Oxford University Press), 8. The Brundtland Commission, however, failed to see the inherent contradiction in its definition with advocating a fivefold increase in the world's gross domestic product.

8. For more details and references on the predicament, see P. Ehrlich and A. Ehrlich, 1990, *The Population Explosion* (New York: Simon & Schuster); and 1991, *Healing the Planet* (Boston: Addison-Wesley).

Chapter One

1. R. McNamara, 1991, speech to the African Leadership Forum in Lagos (June 1990), quoted in *People* (International Planned Parenthood Federation) 18 (1):8.

2. Co-author Gretchen C. Daily was not a member of the group. Here and later in the book, we use "we" for any combination of the authors.

3. E.g., C. Holdren and A. Ehrlich, 1984, "The Virunga volcanoes: Last redoubt of the mountain gorilla," *Not Man Apart* 8 (June):8–9.

4. J. Gasana, 1991, "A very tough challenge for us," *International Agricultural Development,* 11 (Sept./Oct.):8.

5. J. Hammond, 1994, "Land pressure adds to Rwanda conflict," *People and Planet* 3(4):9.

6. T. Conover, 1993, "Trucking through the AIDS belt," *New Yorker,* 16 Aug., 56–75.

7. For an overview of Africa's problems, see T. J. Goliber, 1989, "Africa's expanding population: Old problems, new policies," *Population Bulletin* (Population Reference Bureau, Washington D.C.) 44(3).

8. See, for instance, Robert Kaplan, 1994, "The coming anarchy," *Atlantic Monthly,* Feb., 10–36; T. Homer-Dixon, J. Boutwell, and G. Rathjens, 1993, "Environmental change and violent conflict," *Scientific American,* Feb., 38–45.

9. Population projections are at best educated guesses, based on expected trends in birth or death rates and information about existing population structure. For instance, the number of newborn females in a pop-

ulation can tell a demographer a lot about the number of babies likely
to be born in the next generation. But birth and death rates can
change in response to many factors, so projections cannot be expected
to be entirely accurate (especially for far into the future), and they
ordinarily undergo frequent revision. See J. M. Stycos, 1994, "Population, projections, and policy: A cautionary perspective," EPAT/
MUCIA Working Paper no. 12 (June); *see also* W. Lutz, 1994, "The
future of world population," *Population Bulletin* 49 (June):1; and
C. Haub, "Understanding population projections," *Population Bulletin*
(Population Reference Bureau, Washington D.C.) 42 (Dec.):4.

10. C. Haub and M. Yanagishita, 1994, *1994 World Population Data Sheet*
(Washington, D.C.: Population Reference Bureau, 1875 Connecticut
Ave., NW, Suite 520, Washington DC 20009 [202] 483-1100). This
annual publication is a gold mine of information on one sheet and the
source of current demographic information in this volume, unless otherwise noted.

11. The total fertility rate is calculated as the number of children an average woman would have if current age-specific birth rates did not
change during her childbearing years (normally considered ages fifteen
through forty-nine), without considering mortality during those years.
One can think of it as roughly the average completed family size.

12. J. C. Caldwell and P. Caldwell, 1990, "High fertility in sub-Saharan
Africa," *Scientific American,* May.

13. C. H. Bledsoe, D. C. Ewbank, and U. C. Isiugo-Abanihe, 1988, "The
effect of child fostering on feeding practices and access to health ser-
vices in Sierra Leone," *Social Science and Medicine* 27(6):627–636;
H. Ware, 1977, "Economic strategy and the number of children," in
J. C. Caldwell (ed.), *The Persistence of High Fertility, Part 2* (Canberra:
Australian National University), 469–590.

14. S. Sonko, 1994, "Fertility and culture in sub-Saharan Africa: A review," *International Social Science Journal* 46(3):397–411.

15. J. Mbiti, 1973, "The Kamba of central Kenya," in A. Molnos (ed.),
Cultural Source Materials for Population Planning in East Africa, vol. 3 of
Beliefs and Practices (Nairobi: East African Publishing House), 97–113.

16. R. Winterbottom, 1991, "Environment action plans for the greening
of Africa," *People* 18(1):9–13.

17. Population Reference Bureau, Washington, D.C. *World Population Data
Sheets, 1980* and *1994.* Indeed, in sub-Saharan Africa as a whole, the
average annual growth rate of GNP has only been 0.2 percent per year
since 1965 and has declined in seventeen of thirty-two nations; see
World Bank, 1993, *Social Indicators of Development 1993* (Baltimore:
Johns Hopkins Press).

18. World Bank, 1992, *World Development Report 1992* (Washington,
D.C.: World Bank); World Resources Institute, 1992, *World Resources
1992–93,* New York: Oxford University Press); L. R. Brown and
E. C. Wolf, 1985, "Reversing Africa's decline," *Worldwatch Paper 65*
(Worldwatch Institute, Washington D.C.), June; A. B. Durning, 1989,

"Poverty and the environment: Reversing the downward spiral," *Worldwatch Paper 92* (Worldwatch Institute, Washington D.C.), Nov.

19. R. McNamara, 1991, "A blueprint for Africa," *People* 18(1):3–6.

20. L. R. Brown, 1994, "Who will feed China?," *World Watch,* Sept./ Oct., 10–19.

21. Quoted in Brown, 1994, p. 17.

22. Winterbottom, 1991; World Resources Institute, 1994, *World Resources, 1994–95* (New York Oxford University Press), ch. 3. The exact dimensions of the fuelwood shortage in Africa and elsewhere are very difficult to estimate—see G. Leach and R. Mearns, 1988, *Beyond the Fuelwood Crisis* (London: Earthscan).

23. S. Okie, 1994, "The deepening shadow of AIDS over Africa," *Washington Post Weekly,* 17–23 Oct., 20.

24. S. Kalish, 1992, "New UN projections include local effects of AIDS," *Population Today* 20 (Oct.):1–2; W. H. Mosley and P. Cowley, 1991, "The challenge of world health," *Population Bulletin* (PRB) 46 (Dec.): 20–21. In a separate study, the U.S. Census Bureau estimated that child mortality rates could triple in sub-Saharan Africa and some other heavily infected regions by 2010; see S. Holmes, 1994, "Child death rate for AIDS expected to triple by 2010," *The New York Times,* 29 April. By 2020, the study indicated, the populations of Uganda and Zambia, two of the most affected nations, could have population sizes more than 45 percent smaller than they would be with no AIDS. Even so, their populations will continue to grow, if slowly. In a region whose population could by then be more than a billion, these reductions will hardly be noticed; the disruptive effects on family and social structures in affected communities are likely to be far greater. Indeed, the possibility exists that widespread AIDS and awareness of impending mortality among some of Rwanda's leaders and elites may have contributed to the political crisis in 1994. For a vivid description of the attitudes of Central African men toward AIDS and their limited understanding of it, see T. Conover, 1993.

25. Quoted by Okie, 1994.

26. World Bank, 1989, *Sub-Saharan Africa: From Crisis to Sustainable Growth* (Washington, D.C.: World Bank). Notice that the World Bank, typically, did not recognize the need to *stop* population growth and included the oxymoron "sustainable growth" in the report's title.

27. The U.S. population has not grown that fast since the last century, and much of the growth then was from immigration. The population has doubled since 1940 (fifty-five years), and many facilities and amenities to support the population have more than doubled in that period. Increases were very uneven: gross domestic product (in constant dollars) doubled between 1960 and 1984; housing units roughly doubled between 1960 and 1994, as did the number of physicians. Miles of highway jumped nearly 50 percent in the 1960s and increased only modestly after that, although their condition generally deteriorated, and the number of motor vehicles nearly tripled. Airline passenger miles much more than doubled, and the number of hospitals and hos-

pital beds declined; see *Statistical Abstract of the United States (American Almanac)*, (Austin, Tex.: Reference Press, 1993–1994); and U.S. Bureau of the Census, 1961, *Historical Statistics of the United States, Colonial Times to 1957* (Washington D.C.: USGPO). Unlike most African populations, the American population is well educated (three out of four Americans today have more than twelve years of education); see World Resources Institute, 1994, ch. 3; and *American Almanac 1993–94*.

Even ignoring environmental constraints, and with all the advantages of economic and military power, resources, infrastructure, education, capital, and scientific expertise possessed by the United States, whether another doubling of infrastructure and amenities could be accomplished is problematic. And the environmental constraints could not be ignored in reality. Still, the United States might manage. The task would be even more difficult if the nation were burdened with a much higher rate of population growth and the high proportion of unproductive young people that rapid growth would generate.

28. The total percentage of dependents—those under 15 or over 65—is around 36 in most rich countries, and pushes 50 percent in the most rapidly growing poor nations (in East Africa it was 51 percent in 1994).

29. World Bank, 1993, *Social Indicators of Development 1993* (Baltimore: Johns Hopkins Press). For comparison, India has about one doctor for every 2,500 people, and the United States has one for every 420. The education figure is mean number of years of school for women over the age of 25 in 1990 (from World Resources Institute, 1994, ch. 3)—a figure that more accurately represents the human capital available *now* than does the percentage of the school-age population enrolled in secondary school. For the countries mentioned in the next paragraph, the percentage of school-age girls in secondary education in Burkina Faso is 5; in Chad, 3; Ethiopia, 12; Tanzania, 4; Uganda, 13; and Zaire, 16.

30. Zaire's situation is hardly unique; Ethiopia, with 49 percent of its people under 15 and 8 months average schooling per woman, has over 32,000 people per physician; Tanzania, with 47 percent under 15, 16 months of schooling for women, has 33,000 potential patients for each doctor. For Mali, the equivalent numbers are 46 percent, 1 month, and 19,000; for Burkina Faso, 48 percent, 1 month, and 57,000.

31. P. R. Ehrlich, 1980, "Variety is the key to life," *Technology Today* 82(5):58–68. For example, traditional knowledge of local medicinal plants and alternative food crops is greatly undervalued and disappearing rapidly worldwide.

32. A recent sample survey suggests it might have been as low as 5.4 in 1990–1992; see Population Council, 1994, "Kenya 1993: Results from the demographic and health survey," *Studies in Family Planning* 25:310–314.

33. Botswana and Zimbabwe had TFRs of 4.6 and 5.3 respectively in 1994, also representing significant declines from 6.5 and 6.6 in 1984;

see Population Reference Bureau, 1984, *1984 World Population Data Sheet* (Washington, D.C.: PRB); B. Robey, S. O. Rutstein, and L. Morris, 1993, "The fertility decline in developing countries," *Scientific American,* Dec., 60–67). Of course, these are still very high, generating population growth rates of 2.7 percent (Botswana) and 3.0 percent (Zimbabwe) and doubling times of 23 to 26 years.

34. United Nations Population Fund (UNFPA), 1993, *The State of World Population 1993* (New York: United Nations).
35. Doubling times are the time it would take these populations to double in size if their growth rates did not change. Growth rates of course do change, but doubling times help us understand the meaning of those rates.
36. A 1994 revision of the United Nations' projections indicated that growth had fallen to 1.57 percent per year from 1990 to 1994, the lowest rate since World War II; see United Nations, 1994, "The 1994 revision of the population estimates and projections," *Population Newsletter;* June.
37. Demographers project another doubling or more before growth ends because of the momentum of population growth. Since 35 to 50 percent of the people in a rapidly growing population are youngsters (future parents), even if birth rates are sharply reduced, the population will continue to increase for many decades until the first generation of parents of small families reaches old age and begins to die off.
38. The medium demographic estimates made in 1991 by the United Nations projected that the world population will pass 8.5 billion by 2025, reach 10 billion near midcentury, and ultimately stop growing at 11.6 billion; see United Nations Population Fund (UNFPA), 1992, *State of the World Population, 1992* (New York: United Nations). UNFPA based its summary on United Nations Population Division, 1991, *Long-Range World Population Projections,* ST/SEA/SER.A/125 (New York: United Nations). *See also* C. Haub, 1992, "New UN projections show uncertainty of future world," *Population Today* (Population Reference Bureau, Washington, D.C.) 20(2):1, 6–7. Of course, unexpected changes in reproductive rates (or mortality rates) could cause the growth rate to be either considerably higher or lower.

The UN's high projection shows no end to growth, with the population soaring past 28 billion around 2150—a clearly unsustainable scenario. The low projection indicates a peak population size of 8 billion around 2050, followed by a slow decline, passing 5 billion again within a century.
39. In general, the industrialized nations of Europe, North America, Japan, Australia, and New Zealand have annual population growth rates of less than 1 percent: those of most European countries are under 0.5 percent, with some actually shrinking. All of Russia and former Soviet republics Belarus, Ukraine, and Moldova are now considered European nations, according to a new United Nations classification.
40. Legal immigration accounts for about one fourth of U.S. population

growth (about 900,000 per year), but an unknown number of illegal immigrants adds several hundred thousand more each year. For recent details, see P. Martin and E. Midgeley, 1994, "Immigration to the United States: Journey to an uncertain destination," *Population Bulletin* 49 (Sept.):2.

41. G. Daily and P. Ehrlich, 1992, "Population, sustainability, and Earth's carrying capacity," *BioScience* 42:761–771.

42. J. P. Holdren, 1990, "Energy in transition," *Scientific American*, Sept., 157–163.

43. P. R. Ehrlich, A. H. Ehrlich, and G. C. Daily, 1993, "Food security, population, and environment," *Population and Development Review* 19 (March):1–32.

44. We are referring here to resources that can be renewed in theory, but not in practice. For instance, soils are usually renewed at a rate of inches per millennium, but they are being eroded away in many areas at rates of inches per decade.

45. B. Turner II, W. Clark, R. Kates, J. Richards, J. Mathews, and W. Meyer (eds.), 1990, *The Earth as Transformed by Human Action* (Cambridge and New York: Cambridge University Press); P. Ehrlich and A. Ehrlich, 1991, *Healing the Planet* (Boston: Addison-Wesley).

46. See the analysis by T. Dyson, 1994, "Population growth and food production: Recent global and regional trends," *Population and Development Review* 20:397–411. It shows per-capita grain production declining since 1985, but claims, correctly, that there has been, globally, "no abrupt reversal in food production per head" (p. 409).

47. D. Pimentel and C. W. Hall (eds.), 1989, *Food and Natural Resources,* (San Diego: Academic Press); Ehrlich, Ehrlich, and Daily, 1993.

48. United Nations Children's Fund (UNICEF), 1992, *State of the World's Children 1992* (New York: United Nations).

49. Estimates of the numbers of undernourished people range from a conservative estimate of about 550 million—e.g., United Nations, 1993, *Report on the World Social Situation 1993,* ch. 2 (New York: United Nations); World Food Council, 1992, "The global state of hunger and malnutrition," report at 18th Ministerial Session, Nairobi, Kenya, 23–26 June (WFC/1992/12)—to a billion or more; see UNICEF, 1992, and World Bank estimates, cited in R. Kates and V. Haarmaan, 1992, "Where the poor live: Are the assumptions correct?," *Environment* 34(4):5–11, 25–28.

50. Some serious analysts claim that estimates of undernourishment in sub-Saharan Africa are exaggerated. See P. Svedberg, 1991, "Undernutrition in sub-Saharan Africa: A critical assessment of the evidence," in J. Drèze and A. Sen, *The Political Economy of Hunger,* vol. 3, *Endemic Hunger* (Oxford, Eng.: Oxford University Press), 155–193.

51. For details on the problems of estimating how many hungry people there are, see D. Grigg, 1993, *The World Food Problem,* 2nd ed. (Oxford, Eng.: Blackwell), especially ch. 2 and Table 2.10.

52. E.g., S. Budiansky, 1994, "10 billion for dinner, please," *US News and*

World Report, 12 Sept., 57–62, which asserted that "With technology and free trade, the Earth can defy the doomsayers—and feed twice as many people."

53. E.g., Drèze and Sen, 1991.
54. R. Chen (ed.), 1990, *The Hunger Report: 1990* (Providence, R.I.: The Alan Shawn Feinstein World Hunger Program, Brown University).
55. The estimate of a 40 percent overall loss between production and consumption includes a 10 to 15 percent loss after food leaves retail outlets, based on the United Nations Food and Agriculture Organization (FAO) and other sources. Other estimates are more conservative, but not strictly comparable; for instance, M. Greely, 1991, "Postharvest losses—the real picture," *International Agricultural Development,* Sept./Oct., 9–11.
56. For a good, brief discussion of these social factors (which, however, somewhat understates the contribution of population growth to the dilemma), see W. Murdoch, 1990, "World hunger and population," in C. Carroll, J. Vandermeer, and P. Rosset (eds.), *Agroecology* (New York: McGraw-Hill), 3–20.
57. D. Colander and A. Klamer, 1987, "The making of an economist," *Economic Perspectives* 1:95–111.
58. "Externalities" are commonly thought of as negative—for instance, the costs to the public of air pollution caused by automobiles, which are not included in the prices of cars or gasoline. But there can be positive externalities too, such as the benefit a homeowner receives in increased property value if a neighbor paints his house and landscapes his yard, raising local property values. P. Samuelson and W. Nordhaus, 1989, *Economics* (New York: McGraw-Hill), 972.
59. Herman Daly has been a leading economist who knows better: see H. E. Daly and J. B. Cobb Jr., 1989, *For the Common Good* (Boston: Beacon Press).
60. A. Gore, 1992, *Earth in the Balance: Ecology and the Human Spirit* (New York: Houghton Mifflin), 189.
61. Summarized in Hilary F. French, 1990, "Green revolutions: Environmental reconstruction in Eastern Europe and the Soviet Union," *Worldwatch Paper* (Worldwatch Insititute, Washington, D.C.), Nov.; D. J. Peterson, 1993, *Troubled Lands: The Legacy of Soviet Environmental Destruction* (Boulder, Colo.: Westview Press); Murray Feshbach and Alfred Friendly, 1992, *Ecocide in the USSR* (New York: Aurum); Boris Komarov, 1980, *The Destruction of Nature in the Soviet Union* (White Plains, N.Y.: M. E. Sharpe). A recent account is M. Simons, 1994, "Capitalist or communist, the air is still bad," *The New York Times,* 3 Nov.
62. T. Colburn and C. Clement (eds.), 1992, *Chemically-induced Alterations in Sexual and Functional Development: The Wildlife/Human Connection. Advances in Modern Environmental Toxicology,* vol. 21 (Princeton, N.J.: Princeton Scientific Publishing Co.). For a current overview, written

for the layperson, see T. Colborn, D. Dumanoski, and J. Myers, 1995, *And the Cradle Will Fall: The Chemical Assaults on the Unborn* (New York: Dutton).

63. In 1970, the Ehrlichs wrote: "We know that DDT affects the sex hormones of rats and birds, and we know that rat reproductive physiology is very similar to human reproductive physiology. We do not know whether hormonal changes are induced in man, or what their effects will be if they are. . . . Breastfed babies in Sweden get 70% more than what is considered the maximum acceptable amount of DDT, and British and American breastfed babies consume 10 times the recommended maximum amount of dieldrin. . . . At any rate, we shall almost certainly find out what the overall effect of the chlorinated hydrocarbon load will be, since the persistence of these compounds guarantees decades of further exposure, even after their use has been discontinued. The continued release of chlorinated hydrocarbons into our environment is tantamount to a reckless global experiment, and we humans as well as all other animals that live on this globe, are playing the role of guinea pigs," in P. R. Ehrlich and A. H. Ehrlich, 1970, *Population, Resources, Environment: Issues in Human Ecology* (San Francisco: W. H. Freeman & Co.), 129–134. Of course, the hazards of saturating the environment with synthetic organic chemicals were first brought widely to public attention by Rachel Carson in her 1962 classic, *Silent Spring* (Boston: Houghton Mifflin).

64. P. R. Ehrlich and J. P. Holdren, 1971, "The impact of population growth," *Science* 171:1212–1217; J. P. Holdren and P. R. Ehrlich, 1974, "Human population and the global environment," *American Scientist* 62:282–292; P. R. Ehrlich and A. H. Ehrlich, 1990, *The Population Explosion* (New York: Simon & Schuster); Ehrlich and Ehrlich, 1991. The equation is, of course, an identity, and P, A, and T are clearly not independent of one another.

65. Based on per-capita energy data from World Resources Institute, 1994, 334–335.

66. McNamara, 1991.

67. E.g., W. Mauldin and J. Ross, 1994, "Prospects and programs for fertility reduction, 1990–2015," *Studies in Family Planning* 25:77–95.

68. A. Bartlett and E. Lytwak, 1995, "Zero growth of the population of the United States," *Population and Environment* (forthcoming).

69. D. Pirages and P. Ehrlich, 1972, "If all Chinese had wheels," *The New York Times*, 16 March.

70. M. Gadgil and R. Guha, 1992, *This Fissured Land: An Ecological History of India* (London: Oxford University Press).

71. *Joint Statement by Fifty-eight of the World's Scientific Academies,* signed at a "Science Summit" on World Population, New Delhi, India, October 24–27, 1993. The academies included the U.S. National Academy of Sciences; Royal Society of London; Royal Swedish Academy of Sciences; French Academy of Sciences; Russian Academy of Sciences; Polish Academy of Sciences; Federation of Asian Scientific Academies

and Societies; Bangladesh Academy of Sciences; Indian National Scientific Academy; Pakistan Academy of Sciences; National Academy of Science and Technology, Philippines; Kenya National Academy of Sciences; Nigerian Academy of Science; Uganda National Academy of Science and Technology; Brazilian Academy of Sciences; National Academy of Sciences, Mexico; Cuban Academy of Sciences; National Academy of Physics, Mathematics, and Natural Sciences of Venezuela; Royal Scientific Society of Jordan; Israel Academy of Sciences and Humanities; the Third World Academy of Sciences; and those of thirty-eight other nations.

Chapter Two

1. N. Himes, 1936, *Medical History of Contraception* (New York: Gamut Press), 185, quoted in D. Bogue, 1969, *Principles of Demography* (New York: Wiley), 55. The emphasis on early contraception varies greatly among demography texts, but it is always mentioned. Demographers Warren Thompson and David Weeks wrote in 1965: "Possibly no society has ever realized its full fecundity. . . . In some tribal groups these restrictions [on fertility] may have been rather severe"; in *Population Problems,* 5th ed. (New York: McGraw-Hill). John Weeks, in his 1992 textbook of demography, wrote, "It is probable that at least some people in most societies throughout human history have pondered ways to control fertility," in *Population,* 5th ed. (Belmont, Calif.: Wadsworth).
2. In more technical language, there is differential reproduction of genotypes, where the differential is greater than can be accounted for by random processes (genetic drift).
3. This does not necessarily mean maximizing the number of offspring produced. A female rodent that produces, say, eight young may so stress herself that she can't adequately care for them and six die. Worse yet, she may be so weakened she succumbs to the stress of winter. A female that produces five may rear all of them and survive to breed again the next year. It is survival to reproduce in the next generation that counts, not just production of infant offspring.
4. Sperm may be produced either by the same individuals that produce eggs (in simultaneous hermaphrodites such as snails and barnacles) or by separate individuals (males, as in mammals and most other vertebrates).
5. See R. E. Michod and B. R. Levin (eds.), 1988, *The Evolution of Sex* (Sunderland, Mass.: Sinauer). One possible element of the answer is that sex is an adaptive response to environmental uncertainty. Sexual populations do not respond evolutionarily to fluctuating conditions to the same degree as asexual ones must. As a result, sexual populations have higher growth rates on average than asexual ones facing the same environmental conditions, and are thus more likely to persist in the long run. That benefit is generally greater than the cost of producing sperm; asexual populations tend to be evolutionary dead ends that do not persist for long periods. In this view, individual selection favoring

asexual reproduction is generally overwhelmed in evolutionary time by group selection favoring sexually reproducing populations. This review of the issue is necessarily greatly oversimplified; for more detail and a quantitative model of this explanation of the evolution of sex, see J. Roughgarden, 1990, "The evolution of sex," *American Naturalist* 138:934–953.

6. James Reed, 1978, *From Private Vice to Public Virtue: The Birth Control Movement and American Society Since 1830* (New York: Basic Books), ix–x. This naive statement in no way detracts from the value of this fine work as an overview of the history of contraceptive technology, practice, and politics in the United States.

7. This does not mean that males and females will necessarily "desire" the same number of offspring. There is a rich literature on sexual strategies, which we cannot go into here. Those wishing more information should consult P. Bateson (ed.), 1983, *Mate Choice* (Cambridge, Eng.: Cambridge University Press); or T. H. Clutton-Brock (ed.), 1988, *Reproductive Success* (Chicago: University of Chicago Press).

8. This is the fundamental difficulty with group selection as an explanation for the evolution of altruistic behavior, and the basic reason why the explanation of the evolution of sexual reproduction involving group selection remains controversial. See G. C. Williams, 1966, *Adaptation and Natural Selection* (Princeton, N.J.: Princeton University Press); D. J. Futuyama, 1986, *Evolutionary Biology* (Sunderland, Mass.: Sinauer).

9. J. H. Barkow, 1989, *Darwin, Sex, and Status: Biological Approaches to Mind and Culture* (Toronto: University of Toronto Press).

10. For a provocative review of cultural maladaption, see R. B. Edgerton, 1992, *Sick Societies: Challenging the Myth of Primitive Harmony* (New York: Free Press).

11. For a good summary, see D. Symons, 1979, *The Evolution of Human Sexuality* (New York: Oxford University Press). Much of our approach in this chapter would be characterized as "cultural materialism": see especially M. Harris, 1979, *Cultural Materialism: The Struggle for a Science of Culture* (New York: Random House).

12. R. Ornstein and P. R. Ehrlich, 1989, *New World/New Mind: Moving Toward Conscious Evolution* (New York: Doubleday).

13. For the quite distinct and interesting views of a female anthropologist, see M. F. Small, 1994, *Female Choices: Sexual Behavior of Female Primates* (Ithaca, N.Y.: Cornell University Press).

14. See M. Harris, 1979, especially the discussion of elite female infanticide, pp. 137–139. This book also provides insight into some of the baroque arguments about human behavior to be found in anthropological and other social scientific literature, a testimony to the difficulties our species has in studying itself. For a technical discussion of the spread of a cultural trait like birth control, see L. Carotenuto, M. Feldman, and L. Cavalli-Sforza, 1989, "Age structure in models of cultural transmission," *Morrison Institute for Population and Resource Studies*, Stanford University, paper no. 16.

15. A. J. Coale, 1972, *The Growth and Structure of Human Populations: A Mathematical Investigation* (Princeton, N.J.: Princeton University Press).
16. Of course, such a high birth rate in a very poor country no doubt contributed to the pressures leading to the explosive and devastating ethnic conflict of 1994, in which at least 1 million of the 8 million Rwandans perished and 2 million more fled to neighboring countries. While Hutu–Tutsi ethnic hatred obviously played a major role in generating the conflict, this was undoubtedly exacerbated by shortages of land and feared shortages of food. See C. Holdren and A. Ehrlich, 1984, "The Virunga volcanoes: Last redoubt of the mountain gorilla," *Not Man Apart* 8 (June):9, and the statement by the minister of agriculture, livestock, and forests of Rwanda: J. Gasana, 1991, "A very tough challenge for us," *International Agricultural Development* 11 (5):8.
17. E.g., R. N. Ostling, 1993, "Sex and the single priest," *Time,* 5 July.
18. E.g., E. O. Wilson, 1975, *Sociobiology* (Cambridge, Mass.: Harvard University Press).
19. H. F. Harlow, 1958, "The evolution of learning," in A. Roe and G. G. Simpson, *Behavior and Evolution* (New Haven, Conn.: Yale University Press), 169–290.
20. J. V. Neel, 1970, "Lessons from a 'primitive' people," *Science* 170: 816–817.
21. A. M. Carr-Saunders, 1922, *The Population Problem: A Study in Human Evolution* (London: Oxford University Press).
22. V. C. Wynne-Edwards, 1962, *Animal Dispersion in Relation to Social Behavior* (Edinburgh: Olvier and Boyd), 402.
23. See B. Malinowski, s.v. "Anthropology," *Encyclopaedia Britannica,* 1936, 1st supp. vol., 131–140; M. Harris, 1960, "Adaptation in biological and cultural science," *Transactions of the New York Academy of Science* 23:59–65; R. A. Rappaport, 1968, *Pigs for Ancestors: Ritual in the Ecology of a New Guinea People* (New Haven, Conn.: Yale University Press).
24. See review in R. B. Edgerton, 1992, *Sick Societies: Challenging the Myth of Primitive Harmony* (New York: Free Press). On adaptive features of organisms, see S. J. Gould and R. C. Lewontin, 1979, "The spandrels of San Marco and the panglossian paradigm: A critique of the adaptationists' programme," *Proceedings of the Royal Society of London B* 205:581–598.
25. Kahune Medical Papyrus, 1850 B.C.E., cited in J. M. Riddle and J. W. Estes, 1992, "Oral contraceptives in ancient and medieval times," *American Scientist* 80:226–233.
26. Unless otherwise noted, what follows is largely based on Riddle and Estes, 1992.
27. This is a particularly preposterous concern today; see P. R. Ehrlich and A. H. Ehrlich, 1990, *The Population Explosion* (New York: Simon & Schuster), 170.
28. *Silphion* was the Greek name; the Romans called it *silphium.*
29. Reviewed by Riddle and Estes, 1992.

30. E. B. Leacock, 1981, *Myths of Male Dominance* (New York: Monthly Review Press), 289.
31. For a summary of fertility control in preindustrial societies, see M. Harris and E. B. Ross, 1987, *Death, Sex, and Fertility: Population Regulation in Preindustrial and Developing Societies* (New York: Columbia University Press). For an overview of theories of demographic control in historic populations, see R. D. Lee (ed.), 1977, *Population Patterns in the Past* (San Diego: Academic Press).
32. S. B. Hrdy, 1981, *The Woman Who Never Evolved* (Cambridge, Mass.: Harvard University Press).
33. R. Frisch, 1978, "Population, food intake, and fertility," *Science* 199:22–30.
34. E. Wilmsen, 1978, "Seasonal effects of dietary intake on Kalahari San," *Proceedings of Federation American Societies of Experimental Biology* 37:25–32. Suppression of ovulation or implantation failures in lean months are among other possible explanations.
35. Harris and Ross, 1987, p. 27.
36. Unhappily, as noted earlier, the adaptiveness of such taboos and similar cultural mechanisms is very difficult to establish. What is here interpreted as a fertility control device could simply be a case of maladaptation.
37. J. V. Neel, 1968, "Some aspects of differential fertility in two American Indian tribes," *Proceedings of the 8th International Congress of Anthropological and Ethnological Sciences* 1:356–361.
38. R. Wilkinson, 1979, *Poverty and Progress: An Ecological Model of Economic Development* (London: Methuen), 37.
39. E.g., J. T. Noonan, 1966, *Contraception: A History of Its Treatment by the Catholic Theologicans and Canonists* (Cambridge, Mass.: Harvard University Press); M. Potts, P. Diggory, and J. Peel, *Abortion* (Cambridge, Eng.: Cambridge University Press).
40. E.g., D. I. Rubenstein and R. W. Wrangham, 1986, *Ecological Aspects of Social Evolution* (Princeton, N.J.: Princeton University Press).
41. S. B. Hrdy, 1979, "Infanticide among animals: A review, classification and examination of the implications for the reproductive strategies of females," *Ethology and Sociobiology* 1:13–40.
42. For early reports, see G. B. Schaller, 1972, *The Serengeti Lion: A Study of Predator-Prey Relations* (Chicago: University of Chicago Press); and B. C. R. Bertram, 1975, "Social factors influencing reproduction in wild lions," *Journal of Zoology, London* 177:463–482.
43. W. Angst and D. Thommen, 1977, "New data and discussion of infant killing in Old World monkeys and apes," *Folia Primatologia* 27:198–229; D. Fossey, 1983, *Gorillas in the Mist* (Boston: Houghton Mifflin).
44. E.g., summary in D. Finkelhor, 1993, "Epidemiological factors in the clinical identification of child sexual abuse," *Child Abuse and Neglect* 17:67–70.
45. S. Emlen, N. Demong, and D. Emlen, 1989, "Experimental induction of infanticide in female Wattled Jacanas," *Auk* 106:1–7.

46. Wilson, 1975.
47. There is theoretical reason to believe that, in mammals, males and females have different sexual strategies because of the disproportionately greater time and energy that female mammals invest in producing offspring. Some empirical evidence suggests that this theoretical view applies as well to *Homo sapiens,* although some scientists think this is simplistic. This genetic tendency, if it exists, obviously can be overridden by cultural evolution, as is indicated by the significant numbers of males in putatively monogamous societies who do not seek what biologists quaintly call "extra-pair copulations."

    For a fine summary of the "profligate male/choosy female" view, see D. Symons, 1979, *The Evolution of Human Sexuality* (New York: Oxford University Press). For an equally interesting counterargument, emphasizing that the dependency of children has consequences for the reproductive success of both parents, see M. Small, 1993, *Female Choices: Sexual Behavior of Female Primates* (Ithaca, N.Y.: Cornell University Press). That these two books are written by a man and a woman respectively may be more than coincidence. Our team, with representatives of both "sides," is undecided. One of Symons's most persuasive arguments for a genetically based difference in attitudes toward sexual variety comes from different behaviors of male homosexuals and lesbians (pp. 292ff). If you're interested in this topic, a good starting point would be to read both books.
48. Biologists are asked to excuse our teleological wording here, but the circumlocutions required to avoid it are too cumbersome.
49. See summary in P. Ehrlich, D. Dobkin, and D. Wheye, 1988, *The Birder's Handbook* (New York: Simon & Schuster), 307–309. As this source explains, hatching asymmetry is not correlated with brood reduction in all species where it occurs.
50. N. H. H. Graburn, 1987, "Severe child abuse among the Canadian Inuit," in N. Scheper-Hughes (ed.), *Child Survival: Anthropological Perspectives on the Treatment and Maltreatment of Children* (Dordrecht, Holland: D. Reidel), 223.
51. M. Dickemann, 1975, "Demographic consequences of infanticide in man," *Annual Review of Ecology and Systematics* 6:107–137. There appears to be a variety of reasons for which people use infanticide as a fertility control technique; see M. Dickemann, 1984, "Concepts and classification in the study of human infanticide: Sectional introduction and some cautionary notes," in G. Hausfater and S. B. Hrdy, *Infanticide: Comparative and Evolutionary Perspectives* (New York: Aldine), 427–437; and S. C. M. Scrimshaw, 1984, "Infanticide in human populations: Societal and individual concerns," in Hausfater and Hrdy, 1984, pp. 439–462. In general, it appears that infanticide and pedicide are most likely to occur when it will increase the fitness of the parents (i.e., maximize their lifetime reproduction), as suggested by M. Daly and M. Wilson, 1984, "A sociobiological analysis of human infanticide," in Hausfater and Hrdy, 1984, 487–502; and by P. E. Bugos Jr.

and L. M. McCarthy, 1984, "Ayoreo infanticide: A case study," in Hausfater and Hrdy, 1984, 503–520.

52. K. Rasmussen, 1931, *The Netsilik Eskimos,* vol. 8 (Copenhagen, Denmark: Reports of the 5th Thule Expedition), 138.

53. A. Balikci, 1970, *The Netsilik Eskimo* (Garden City, N.Y.: Natural History Press).

54. Ibid., p. 151.

55. E.g., G. Van Den Steenhoven, 1959, *Legal Concepts Among the Netsilik Eskimos of Pelly Bay, N.W.T.* (Ottawa: Northern Co-ordination and Research Centre, Department of Northern Affairs and National Resources), NCRC-59-3.

56. Balikci, 1970, pp. 161–162.

57. Wynne-Edwards, 1962, p. 492.

58. N. Howell, 1979, *Demography of the Dobe !Kung* (San Diego: Academic Press), 120; M. Shostak, 1981, *Nisa: The Life and Words of a !Kung Woman* (Cambridge, Mass.: Harvard University Press), 66.

59. Neel, 1969. In this article, which precedes his *Science* article cited above, he suggests that the transition from prehuman to human occurred at "the point at which parental care evolved to the level of permitting rapid population increase, with the concomitant recognition of the necessity to limit natural fecundity" (p. 358).

60. Quoted in Harris and Ross, 1987, p. 80.

61. J. C. Russell, 1948, *British Medieval Population* (Albuquerque, N.M.: University of New Mexico Press), 168.

62. W. E. H. Lecky, 1869, *A History of European Morals from Augustus to Charlemagne,* vol. 2 (London: Longmans, Green), 27.

63. W. L. Langer, 1972, "Checks on population growth: 1750–1850," *Scientific American,* Feb., 92–99; W. L. Langer, 1974, "Infanticide: A historical survey," *History of Childhood Quarterly* 1:353–365.

64. Langer, 1974, p. 360.

65. Ibid., p. 359.

66. Ibid., p. 360.

67. George Burrington, 1757, "An answer to Dr. William Brakenridge's letter concerning the number of inhabitants within the London bills of mortality" (London: J. Scott).

68. B. Kellum, 1974, "Infanticide in England in the later Middle Ages," *History of Childhood Quarterly* 1:367–388.

69. J. R. McNeil and H. M. Gamer (trans.), 1938, *Medieval Handbooks of Penance* (New York: Columbia University Press), 293, cited in Kellum, 1974, p. 370.

70. Kellum, 1974, pp. 380–381.

71. Dickemann, 1979, p. 328ff.

72. G. Vines, 1993, "The hidden cost of sex selection," *New Scientist,* 1 May, 12–13.

73. H. Yuan Tien et al., 1992, "China's demographic dilemmas," *Population Bulletin* (Population Reference Bureau, Washington, D.C.) 47(June):1.

74. "China on the population question," *China Reconstructs* 23(1974):11; Qu Geping, 1989, "Over the limit," *Earthwatch* (supplement to *People,* International Planned Parenthood Federation) 34:2.

75. N. D. Kristof, 1993, "Peasants of China discover new way to weed out girls," *The New York Times,* 21 July.

76. P. Shenon, 1994, "China's mania for baby boys creates surplus of bachelors," *The New York Times,* 16 August.

77. C. Ford and F. Beach, 1970, *Patterns of Sexual Behavior* (New York: Harper & Row), 143. Female monkeys often mount other females and males other males. While this may be partly a dominance interaction, it also may involve sexual stimulation; see Hrdy, 1981, p. 170.

78. E.g., J. Boswell, 1980, *Christianity, Social Tolerance, and Homosexuality: Gay People in Western Europe from the Beginning of the Christian Era to the Fourteenth Century* (Chicago: University of Chicago Press).

79. D. Werner, 1979, "A cross-cultural perspective on theory and research on male homosexuality," *Journal of Homosexuality* 4:345–362. There is a brief summary on p. 359 of the results of Werner's unpublished master's thesis on this topic.

80. R. C. Kelly, 1976, *Etoro Social Structure: A Study in Structural Contradiction* (Ann Arbor: University of Michigan Press).

81. Harris and Ross, 1987, pp. 84–85.

82. M. Martin and B. Voorhies, 1975, *The Female of the Species* (New York: Columbia University Press).

83. Harris and Ross, 1987, p. 68.

84. S. Lindenbaum, 1979, *Kuru Sorcery: Disease and Danger in the New Guinea Highlands* (Palo Alto, Calif.: Mayfield), 131–133.

85. J. B. Russell, 1972, *Witchcraft in the Middle Ages* (Ithaca, N.Y.: Cornell University Press), 335; Harris, 1974.

86. Harris and Ross, 1987, p. 96; Russell, 1972, pp. 283–284.

87. Harris and Ross, 1987, p. 97.

88. E. A. Wrigley and R. S. Schofield, 1981, *The Population History of England 1541–1871: A Reconstruction* (London: Edward Arnold), 421ff.

89. Tien et al., 1992; "China's fertility patterns closely parallel recent national policy changes," *International Family Planning Perspectives* 17 (June 1991): 75–76.

90. For an overview, see P. Ehrlich, L. Bilderback, and A. Ehrlich, 1981, *The Golden Door: International Migration, Mexico, and the United States* (New York: Wideview Books).

91. J. T. Tanner, 1975, "Population limitation today and in ancient Polynesia," *BioScience* 25:513–516.

92. Ehrlich et al., 1981, pp. 17–27.

93. N. Myers, 1993, "Environmental refugees in a globally warmed world," *BioScience* 43(11): 752–761.

94. United Nations Population Fund (UNFPA), 1993, *State of the World's Population, 1993* (New York: United Nations).

95. Garrett Hardin shot that idea down in flames in a 1957 article ("Interstellar migration and the population problem," *Journal of Heredity*

50:68–70) when he showed that migration to other planets (assuming they were habitable, which they aren't) would fill them Earth-full within a couple of generations at the rate of population growth then prevailing. He also demolished the idea of migrating to the planets of distant stars, because many years would pass before arrival at the nearest possible system, necessitating strict population control during the voyage. Both arguments are reprised in his 1993 book, *Living within Limits* (New York: Oxford University Press), 7–13.

96. Ehrlich and Ehrlich, 1990, p. 62.

97. E.g., J. Caldwell, 1983, "Direct economic costs and benefits of children," in R. A. Bulatao and R. D. Lee, *Determinants of Fertility in Developing Countries* (New York: Academic Press), 458–493; and P. Dasgupta, 1993, *An Inquiry into Well-being and Destitution* (Oxford, Eng.: Oxford University Press).

98. Dickemann, 1975, p. 110. Note that the degree to which demographic control is exercised intentionally is usually unclear.

99. Ibid., p. 111. For a detailed discussion of this myth, see Edgerton, 1992.

100. Carr-Saunders, 1922, p. 214.

101. V. C. Wynne-Edwards, 1962. This classic work is a gold mine of information; it has been somewhat neglected by biologists because of its heavy emphasis on group selection, a process of doubtful potency in most evolutionary situations.

102. For some of the problems with early views, see Mary Douglas, 1966, "Population control in primitive groups," *British Journal of Sociology* 17:263–273. Their basic claim of near-optimal size for primitive groups cannot be easily defended. And the issue of underpopulation (which can present severe problems for small societies) is not addressed. Douglas believed that population control efforts in preindustrial groups were often motivated by "concern for scarce social resources," those that yield social advantage. Thus the Rendille of Kenya are very concerned about the possibility of human overpopulation in relation to the size of their camel herds and adjust their population size to that of their herds. The Rendille require celibacy for men between the ages of twelve and thirty-one. Only half of all men are able to pay the required bride-price, which must be remitted in camels. Camels are, above all, status symbols; see E. F. Moran, 1979, *Human Adaptibility: An Introduction to Ecological Anthropology* (North Scituate, Mass.: Duxbury Press), 229–231.

103. W. Irons, 1979, "Cultural and biological success," in N. A. Chagnon and W. Irons, *Evolutionary Biology and Human Social Behavior: An Anthropological Perspective* (North Scituate, Mass.: Duxbury Press); B. S. Low, 1990, "Occupational status, landownership, and reproductive behavior in 19th-century Sweden: Tuna Parish," *American Anthropologist* 92:457–468; B. S. Low, 1990, "Marriage systems and pathogen stress in human societies," *American Zoologist* 30:325–339.

104. E.g., M. Livi-Bacci, 1989, *A Concise History of World Population* (Cam-

bridge, Mass.: Blackwell). See especially the discussion of "demographic waste," p. 100ff., which makes it seem that women always struggled to produce enough babies to more than replace themselves.

105. Harris and Ross, 1987, p. 85. The importance of the Church and its property in demographic issues is discussed in J. Goody, 1983, *The Development of the Family and Marriage in Europe* (Cambridge, Eng.: Cambridge University Press), 46–47, 81, 216. The reasons for the generally pronatalist policies that evolved in the Church were, of course, very complex and involved much more than considerations of labor supply. Ironically, some Church policies were clearly antinatalist in result if not in intent, especially clerical celibacy (Goody, 1983, p. 190).

106. See discussions in P. Ehrlich, A. Ehrlich, and J. Holdren, 1977, *Ecoscience: Population, Resources, Environment* (San Francisco: W. H. Freeman & Co.), 776–780; Harris and Ross, 1987.

107. Frisch, 1978.

108. M. Mamlouk, 1982, *Knowledge and Use of Contraception in Twenty Developing Countries* (Washington, D.C.: Population Reference Bureau).

109. Wang Yi-fei, 1994, "Men should take an equal responsibility in all reproductive health matters," *Reproduction and Contraception* 5:21–25 (supplement).

110. Xiao Bi-lian, 1994, "Advances of contraception in China," *Reproduction and Contraception* 5:13–19 (supplement).

111. M. Jitsukawa and C. Djerassi, 1994, "Birth control in Japan: Realities and prognosis," *Science* 265:1048–1051.

112. Ibid., p. 1050.

113. W. D. Mosher, 1990, "Contraceptive practice in the United States, 1982–1988," *Family Planning Perspectives* 22:198–205.

114. J. Bongaarts and G. Rodriguez, 1988, "A new method for estimating contraceptive failure rates," in *Measuring the Dynamics of Contraceptive Use,* proceedings of the United Nations Expert Group Meeting on Measuring the Dynamics of Contraceptive Use, New York.

115. E.g., M. Holloway, 1993, "Obstacle course: Funding and policy stifle contraceptive research," *Scientific American,* April, 18–24; C. Djerassi, 1991, "New contraceptives: Utopian or Victorian?," *Science and Public Affairs* 6:5–15.

116. R. Stone, 1992, "Controversial contraceptive wins approval from FDA panel," *Science* 256:1754.

117. K. Fackelmann, 1992, "Sex protection: Balancing the equation," *Science News* 141:168–169.

118. Wu Wei-xiong, Yuan Shi-jin, Chen Zhen-wen, Gu Yi-qun, and Li Shi-qin, 1994, "Preliminary clinical study of intra-vas device," *Reproduction and Contraception* 5:72–78 (supplement).

119. The herb is *Tripterygium wilfordii,* also referred to as Mang Cao.

120. P. Aldhous, 1994, "A booster for contraceptive vaccines," *Science* 266:1484–1486.

121. C. Djerassi, 1989, "The bitter pill," *Science* 245:356–361.

122. There are other problems as well, having to do with regulation of re-

search on human embryos—see R. Service, 1994, "Barriers hold back new contraception strategies," *Science* 226:1489.

123. Norplant contains no estrogen, and thus avoids the complications it sometimes causes. See M. Frank, A Poindexter, M. Johnson, and L. Bateman, 1992, "Characteristics and attitudes of early contraceptive implant acceptors in Texas," *Family Planning Perspectives* 24:208–213.

124. Gu Sujuan, Du Mingkun, Zhang Ling-De, Liu Yinglin, Wang Shuhua, and I. Sivin, 194, "A five-year evaluation of Norplant II implants in China," *Contraception* 50:27–34.; Gu Sujuan, Du Mingkun, Zhang Ling-De, Liu Yinglin, Wang Shuhua, and I. Sivin, 1994, "A five-year evaluation of Norplant contraceptive implants in China," *Obstetrics and Gynecology* 83:673–678.

125. C. Pies, M. Potts, and B. Young, 1994, "Quinacrine pellets: An examination of nonsurgical sterilization," *International Family Planning Perspectives* 20(4):137–141.

126. R. Berkow (ed.), 1992, *The Merck Manual of Diagnosis and Therapy* (Rahway, N.J.: Merck Research Laboratories).

127. D. T. Hieu et al., 1993, "31,781 cases of non-surgical female sterilization with quinacrine pellets in Vietnam," *Lancet* 342:213–217.

128. M.-T. Feuerstein, 1993, "Family planning in Vietnam," *Lancet* 342:188.

129. Dr. Stephen Mumford, 13 January 1994.

130. Center for Research on Population and Security, 1994, "Major advance in non-surgical sterilization," press release, 15 October 1993.

131. L. Lader, 1991, *RU-486: The Pill That Could End the Abortion Wars and Why American Women Don't Have It* (Reading, Mass.: Addison-Wesley).

132. C. Djerassi, 1991.

133. C. Djerassi and S. Leibo, 1994, "A new look at male contraception," *Nature* 370:11–12.

134. D. Lodge, 1980, *Souls and Bodies* (New York: Penguin).

135. Douglas, 1966, p. 273.

Chapter Three

1. E.g., G. Moffett, 1994, *Critical Masses: The Global Population Challenge* (New York: Viking). While it ignores the major problems generated by overpopulation in rich nations, this book is a good journalistic introduction to population problems in poor countries.

2. For a recent example of this view, see A. Sen, 1994, "Population: Delusion and reality," *New York Review of Books*, 22 Sept. The treatment of food production and the environment in this piece (and especially the failure to see the environmental constraints on production) gives new force to the word *delusion*—and maintains a virtually unbroken record of incompetent treatment of environmental issues in this influential journal.

3. Here and elsewhere, unless otherwise specified, we use *crude birth rates* and *crude death rates,* and both are known as *vital rates.* These are simply the number of births or deaths per 1,000 in the population in a year,

divided by an estimate of the midyear population size. The growth rate is the difference between birth and death rates, plus or minus any migration. The rate of natural increase is the crude birth rate minus the crude death rate. While crude vital rates are measured as births or deaths per 1,000 in the population, growth rates (just to confuse the unwary) are usually stated as a percent. Thus (to take the example of Algeria in 1994) a birth rate of 32 and a death rate of 7 yielded a growth rate of 2.5 percent, corresponding to a doubling time of 28 years.

More refined estimates of fertility or mortality take into account the age composition of the population. For instance, the crude death rate may be higher in a rich country with superb health care but an older population than that in a poor nation with poor health care but a population with up to half its people younger than 15. Thus, in 1994, the crude death rate was 11 per 1,000 in Sweden, where only 18 percent of the population was under 15 years of age, and 8 in Namibia with 45 percent of its population under 15. Age-specific mortalities, however, were lower in Sweden; the higher crude rate was simply a result of an older population.

4. For a more complete overview of the history of population growth, see P. Ehrlich and A. Ehrlich, 1990, *The Population Explosion* (New York: Simon & Schuster). For much more extensive coverage and references, see P. R. Ehrlich, A. H. Ehrlich, and J. P. Holdren, 1977, *Ecoscience: Population, Resources, Environment* (San Francisco: W. H. Freeman & Co.).

5. The exception was North America, where a high rate of natural increase was swelled by high immigration rates; in Europe, population growth was correspondingly slowed by the departure of tens of millions for the New World; see U.S. Census Bureau, 1961, *Historical Statistics of the United States: Colonial Times to 1957* (Washington, D.C.: U.S. Government Printing Office); P. R. Ehrlich, L. Bilderback, and A. H. Ehrlich, 1981, *The Golden Door: International Migration, Mexico, and the United States* (New York: Wideview Books).

6. Replacement-level fertility is the rate at which parents are just replacing themselves in the next generation—slightly more than an average of two children per woman in her lifetime. (The extra fraction compensates for children who don't survive to reproduce.) Technically, with the low mortality rates prevailing in developed nations today, replacement is accomplished with a total fertility rate (TFR) of about 2.1 children per woman. In less developed countries, where mortalities are still somewhat higher, replacement fertility might be 2.2 or more.

7. E. van de Walle and J. Knodel, 1967, "Demographic transition and fertility decline: The European case," paper presented at the 1967 conference of the International Union for the Scientific Study of Population, Sydney, Australia; E. van de Walle and J. Knodel, 1980, "Europe's fertility transition: New evidence and lessons for today's developing world," *Population Bulletin* 34 (Feb.):6.

8. C. Rollet-Echalier, 1990, *La Politique à l'Egard de la Petite Enfance sous*

la IIIe Republique (Paris: Presses Universitaires de France). Anthropologist Virginia Abernethy has a nice essay questioning this widely held notion; see V. Abernethy, 1993, "The demographic transition revisited: Lessons for foreign aid and U.S. immigration policy," *Ecological Economics* 8:235–252.

9. Typical growth rates in 1969 were: Malaysia, 3.1 percent; Thailand, 3.1; Philippines, 3.5; Mexico, 3.4; Costa Rica, 3.8; Panama, 3.2; Colombia, 3.4; and Venezuela, 3.3; see Population Reference Bureau, 1969, *1969 World Population Data Sheet* (Washington D.C.: PRB).

10. L. Tabah, 1989, *World Population at the Turn of the Century*, Population Studies no. 111, ST/ESA/SER.A/111 (New York: United Nations).

11. The momentum of population growth causes populations to continue growing for many decades after reaching replacement fertility because of a preponderance of people in younger age groups resulting from previous high fertility. Since human beings live a long time after reproducing, it takes three generations (about seventy years) to wring out the momentum once replacement fertility is initiated. If fertility is significantly *below* replacement, of course, growth can be stopped sooner and a slow decline in population size will begin.

12. World Bank, 1980, *World Development Report 1980* (New York: Oxford University Press); N. Birdsall, 1980, "Population growth and poverty in the developing world," *Population Bulletin* (Population Reference Bureau, Washington, D.C.) 35 (Dec.):5.

13. Summarized in ch. 8 of Ehrlich, Ehrlich, and Holdren, 1977; *see also* a reprint of B. Berelson's well-known 1975 summary of the controversy, "The great debate on population policy: An instructive entertainment" (together with several articles updating the debate), in "The great debate on population policy revisited," *International Family Planning Perspectives* 16 (Dec. 1990):4.

14. See, for instance, World Resources Institute, 1992, *World Resources 1992–93,* and, in 1994, *World Resources 1994–95* (New York: Oxford University Press); L. R. Brown, et al., 1994, *State of the World 1994* (New York: Norton); or World Bank, 1992.

15. W. Rich, 1973, *Smaller Families through Social and Economic Progress,* Monograph no. 7 (Washington, D.C.: Overseas Development Council).

16. J. E. Kocher, 1973, *Rural Development, Income Distribution and Fertility Decline* (New York: The Population Council).

17. For a discussion of the fertility declines in East Asian nations (specifically Japan, Taiwan, South Korea, and China), see G. Feeney, 1994, "Fertility decline in East Asia," *Science* 266:1518–1523. Feeney maintains that the increased survival of children was an important factor in motivating family limitation in the four countries he analyzed.

18. E.g., Abernethy, 1993, "The demographic transition revisited," in V. Abernethy, 1993, *Population Politics: The Choices that Shape Our Future* (New York: Plenum Press).

19. Abernethy, 1993, "The demographic transition revisited," 256.

20. Abernethy, 1993, *Population Politics,* 83.

21. J. Knodel, A. Chamratrithirong, and N. Debavalya, 1987, *Thailand's Reproductive Revolution* (Madison: University of Wisconsin Press), 124–142; A. Bennet, C. Frisen, P. Kamnuansilpa, and J. McWilliam, 1990, *How Thailand's Family Planning Program Reached Replacement Level Fertility: Lessons Learned,* Population Technical Assistance occasional paper no. 4 (Dual Associates).

22. K. Holl, G. Daily, and P. Ehrlich, 1993, "The fertility plateau in Costa Rica: A review of causes and remedies," *Environmental Conservation* 20:317–323.

23. Holl et al., 1993.

24. In technical terms, scientists often point out that "correlation does not mean causation." For instance, the price of IBM stock in the United States was going up while the Costa Rican TFR was dropping. This does not mean that the rise in IBM stock prices was causing fewer Tico babies, or vice versa.

25. Holl et al., 1993.

26. For the reasons behind this statement, see the discussion in Chapter 1; for more details, see P. R. Ehrlich and A. H. Ehrlich, 1991, *Healing the Planet* (New York: Addison-Wesley), 32–35.

27. M. Ballara, 1991, *Women and Literacy* (London: Zed Books); UNFPA, 1990, *State of the World Population 1990* (New York: United Nations).

28. Whether the lower fertility associated with education is a result of deferring marriage and childbirth in favor of education or the arrival of the first child in adolescence cuts education short is not always clearcut, especially in very poor regions such as sub-Saharan Africa; see, e.g., C. H. Bledsoe and B. Cohen, 1993, *Social Dynamics of Adolescent Fertility in Sub-Saharan Africa* (Washington, D.C.: National Academy Press), 114–115.

29. See J. Drèze, 1990, "Famine prevention in India," in J. Drèze and A. Sen, *The Political Economy of Hunger,* pp. 13–99, in vol. 2 of *Famine Prevention* (Oxford, Eng.: Oxford University Press).

30. World Resources Institute, 1994.

31. Expert Group on Population Policy, 1994, *Draft National Population Policy,* submitted to Ministry of Health and Family Welfare, Government of India, New Delhi, May 21.

32. *Travancore Education Reforms Committee Report,* 1933 (Trivandrum, India: Government Press), 65–66.

33. Ministry of Home Affairs, 1993, *Sample Registration System: Fertility and Mortality Indicators 1991* (New Delhi); P. Mari Bhat and S. Rajan, 1990, "Demographic transition in Kerala revisited," *Economic and Political Weekly* 25:1957–1990; M. S. Swaminathan, personal communication, 10 November 1994.

34. R. Jeffrey, 1987, "Governments and culture: How women made Kerala literate," *Pacific Affairs* 60:467–479.

35. Ibid., p. 471.

36. Bhat and Rajan, 1990, p. 1979.

37. Knodel et al., 1987; Bennet et al., 1990.

38. L. Summers, 1993, "The most influential investment," *People and the*

Planet (International Planned Parenthood Federation, London) 2(1):10. This special issue was devoted to the subject of educating girls.

39. D. Taylor, 1993, "Meeting the need," *People and the Planet* 2(1):7.

40. M. A. Hill and E. M. King, 1992, "Women's education in developing countries: An overview," in M. A. King and E. M. Hill (eds.), *Women's Education in Developing Countries: Barriers, Benefits, and Policies* (Baltimore: Johns Hopkins Press, 1992), 1–50; *see also* K. Subbarao and L. Raney, 1993, "Social gains from female education: A cross-national study," *World Bank Discussion Paper* no. 194 (Washington, D.C.: World Bank).

41. Hill and King, 1992.

42. J. L. Jacobson, 1991, "Women's reproductive health: The silent emergency," *Worldwatch Paper 102* (Worldwatch Institute, Washington, D.C.), June.

43. At this rate, abortion–related deaths in the U.S. would amount to three or less per year; there are roughly 2 million abortions performed each year and about 4 million births.

44. A. Starrs, 1987, "Preventing the tragedy of maternal deaths: A report on the International Safe Motherhood Conference," Nairobi, Kenya (cited in Jacobson, 1991).

45. Worldwide, the annual number of maternal deaths in the mid-1980s was conservatively estimated to be a half million; three fifths of them in Asia, 150,000 in Africa, 34,000 in Latin America, and 6,000 in developed countries; see J. Walsh et al., "Maternal and perinatal health," in D. T. Jamison and W. H. Mosley (eds.), 1992, *Disease Control Priorities in Developing Countries* (New York: World Bank/Oxford University Press); World Health Organization, 1986, *Maternal Mortality Rates: A Tabulation of Available Information* (Geneva, Switz.: WHO); Starrs, 1987; S. Armstrong, 1990, "Labour of death," *New Scientist,* 31 March, 50–55. Since about 140 million infants were being born each year, one woman died for every 280 births (and about 20 of the infants failed to reach their first birthdays). Of course, deaths from reproductive causes, especially abortion, are underreported even in nations like the United States. In developing nations, maternal deaths are estimated to be underreported by at least 70 percent; see Pan American Health Organization, 1990, *Regional Plan of Action for the Reduction of Maternal Mortality in the Americas* (Washington, D.C.) 7 August. Thus the actual number of deaths each year could be upward of 750,000. And the number of women who suffer illness and disability but survive is many times greater; see Walsh, 1992.

46. W. H. Mosley and P. Cowley, 1991, "The challenge of world health," *Population Bulletin* (Population Reference Bureau, Washington, D.C.) 46(Dec.):7–8.

47. E. DeMaeyer and M. Adiels-Tegman, 1985, "The prevalence of anemia in the world," *World Health Statistics Quarterly* 38, cited in Jacobson, 1991.

48. Jacobson, 1991: various sources cited therein.

49. Mosley and Cowley, 1991; Jacobson, 1991. Besides syphilis and gon-

orrhea, several newer diseases have been spreading rapidly that pose particular dangers to women: genital herpes, genital warts, chancroid, bacterial vaginosis, trichomoniasis, and chlamydia; see World Health Organization, 1990, "Sexually transmitted infections increasing by 250 million new infections annually," press release, 20 Dec., Geneva. Many also are associated with AIDS, either facilitating or accompanying its transmission.

Some diseases initially have few or subtle symptoms that go unrecognized in women, but if untreated they can be serious or even fatal; see R. Dixon-Mueller and J. Wasserheit, 1991, *The Culture of Silence* (New York: International Women's Health Coalition); Jacobson, 1991. The least serious, trichomoniasis, is the commonest, with about 120 million new cases each year around 1990. Genital herpes and chancroid are often painful and can facilitate AIDS infections. Genital warts can lead to cervical cancer, of which nearly 500,000 cases per year were reported in the early 1990s; see Mosley and Cowley, 1991, p. 20. Both gonorrhea and syphilis can develop serious complications including infertility, heart disease, and dementia, and can be involved in AIDS transmission. Bacterial vaginosis and chlamydia each can lead to pelvic inflammatory disease, ectopic pregnancy, and infertility; chlamydia has an AIDS connection too. An estimated 50 million new cases of chlamydia occur annually worldwide, as do tens of millions of new cases of genital warts, gonorrhea, and herpes.

AIDS had infected an estimated 17 million people by 1994, and the rate of spread was alarmingly rapid—up to 2 million new cases per year. The problem seemed most acute in Africa, where the occurrence of other STDs is also rampant. More than half of the infections to date are believed to be in Africa, although the infection rate in parts of Asia is rising fast. Also associated with AIDS in Africa and elsewhere is resurgent, often drug-resistant tuberculosis.

50. Mosley and Cowley, 1991.
51. L. A. Mtimavalye and M. A. Belsey, 1987, *Infertility and Sexually Transmitted Disease: Major Problems in Maternal and Child Health and Family Planning* (New York: Population Council); A. DeSchryver and A. Meheus, 1990, "Epidemiology of sexually transmitted diseases: The global picture," *Bulletin of the World Health Organization* 68:5.
52. See Ehrlich, Ehrlich, and Holdren, 1977, ch. 7.
53. Jacobson, 1991; R. Dixon-Mueller, 1993, "The sexuality connection in reproductive health," *Studies in Family Planning* 24 (Sept./Oct.):269–281.
54. United Nations Population fund (UNFPA), 1994, "ICPD '94," *Populi* (special issue, UNFPA, New York) 21 (Oct.).
55. D. Symons, 1979, *The Evolution of Human Sexuality* (New York: Oxford University Press); M. Small, 1993, *Female Choices: Sexual Behavior of Female Primates* (Ithaca, N.Y.: Cornell University Press).
56. K. Mason and A. Taj, 1987, "Differences between women's and men's reproductive goals in developing countries," *Population and Develop-*

ment Review 13:611–638; Anonymous, 1986, "Men—new focus of family planning programs," *Population Reports,* Series J 33 (Nov./Dec.).

57. S. Greenhalgh, 1991, "Women in the informal enterprise: Empowerment or exploitation?," *Research Division Working Papers* no. 33 (New York: Population Council).

58. S. Desai and D. Jain, 1994, "Maternal employment and changes in family dynamics: The social context of women's work in rural South India," *Population and Development Review* 20(1):115–136.

59. C. Lloyd, 1994, "Family and gender issues for population policy," in L. Mazur (ed.), 1994, *Beyond the Numbers: A Reader on Population, Consumption, and the Environment* (Washington, D.C.: Island Press), 242–256.

60. W. Parish and R. Willis, 1992, "Daughters, education, and family budgets: Taiwan experiences," *Discussion Paper Series* (Economic Research Center, NORC, University of Chicago).

61. Vividly described in J. W. Anderson and M. Moore, 1993, "The burden of womanhood," *Washington Post Weekly,* 22–28 March, 6–7; and C. Murphy, 1993, "Pulling aside the veil," *Washington Post Weekly,* 12–18 April, 10–11.

62. M. Hornblower, 1993, "The skin trade," *Time,* 21 June, 45–55; M. Simons, 1994, "The littlest prostitutes," *New York Times Magazine,* 16 January, 30–35; S. Greenhouse, 1994, "State Dept. finds widespread abuse of world's women," *The New York Times,* 3 February, 1.

63. J. C. Caldwell, 1980, "Mass education as a determinant of the timing of fertility decline," *Population and Development Review* 6:225–255; J. C. Caldwell, 1982, *The Theory of Fertility Decline* (New York: Academic Press).

64. 1991 data from Expert Group on Population Policy, 1994.

65. S. Ramasundaram, 1994, "Population control in India—the Tamil Nadu experience and policy options for the future," paper presented at the Southern India Seminar on Population Control at the M. S. Swaminathan Foundation, Madras, Feb. 13.

66. December 15, 1994. This visit and those subsequently described were arranged through the courtesy of Dr. M. S. Swaminathan. Our notes were taken hurriedly during fascinating but all-too-brief visits, and we apologize for any misspelling of names.

67. *See also* District Family Welfare Bureau, Corporation of Madras, 1994, "Brief report on information, education and communications" (I.E.C.), World Bank, India Population Project-V.

68. District ICDS Cell, Chengalpattu MGR District.

69. Government of Tamil Nadu, 1993, *State Plan of Action for the Child in Tamil Nadu* (Madras), Nov., 90.

70. District ICDS Cell, 1994, *Status Report Chengalpattu M.G.R. District* (Tambaram, Madras-45), Nov.

71. S. Ramasundaram, 1994, "Tamil Nadu experience," *The Hindu,* 25 October; Government of Tamil Nadu, Department of Social Welfare and Nutritious Meal Programme, 1993, *Dr. J. Jayalalitha, 15-point*

Programme for Child Welfare. The fifteen points are: (1) increase average birth weight to 3 kg.; (2) eliminate vaccine-preventable diseases; (3) reduce infant mortality to less than 30 per 1,000; (4) reduce severe and moderate malnutrition among children; (5) eliminate deficiencies of micronutrients; (6) make hospitals and maternity centers "baby friendly"; (7) liberate women from early and frequent childbearing; (8) provide universal compulsory primary education, ensuring five years of primary education for every child; (9) raise women's literacy and status; (10) provide safe drinking water and better access to sanitary facilities at all children's centers; (11) eliminate child labor; (12) popularize girl-child protection scheme; (13) eradicate female infanticide in Tamil Nadu; (14) prevent childhood disability and provide early detection for rehabilitation; (15) identify congenital heart diseases early and provide free surgery for children. The program gives special attention to the needs of the poorest children and others (women and those working in hazardous industries) most in need of protection.

72. Once every two weeks, eggs are served.
73. District ICDS Cell, 1994.
74. S. Ramasundaram, 1994, "Population control in India: Policy issues and programme strategies," paper presented August 20 at the Family Welfare Conference at JIPMER, Pondicherry-605006.
75. W. P. Handwerker, 1989, *Women's Power and Social Revolution: Fertility Transition in the West Indies* (Newbury Park, Calif.: Sage Publications).
76. For a historical overview, see M. Harris and E. B. Ross, 1987, *Death, Sex, and Fertility* (New York: Columbia University Press). *See also* J. Caldwell, 1983, "Direct economic costs and benefits of children," in R. A. Bulatao and R. D. Lee, *Determinants of Fertility in Developing Countries* (New York: Academic Press), 458–493.
77. World Resources Institute, 1994, ch. 3.
78. Partha Dasgupta, 1993, *An Inquiry into Well-being and Destitution* (Oxford, Eng.: Oxford University Press).
79. K. Smith, Gu Shuhua, Huang Kun and Qiu Daxiong, 1993, "One hundred million improved cookstoves in China: How was it done?," *World Development* 21(6):942–963.
80. P. Warshall, personal communication, November 1994.
81. Harris and Ross, 1987, p. 124.
82. M. Cain, 1981, "Risk and insurance: Perspectives on fertility and agrarian change in India and Bangladesh," *Population and Development Review* 7:435–474.
83. Ibid., p. 50.
84. For a general discussion of the urban factor in family-size decisions, see J. Caldwell, 1983, "Direct economic costs and benefits of children," in R. A. Bulatao and R. D. Lee, *Determinants of Fertility in Developing Countries* (New York: Academic Press), 458–493; and other articles in the same book.
85. E.g., E. Linden, 1993, "Megacities," *Time,* 11 January, 28–38.
86. S. WuDunn, 1993, "As China leaps ahead, the poor slip behind," *The New York Times,* 23 May *(Week in Review* section), 3; S. WuDunn,

1994, "China's rush to riches," *The New York Times Magazine,* 4 Sept., 38–41, 46–48, 54.

87. K. Wakabayashi, 1990, "Migration from rural to urban areas in China," *The Developing Economies* 28(4):503–523; Zeng Yi, Tu Peng, Guo Liu, and Xie Ying, 1991, "A demographic decomposition of the recent increase in crude birth rates in China," *Population and Development Review* 17 (Sept.): 435–458; Minja Kim Choe, Wu Jianming, and Zhang Ruyue, 1992, "Progression to second and third births in China: Patterns and covariates in six provinces," *International Family Planning Perspectives* 18 (Dec.): 130–136; Digest section, 1991, "Women in rural China want two children despite adherence to government one-child family policy," *International Family Planning Perspectives* 17 (Dec.):152–153.

88. Gita Sen, personal communication, 11 October 1994.

89. Ehrlich, Ehrlich and Holdren, 1977, ch. 8.

90. M. Cain and G. McNicoll, 1988, "Population growth and agrarian outcomes," in R. D. Lee, W. B. Arthur, A. C. Kelley, G. Rodgers, and T. N. Srinivasan (eds.), *Population, Food, and Rural Development* (Oxford, Eng.: Clarendon Press), 106–108.

91. Handwerker, 1989.

92. A. M. Basu, 1992, *Culture, the Status of Women, and Demographic Behavior* (Oxford, Eng.: Clarendon Press).

93. The northerners had smaller households (fewer members of the extended family living together) than the southerners, even though the northerners had more children.

94. Basu, 1992, p. 10.

95. E. van de Walle and J. Knodel, 1980.

96. M. Harris, 1987, *The Sacred Cow and the Abominable Pig: Riddles of Food and Culture* (New York: Simon & Schuster).

97. P. J. Donaldson and A. O. Tsui, 1990, "The international family planning movement," *Population Bulletin* 45 (Nov.):3; B. Robey, S. O. Rutstein, and L. Morris, 1992, "The reproductive revolution: New survey findings," *Population Reports* (Population Information Program, Johns Hopkins University, Baltimore), Special Topics, Series M, 11 (Dec.).

98. R. Gillespie, reported in Anonymous, 1992, "The demographic transition and the technological transition," *Engineering and Science* 55(3):25–30.

99. District Family Welfare Bureau, 1994.

100. T. Conover, 1993, "Trucking through the AIDS belt," *New Yorker,* 16 August, 56–75.

101. See G. Narayana and J. Kantner, 1992, *Doing the Needful: The Dilemma of India's Population Policy* (Boulder, Colo.: Westview Press), for an analysis of the social and cultural problems involved in population control in India.

102. S. Ramasundaram, 1994, "Interpreting demographic data: What do the latest S.R.S. figures indicate?," November, Health and Welfare Department, Government of Tamilnadu, Madras.

103. K. Mahadevan (ed.), 1989, *Women and Population Dynamics: Perspectives from Asian Countries* (Delhi, India: Sage Publications); R. Repetto, 1994, *The "Second India" Revisited: Population, Poverty, and Environmental Stress Over Two Decades* (Washington, D.C.: World Resources Institute).
104. P. Nayar, 1989, "Kerala women in historical and contemporary perspective," in K. Mahadevan (ed.), *Women and Population Dynamics* (New Delhi: Sage Publications, 1989).
105. Anonymous, 1994, "South ahead in population control," *The Hindu* (Madras), 14 February; Anonymous, 1994, "Not by contraception alone," *Indian Express* (Madras), 14 Feb.; I. V. Antony, former Tamil Nadu Chief Secretary, personal communication, 14 Dec. 1994, Madras.
106. Dr. B. Surykum, personal communication, 14 Dec. 1994.
107. B. Robey, S. Rutstein, and L. Morris, 1993, "The fertility decline in developing countries," *Scientific American,* Dec., 60–67.
108. H. Suyono, 1994, address to the meeting of the Population Association of America, 5 May, 1994, Office of the Ministry of Population, Jakarta.
109. W. Brass and C. Jolly (eds.), 1993, *Population Dynamics of Kenya* (Washington, D.C.: National Research Council, National Academy Press), 3.
110. J. Phillips, R. Simmons, M. Koenig, and J. Chakraborty, 1988, "Determinants of reproductive change: Evidence from Matlab, Bangladesh," *Studies in Family Planning* 19:313–334.
111. Phillips et al., 1988, p. 322.
112. United Nations Population Fund (UNFPA), 1992, *State of the World's Population 1992* (New York: United Nations).
113. Demographic projections traditionally have taken no account of potential physical or biological limits to population growth. The standard assumption has been that the twentieth-century improvement of health standards and the associated decline in death rates will continue everywhere, especially in less developed nations. The possibility of massive die-offs was not considered. To give demographers their due, such dire eventualities are impossible to predict and therefore could not reasonably be included in projections. In 1992, this practice was changed when the United Nations included an assessment of the impact of AIDS in Africa, in which higher death rates and lower growth rates would prevail in several hard-hit nations, especially Uganda, Zaire, Tanzania, and Zambia, each of which was projected to suffer 1.1 to 1.8 million extra deaths by 2005; see United Nations, 1992, "The 1992 revision of world population prospects," *Population Newsletter* 54 (Dec.). But until the relatively predictable increases in deaths from AIDS loomed, such possibilities were never even mentioned. Past trends in fertility and mortality were blithely expected to continue into the future without a hitch.

While all projections should be interpreted with caution because so much is not considered, they do provide useful information on how a population is likely to change in future decades. And over the years, projections have become much more sophisticated and much more accurate, in part because information gathering about populations has improved greatly.

114. P. Brown, 1994, "Condoms before economics," *New Scientist,* 17 September, 11.

115. The number is somewhat controversial, although most family planning professionals subscribe to it. See J. Bongaarts, 1991, "The KAP-gap and the unmet need for contraception," *Population and Development Review* 17 (June):293–313; Robey et al., 1993; S. W. Sinding, J. A. Ross, and A. G. Rosenfield, 1994, "Seeking common ground: Unmet need and demographic goals," *International Family Planning Perspectives* 20 (March):23–27ff; R. Dixon-Mueller and A. Germain, 1992, "Stalking the elusive 'unmet need' for family planning," *Studies in Family Planning* 23 (Sept./Oct.):330–335.

116. F. Pearce, 1994, "Welcome to the contraception cafe," *New Scientist,* 1 October, 12–13.

117. R. McNamara, 1994, "A global population policy to advance human development in the 21st century," paper presented Jan. 26 at the Meeting of Eminent Persons on Population and Development (a preparatory meeting for the International Conference on Population and Development, 1994).

118. M. Potts, 1990, "Meeting the challenge of a growing population," *People* (International Planned Parenthood Federation, London) 17(3):5–7. "Modern contraceptive usage" excludes people who use traditional methods such as withdrawal and the rhythm method; use of all methods in developing nations excluding China in the early 1990s was about 42 percent.

119. *Populi,* October 1994; A. Cowell, 1994, "Conference on population has hidden issue: Money," *The New York Times,* 12 September.

120. Calculated from data on development assistance from World Bank, 1994, *World Development Report 1994* (New York: Oxford University Press).

Chapter Four

1. *Time,* 22 February 1992. For a detailed account of Vatican involvement in U.S. population policy, see S. D. Mumford, 1994. *The Life and Death of NSSM 200* (Research Triangle Park, N.C.: Center for Research on Population and Security).

2. Claiming as an excuse that the international agencies were supporting abortions, Reagan first reduced U.S. contributions, then totally cut off funds to the International Planned Parenthood Federation and the United Nations Population Fund (UNFPA). Funding for family planning continued through other channels, mainly bilateral programs through US AID (Agency for International Development), but at diminished levels.

3. M. S. Teitelbaum, 1993, "The population threat," *Foreign Affairs,*

Spring, 63–78; L. R. Brown and J. Jacobson, 1986, "Our demograph-ically divided world," *Worldwatch Report 74* (Worldwatch Institute, Washington, D.C.).

4. P. R. Ehrlich and A. H. Ehrlich, 1990, *The Population Explosion* (New York: Simon & Schuster), 195.

5. P. J. Donaldson and A. O. Tsui, 1990, "The international family plan-ning movement," *Population Bulletin* 45 (Nov.):3.

6. R. McNamara, 1994, "A global population policy to advance human development in the 21st century," paper presented Jan. 26 at the Meeting of Eminent Persons on Population and Development (a pre-paratory meeting for the International Conference on Population and Development, 1994).

7. S. Horiuchi, 1992, "Stagnation in the decline of the world population growth rate during the 1980s," *Science* 257 (7 August):761–765; Digest section, 1991, "China's fertility patterns closely parallel recent national policy changes," *International Family Planning Perspectives* 17 (June): 75–76. The 1980s fertility rise in the United States was partly due to a "baby boom echo" as millions of boomers began reproducing. Many of them had deferred earlier childbearing as more women sought ca-reers. In addition, increasing numbers of immigrants who preferred larger families were contributing to the nation's birth rate.

8. S. A. Holmes, 1993, "Clinton ready to back global birth controls," *The New York Times,* 12 May; United States Mission to the United Nations press release #63-(93), 11 May 1993; A. de Sherbinin and S. Kalish, 1993, "Clinton administration charting its own course on population issues," *Population Today* (Population Reference Bureau, Washington, D.C.) 21 (Dec.):12.

9. J. Tagliabue, 1992, "Clinton is warned by Vatican paper," *The New York Times,* 8 November.

10. What amounted to a campaign against provision of legal abortion in the United States was launched by Reagan and pursued by his succes-sor, George Bush, both of whom vetoed or administratively rescinded any provision of federal funding for abortions for welfare recipients or for members of the armed services in service hospitals (even at their own expense). They also transparently chose Supreme Court justices with an eye to overturning the 1973 decision that had legalized abor-tion in the United States.

11. For details on the political and medical history and epidemiology of abortion in the United States and Europe, see M. Potts, P. Diggory, and J. Peel, 1977, *Abortion* (Cambridge, Eng.: Cambridge University Press).

12. E.g., A. Gore, 1992, *Earth in the Balance* (Boston: Houghton Mifflin).

13. P. Martin and E. Midgley, 1994, "Immigration to the United States: Journey to an uncertain destination," *Population Bulletin* (Population Reference Bureau, Washington, D.C.) 49 (Sept.):2.

14. P. R. Ehrlich, L. Bilderback, and A. H. Ehrlich, 1981, *The Golden Door; International Migration, Mexico and the United States* (New York: Wideview Books).

15. Potts et al., 1977; L. Lader, 1973, *Abortion II: Making the Revolution* (Boston, Beacon); L. Lader, 1991, *RU-486* (Reading, Mass.: Addison-Wesley).

16. E.g., M. McGrory, 1993, "What to do about parents of illegitimate children," *Washington Post Weekly,* 15–21 February, 25.

17. S. A. Holmes, 1994, "Birthrate for unwed women up 70% since '83, study says," *The New York Times,* 20 July, 1.

18. B. D. Whitehead, 1993, "Dan Quayle was right," *Atlantic Monthly,* April, 47–84; S. J. Zuravin, 1991, "Unplanned childbearing and family size: Their relationship to child neglect and abuse," *Family Planning Perspectives* 23:155–161.

19. Obviously, the best approach is prevention, which is one focus of the Clinton administration's reform proposals, although limitation of time on welfare was another component; See J. Klein, "Learning how to say no," *Newsweek,* 13 June, 29.

20. F. Pearce, 1994, "Welcome to the contraceptive cafe," *New Scientist,* 1 October, 12–13.

21. See P. R. Ehrlich, A. H. Ehrlich and J. P. Holdren, 1977, *Ecoscience: Population, Resources, Environment* (San Francisco: W. H. Freeman & Co.), ch. 13.

22. See ibid., ch. 10 and references therein.

23. J. Parsons, 1971, *Population Versus Liberty* (London: Pemberton Books).

24. For more on these issues, see G. Daily, A. Ehrlich, and P. Ehrlich, 1994, "Optimum human population size," *Population and Environment* 15(6):469–475.

25. Joint Statement by Fifty-eight of the World's Scientific Academies, New Delhi, India, October 24–27, 1993.

26. Interview on 19 December 1994.

27. L. Orleans, 1973, "Family planning developments in China, 1960–1966; Abstracts from medical journals," *Studies in Family Planning* 4:8; L. Orleans, 1971, "China: Population of the People's Republic," *Population Bulletin* 27:6.

28. C. Walford, 1878, "The famines of the world: Past and present," *Royal Statistical Society Journal* 41:433–526.

29. Digest section, 1991, "China's fertility patterns closely parallel recent national policy changes," *International Family Planning Perspectives* 17(June): 75–76; L. Tabah, 1989, "World population at the turn of the century," *Population Studies,* no. 111, ST/ESA/SER.A/111 (New York: United Nations); G. Feeney, 1994, "Fertility decline in East Asia," *Science* 266:1518–1523.

30. For a description of that bureaucracy and its operations, see Mi Guo-qing, 1994, "Family planning services," *Reproduction and Contraception* 5:6–12 (supplement).

31. Zhu Yao-hua, 1994, "Family planning in China," *Reproduction and Contraception* 5:2 (supplement). Zhu is in the Department of Science and Technology of the State Family Planning Commission, so this represents an official view.

32. Anonymous, 1994, "Population and development in China: Facts

and figures," *China Population Today* 11 (July), special issue for ICPD.

33. Digest section, 1991; Zeng Yi, Tu Ping, Guo Liu, Xie Ying, 1991, "A demographic decomposition of the recent increase in crude birth rates in China," *Population and Development Review* 17(1):415–438.

34. Gita Sen, personal communication, 14 October 1994.

35. Interviews in and around Beijing conducted in December 1994; our informants wished to remain anonymous.

36. Xiao Bilian, 1994, "Advances in contraception in China," *Reproduction and Contraception* 5:17 (supplement); Anonymous, 1994, "Husbands share responsibility," *China Population Today* 11(3):19.

37. S. Ramasundaram, 1994, "A positive approach to population control," *The Hindu* (Madras), 6 Sept.; Anonymous, 1994, "Not by contraception alone," *Indian Express* (Madras), 14 Feb.; S. Ramasundaram, 1994, "Tamil Nadu Experience," *The Hindu* (Madras), 25 Oct.; Expert Group on Population Policy, 1994, *Draft National Population Policy,* submitted May 21 to Ministry of Health and Family Welfare, Government of India, New Delhi.

38. Mi Guo-qing, 1994, "Family planning services," *Reproduction and Contraception* 5:12 (supplement).

39. S. WuDunn and N. D. Kristof, 1994, "China's rush to riches," *The New York Times Magazine,* 4 September, 38–41ff.

40. Vice Minister Jiang of the State Family Planning Commission, personal communication, April 1994.

41. J. Kahn, 1994, "Creating a wonder for China: Family car," *The Wall Street Journal,* 21 November.

42. E.g., D. Pirages and P. Ehrlich, 1972, "If all Chinese had wheels," *The New York Times,* 16 March.

43. R. Ornstein and P. R. Ehrlich, 1989, *New World/New Mind* (New York: Doubleday).

44. Pirages and Ehrlich, 1972.

45. L. Grant (ed.), 1992, *Elephants in the Volkswagen: Facing the Tough Questions about our Overcrowded Country* (New York: W. H. Freeman & Co.).

46. C. Haub and M. Yanagishita, 1994, *1994 World Population Data Sheet* (Washington, D.C.: Population Reference Bureau). As in other chapters, this is the source of all current population data unless otherwise noted.

47. Martin and Midgley, 1994.

48. D. A. Ahlburg, 1993, "The Census Bureau's new projections of the US population," *Population and Development Review* 19 (March): 159–174.

49. D. J. van de Kaa, 1987, "Europe's second demographic transition," 42 (March):1, Population Reference Bureau, Washington, D.C.

50. Haub and Yanagishita, 1994; M. Delgado Perez and M. Livi-Bacci, 1992, "Fertility in Italy and Spain: The lowest in the world," *Family Planning Perspectives* 24 (July/Aug.):162–171; *The Economist,* 1991, "The missing children," 3 August, 43–44.

51. M. Specter, 1994, "Climb in Russia's death rate sets off population implosion," *The New York Times,* 6 March, 1; L. Hockstader, 1994, "Faltering vital signs," *Washington Post Weekly,* 14–20 March, 8.

52. Digest section, 1993, "Low national fertility a concern to Japanese couples; half have two children and most want no more," *Family Planning Perspectives* 25 (Dec.):276–277.

53. News section, 1994, "The demographic slant," *Populi* 21 (Sept.):5.

54. The other major exceptions are the oil-rich Arab nations.

55. E.g., J. Kavanaugh, 1967, *A Modern Priest Looks at His Out-dated Church* (New York: Pocket Books); B. G. Harrison, 1994, "Arguing with the Pope," *Harper's Magazine,* April, 50.

56. The latter commandments, if obeyed, would generally reduce fertility. They made sense in the Middle Ages, since the Church was interested in acquiring the property of people who lacked male heirs and would likely leave their worldly goods to the Church; see J. Goody, 1983, *The Development of Marriage and the Family in Europe* (New York: Cambridge University Press). Priestly celibacy also helped retain wealth within the Church, since celibate priests had no legitimate children who could inherit their possessions; see M. Harris and E. B. Ross, 1987, *Death, Sex, and Fertility* (New York: Columbia University Press), 86. Some relatively antinatalist positions of the Church may have been compromises with the desires of parishioners to avoid unwanted offspring. Abortion up to the time of "quickening" (sixteenth to eighteenth week of pregnancy, when movement can be detected) was not condemned until well into the 1800s, although contraception was banned in the thirteenth century; see J. T. Noonan, 1965, *Contraception: A History of Its Treatment by the Catholic Theologians and Canonists.* (Cambridge, Mass: Harvard University Press), 231.

57. Quoted in J. Updike, 1993, "Even the Bible is soft on sex," *New York Times Book Review,* 20 June.

58. E. O. Wilson, quoted in J. Horgan, 1994, "Revisiting old battlefields," *Scientific American,* April, 36–41.

59. E. Eckholm, 1994, "Wellspring of priests dries, forcing parishes to change," *The New York Times,* 30 May; P. Steinfels, 1993, "The Church faces the trespasses of priests," *The New York Times,* 27 June; H. Shulruff, 1993, "His specialty: Sex abuse suits against priests," *The New York Times,* 25 June; R. Ostling, 1993, "Sex and the single priest," *Time,* 5 July, 48–49; K. Woodward, 1993, "The sins of the fathers . . . ," *Newsweek,* 12 July, 57.

60. E.g., C. F. Westoff and L. L. Bumpass, 1973, "The revolution of birth control practices of U.S. Roman Catholics," *Science* 179:41–44; C. F. Westoff and E. F. Jones, 1979, "The end of 'Catholic' fertility," *Demography* 16:209–217. For a somewhat contrasting view, see W. D. Mosher and G. E. Hendershot, 1984, "Religion and fertility," *Demography,* May.

61. Judith Blake, 1984, "Catholicism and fertility: On attitudes of young Americans," *Population and Development Review* 10:329–340.

62. W. D. Mosher, L. B. Williams, and D. P. Johnson, 1992, "Religion and

fertility in the United States: New patterns," *Demography* 29: 199–220; F. Althaus, 1992, "Differences in fertility of Catholics and Protestants are related to timing and prevalence of marriage," *Family Planning Perspectives* 24:234–235.

63. C. Goldscheider and W. D. Mosher, 1991, "Patterns of contraceptive use in the United States: The importance of religious factors," *Studies in Family Planning* 22:102–106.

64. New York Times/CBS News Poll, published in *The New York Times,* 1 June 1994.

65. S. K. Henshaw and J. Silverman, 1988, "The characteristics and prior contraceptive use of U.S. abortion patients," *Family Planning Perspectives* 20:158–168.

66. Haub and Yanagishita, 1994.

67. The Vatican found the draft document unacceptable because it "linked family planning to contraception." The reactions of Italian newspapers were enlightening. Rome's *La Repubblica* said, "The Pope's war against the United Nations goes on non-stop." Milan's *Il Giornale* wrote that "on Papal orders the Vatican has launched a real offensive against the United Nations." The Pope claimed the UN was destroying the family and called for a "Maginot Line" against the conference, but insisted that it must be "more effective" than the original French fortifications, which were flanked by the Germans in 1940. (The quotes are from the following sources: Reuters, 1994, "Pope hits out at U.N. population conference plans" 14 April; Reuters, 1994, "Angry Pope criticises U.N. population conference," 17 April; Reuters, 1994, "Pope takes off gloves against U.N. conference," 18 April; Reuters, 1994, untitled, 21 April; Reuters, 1994, "Vatican pressure felt at population meeting. Pope says U.N. population conference destructive," 25 April; M. L. Usdansky, 1994, "Catholic leaders balk at birth control, abortion," *USA Today,* 18 July.)

68. *San Francisco Chronicle,* 25 April 1994.

69. *USA Today,* 18 July, 1994.

70. "A heavyweight contest," *Newsweek,* 6 June 1994.

71. G. Niebuhr, 1994, "Forming earthly alliances to defend God's kingdom," *The New York Times,* 28 August.

72. In the United States and some other nations, fear of change is centered in the "Catholic right," especially in a series of conservative Catholic organizations; see *Conscience: A Newsjournal of Prochoice Catholic Opinion* 15 (Spring 1994). In the U.S., the Catholic right consists of less than 0.5 percent of Catholics. Nevertheless, these conservatives have become disproportionately influential in both Church affairs and lay politics. Their strength comes from their "rapidly-deepening influence over two centers of power and resources: the clerical authority of the US Catholic Conference/National Conference of Catholic Bishops and the enormous wealth of the Knights of Columbus." The right is embodied in a series of organizations, including the Knights of Malta, Opus Dei, Tradition, Family and Property, the Cardinal Mindszenty Foundation, Catholics United for the Faith, the National Committee

of Catholic Laymen, Sword of the Spirit, Word of God, and a series of right-wing antiabortion organizations.

In the United States and elsewhere, organizations like Opus Dei hope to turn back the clock to the good old days of clerical states and absolute obedience. Tradition, Family, and Property is "a literally reactionary movement that views even Pope John Paul II as a left-wing apostate, [and] harks back to the 'Holy Inquisition' as Catholicism's last great moment"; in Steve Askin, 1994, *Conscience* 15(1):6.

These groups want, among other things, to reverse the Vatican II reforms and stop the movement to give women the same rights as men within the Church. Their activities are worldwide. For instance, when one of us (PRE) gave an invited lecture to the newly formed Costa Rican National Academy of Sciences, a venomous attack from Opus Dei was immediately elicited in a local newspaper; see Enrique Vargas Soto, 1994, "Paul Ehrlich y el Tecnológico," *La Nación*, 23 March. Besides blatantly misrepresenting what Ehrlich was doing in Costa Rica and the content and purpose of his lecture, Vargas relied on the authority of ancient and discredited Catholic sources, especially demographer Colin Clark (notable mostly for predicting in 1969 that, because of its huge population size and rapid growth, India would be the most powerful country in the world within a decade!). Catholic members of the Costa Rican Academy responded to the personal attack; see Eugenia M. Flores y Pedro Morera, 1994, "Paul Ehrlich y el Tecnológico," *La Nación*, 26 March.

73. S. D. Mumford, 1984, *American Democracy and the Vatican: Population Growth and National Security* (Amherst, N.Y.: Humanist Press).
74. G. Sen, at Royal Swedish Academy/Volvo Environmental Prize Symposium, 14 October 1994.
75. *National Catholic Register*, 8 November 1992.
76. K. D. Holl, G. C. Daily, and P. R. Ehrlich, 1993, "The fertility plateau in Costa Rica: A review of causes and remedies," *Environmental Conservation* 20:317–323; *Repubblica*, 25 January 1992.
77. V. M. Gómez, 1989, "Fertility change in Costa Rica, 1964–1986," Ph.D. thesis, University of Wisconsin, Madison; J. Madrigal, D. Sosa, and M. Gómez, 1992, *El embarazo no deseado en Costa Rica* (San José, Costa Rica: Asociación Demográfica Costarricense).
78. "The Cardinal fights again," *The Economist*, 20 August 1994.
79. *The Economist*, 20 August 1992, p. 32.
80. H. Rowen, 1994, "It's about money," *Washington Post Weekly*, 25–31 July.
81. Quoted in Rowen, 1994. The material in the rest of the paragraph is based on this source.
82. R. Smothers, 1994, "Abortion doctor and bodyguard slain in Florida," *The New York Times*, 30 July. This article reported that a Roman Catholic priest in Mobile, Alabama, David Trosch, sought to place an advertisement in newspapers calling the killing of abortion doctors "justifiable homicide." He was suspended by the Church, and the National Conference of Catholic Bishops said the shooting "makes a

mockery of everything we stand for." But, of course, the doctrine that aborting a conceptus is the same as murdering an independent individual is a root cause of the violence against those involved in abortion. The media attention Trosch has received may lead to future violence, since it has given him a platform to spread his views, which are quite logical if one accepts the preposterous premise that the life of a human zygote is equivalent to the life of a child or an adult individual. See also L. Belkin, 1994, "Kill for life?," *The New York Times Magazine,* 30 Oct., 46–51ff; J. Kifner, 1994, "Gunman kills 2 at abortion clinics in Boston suburb," *The New York Times,* 31 Dec. Five others were wounded in that attack, and the suspect made another attempt the next day in Norfolk, Virginia, before being arrested.

83. Mumford, 1984, ch. 10.
84. H. P. David and S. Pick de Weiss, 1992, "Abortion in the Americas," in *Reproductive Health in the Americas* (Washington, D.C.: Pan American Health Organization).
85. J. M. Paxman, A. Rizo, L. Brown, and J. Benson, 1993, "The clandestine epidemic: The practice of unsafe abortion in Latin America," *Studies in Family Planning* 24:205–226.
86. M. Weisner, 1990, "Induced abortion in Chile, with references to Latin American and Caribbean countries," paper presented at the annual meeting of the Population Association of America, Toronto, 3–5 May.
87. Paxman et al., 1993.
88. S. Singh and D. Wulf, 1991, "Estimating abortion levels in Brazil, Colombia, and Peru, using hospital admissions and fertility survey data," *International Family Planning Perspectives* 17(1):8–13.
89. O. Paz, 1982, "Impresiones sobre la realidad," *El Comercio,* 17 July, 11.
90. Sperm and eggs are every bit as much human life as zygotes. Each sperm and egg contains a complete set of chromosomes, each zygote two sets. There is no reason to consider the single-set stage of the continuous life cycle as any less "alive" or "human" than the two-set stage. In mosses, the plant you see is the one-set stage; the tiny reproductive cells have two sets. When a "person" begins is purely and simply an issue of cultural choice and legal decision.
91. Frances Kissling and her colleagues do not stand alone. For instance, we were sent a copy of a letter dated 25 July 1994 to John Paul II from the Italian branch of the World Wide Fund for Nature (Fondo Mondiale per la Natura). The letter urged the Pope to consider the seriousness of the population problem and to permit couples to practice "safe and effective contraception." There is considerable dissent within the hierarchy itself, and some elements within that hierarchy are not as pronatalist as John Paul II's group in the Vatican. In November 1991, the U.S. Catholic Bishops published a paper that stated: "Even though it is possible to feed a growing population, the ecological costs of doing so ought to be taken into account. . . . Our mistreatment of the natural world diminishes our own dignity and sacredness, not only

because we are destroying resources that future generations of humans need, but because we are engaging in actions that contradict what it means to be human"; quoted in R. Beck, 1992–1993, "Religions and the environment: Commitment high until U.S. population issue raised," *Social Contract* 3:75–89.

92. Pontificia Academia Scientiarum, *Population and Resources: A Report* (Milan, Italy: Vita e Pensiero—Largo A. Gemelli). Reported in *The New York Times*, 16 June 1994.

93. The author is Edward B. Grothus of Los Alamos; a version was published in *Playboy* 40(6), 1993. The version reproduced here was sent to us by the author on 6 February 1994.

94. Positions of U.S. religious leadership from Beck, 1992–1993.

95. PBS, 25 August 1994.

96. J. Weeks, 1988, "The demography of Islamic nations," *Population Bulletin* 43:4. TFRs in Middle Eastern countries around 1990 ranged between 4.2 and 7.6, and contraceptive usage rates exceeded 35 percent only in Turkey, Egypt, and Bahrain; see A. Omran and F. Roudi, 1993, "The Middle East population puzzle," *Population Bulletin* 48 (July):1.

97. Quoted in Beck, 1992–1993.

98. N. el-Saadawi, 1990, "Woman must own her body," *People* (IPPF) 17(4):16.

99. For example, Betsy Hartmann, 1992, "Population control in the new world order," speech presented June 6 at the Forum on Population Policies, Women's Health and Environment, UNCED 92 Global Forum, Rio de Janeiro, Brazil; and Hartmann's presentation Feb. 20, 1993, in the 7th Annual Stanford Health Policy Forum on *Population Growth and the Environment: A Dangerous Trend?*, Stanford University Medical School, Stanford, Calif. Hartmann persists in spreading out-of-date negative information about the family planning program in Bangladesh, which she witnessed in the early 1970s. According to Edson Whitney of Dhaka ("Bangladesh and population," *The New York Times*, Letters, 11 Oct. 1994), "It is unconscionable to suggest that the priority given to family planning is somehow misdirected. The Bangladesh program is a success story, it is voluntary, and it continues to endeavor to meet the demands of those who request its services."

100. M. Mies and V. Shiva, 1993, *Ecofeminism* (London: Zed Books).

101. J. DeParle, 1994, "Census report sees incomes in decline and more poverty," *The New York Times*, 7 Oct.

102. Quoted in *Time*, 17 Oct. 1994.

103. Alan Guttmacher Institute, 1994, *Sex and America's Teenagers* (New York and Washington, D.C.).

104. *The American Almanac 1993–94* (Austin, Tex.: The Reference Press), 469.

105. Centered around a book by Charles Murray and Richard Herrnstein entitled *The Bell Curve* (New York: Free Press, 1994). For a detailed refutation of the sort of argument advanced in this book, see

P. Ehrlich and S. Feldman, 1977, *The Race Bomb: Skin Color, Prejudice, and Intelligence* (New York: New York Times Books).

106. P. R. Ehrlich, L. Goulder, and G. C. Daily, 1992, "Population growth, economic growth, and market economies," *Contention* 2:17–35.

107. Ehrlich, Bilderback and Ehrlich, 1981; G. Daily, A. Ehrlich, and P. Ehrlich, 1995, "Population and immigration policy in the U.S.," *Population and Environment* (in press).

108. We emphasize that controlling overconsumption is every bit as important as controlling population growth. More details can be found in P. R. Ehrlich and A. H. Ehrlich, 1991, *Healing the Planet* (Reading, Mass.: Addison-Wesley).

Chapter Five

1. The first human beings by our definition were the small-brained but upright australopithecines who lived almost 4 million years ago.

2. We would trace the antiquity of "humanity" back to *Australopithecus afarensis*, the first fully upright of our ancestors, the earliest known of which lived some 3.8 million years ago; see L. C. Aiello, 1994, "New Hadar fossil finds announced," *Institute of Human Origins Newsletter* 12(1):1–3. The recently discovered *A. ramidus* (see J. Wilford, 1994, "New fossils take science close to dawn of humans," *The New York Times*, 22 Sept.) may not have been fully upright, but if it were, it would extend humanity's history back to some 4.4 million years ago. Australopithecines were relatively small-brained, but to exclude them from humanity for that reason would necessarily leave out many university professors and politicians (among others) today and fail to recognize the substantial intellectual talents of relatively small-brained nonhuman primates today.

3. R. Lee and I. DeVore, 1968, *Man the Hunter* (Chicago: Aldine).

4. E.g., W. Tanner and A. Zihlman, 1976, "Women in evolution. Part 1: Innovations and selection in human origins," *Signs* 1:585–608.

5. J. Briggs, 1974, "Eskimo women: Makers of men," in C. Matthiasson (ed.), *Many Sisters: Women in Cross-cultural Perspective* (New York: Free Press) 261–304.

6. M. Harris, 1986, *Good to Eat: Riddle of Food and Culture* (New York: Simon & Schuster); M. Harris and E. B. Ross, 1987, *Death, Sex, and Fertility: Population Regulation in Preindustrial and Developing Societies* (New York: Columbia University Press), 22.

7. E. Deevey, 1960, "The human population," *Scientific American*, Sept.; P. R. Ehrlich, A. H. Ehrlich, and J. P. Holdren, 1977, *Ecoscience: Population, Resources, Environment* (San Francisco: W. H. Freeman & Co.); J-N. Biraben, 1979, "Essai sur l'évolution du nombre des hommes," *Population* 34:13–25.

8. Harris and Ross, 1987, p. 23.

9. M. Sahlins, 1972, *Stone Age Economics* (Chicago: University of Chicago Press); H. Kaplan et al., 1984, "Food sharing among the Ache hunter–gatherers of eastern Paraguay," *Current Anthropology* 25:113–115.

10. For more details, see J. Diamond, 1991, *The Rise and Fall of the Third Chimpanzee* (London: Radius). This book is a great read on the possible animal origins of human behaviors; highly recommended.

11. Ibid.

12. Ibid., p. 169.

13. Ibid.

14. R. Lee and I. DeVore, 1968, "Problems in the study of hunters and gatherers," in Lee and DeVore, *Man the Hunter*, p. 3.

15. Otto and Dorothy Solbrig, 1994, *So Shall You Reap* (Washington, D.C.: Island Press).

16. M. Cohen, 1977, *The Food Crisis in Prehistory: Overpopulation and the Origins of Agriculture* (New Haven: Yale University Press).

17. Similarly, groups that learned to cultivate plants or herd animals to supplement their gathering and hunting would be able to live in denser populations—a shift to which our highly social species seems predisposed. Furthermore, if people could be fed with less effort growing crops than gathering–hunting (as they might in some circumstances, especially where food was sufficiently sparse to require the group to move frequently), the shift to agriculture might have occurred without population growth simply on a cost-benefit basis.

18. J. D. Durand, 1967, "A long-range view of world population growth," *Annals of the American Academy of Political Science* 369:1–8.

19. Harris and Ross, 1993, pp. 50–55.

20. Data on comparison of work in contemporary simple and intensive agricultural groups from W. Minge-Klevana, 1980, "Does labor time decrease with industrialization? A survey of time allocation studies," *Current Anthropology* 21:279–287; and C. Ember, 1983, "The relative decline in women's contribution to agriculture and intensification," *American Anthropologist* 85:285–304.

21. The caste system played an interesting role in fostering this coexistence in India; see M. Gadgil and R. Guha, 1992, *This Fissured Land* (London: Oxford University Press).

22. W. Peters and L. Neuenschwander, 1988, *Slash and Burn: Farming in the Third World Forest* (Moscow, Idaho: University of Idaho Press).

23. The disease is caused by *Trypanosoma brucei* and other relatives of the organisms that cause sleeping sickness in human beings and, like them, is transmitted by tsetse flies *(Glossina* species). The trypanosomes cause a wasting sickness of cattle and kill horses, and have effectively barred domestic animals from much of equatorial Africa. A positive side effect of this, however, has been to prevent the decimation of biodiversity in wide areas.

24. P. Pingali, Y. Bigot, and H. Binswanger, 1987, *Agricultural Mechanization and the Evolution of Farming Systems in Sub-Saharan Africa* (Baltimore: Johns Hopkins University Press).

25. Some polycultural cropping may be prolonged for five years or more. See D. Harris, 1971, "The ecology of swidden cultivation in the upper Orinoco rain forest, Venezuela," *The Geographical Review* 61:475–495;

H. Ruthenberg, 1971, *Farming Systems in the Tropics* (Oxford: Clarendon Press).

26. W. Peters and L. Neuenschwander, 1988.

27. I. Wallerstein, 1974, *The Modern World System I: Capitalist Agriculture and the Origins of the European World Economy in the Sixteenth Century* (New York: Academic Press).

28. F. Bray, 1994, "Agriculture for developing nations," *Scientific American,* July, 30–37.

29. Ibid., p. 32.

30. National Research Council, 1989, *Alternative Agriculture* (Washington, D.C.: National Academy Press); D. Pimentel et al., 1978, "Benefits and costs of pesticide use in the United States," *BioScience* 28:778; D. Pimentel et al., 1989, "Environmental and economic impacts of reducing U.S. agricultural pesticide dependence" (manuscript).

31. D. Pimentel and W. Dazhong, 1990, "Technological changes in energy use in U.S. agricultural production," in C. Carroll, J. Vandermeer, and P. Rosset (eds.), 1990, *Agroecology* (New York: McGraw-Hill), 147–164; D. Pimentel and C. Hall (eds.), 1984, *Food and Energy Resources* (Orlando, Fla.: Academic Press). The numbers reported as the proportion of U.S. energy that goes into the agricultural enterprise depend heavily on where one places the boundaries of that enterprise (e.g., should the manufacture of aluminum cans for beer be included?). E. Heady and D. Christiansen estimate that about 1 percent is actually used in agricultural production itself; see "Potentials in producing alcohol from corn grain and residue in relation to prices, and use, and conservation," in Pimentel and Hall, 1984, p. 237.

32. H. Odum, 1971, *Environment, Power, and Society* (New York: Wiley), 115–116.

33. D. Grigg, 1993, *The World Food Problem,* 2nd ed. (Oxford, Eng., and Cambridge, Mass.: Blackwell).

34. The plan was named for the city where it was organized: Colombo, Ceylon.

35. V. Shiva, 1991, *The Violence of the Green Revolution* (London: Zed Books).

36. For a detailed summary and references on the material on the early history of the green revolution, see Ehrlich, Ehrlich, and Holdren, 1977, ch. 7.

37. No one knows how many people starved to death in this famine, since people normally die of other causes when weakened by hunger. For insight into the Bihar famine and India's overall food situation, see J. Drèze, 1990, "Famine prevention in India," in *The Political Economy of Hunger,* vol. 2; J. Drèze and A. Sen. (eds.), *Famine Prevention* (Oxford, Eng.: Oxford University Press), 13–122; for the question of how many died, see p. 55ff.

38. D. Tribe, 1994, *Feeding and Greening the World: The Role of International Agricultural Research* (Oxon, Eng.: CAB International).

39. L. R. Brown, H. Kane, and D. M. Roodman, 1994, *Vital Signs* (New York: Norton), 41.
40. J. Cohen, J. Williams, D. Plunknett, and H. Shands, 1991, "Ex-situ conservation of plant genetic resources: Global development and environmental concerns," *Science* 253:866–872.
41. Readers who are interested in more detail on the differences in agricultural systems among continents should consult the second edition of David Grigg's *World Food Problem*. This fine book contains a detailed account of the history of hunger and food production.
42. A fine description of climate and how it influences human affairs can be found in S. Schneider and R. Londer, 1984, *The Coevolution of Climate and Life* (San Francisco: Sierra Club Books). Stephen Schneider is an outstanding climatologist, and has done more than any other scientist to bring to the general public a balanced view of climatic issues such as global warming.
43. Food and Agriculture Organization, 1990, *Production Yearbook 1989* (Rome: FAO).
44. R. Montgomery and T. Sugito, 1980, "Changes in the structure of farms and farming in Indonesia between censuses, 1963–1973: The issues of inequality and near landlessness," *Journal of South East Asian Studies* 11:348–365.
45. Grigg, 1993.
46. Ehrlich, Ehrlich, and Holdren, 1977. For a more recent history of the green revolution and an account of its limitations, see Lester R. Brown and Hal Kane, 1994, *Full House* (New York: Norton).
47. G. Higgins, A. Kassam, L. Maiken, and M. Shah, 1981, "Africa's agricultural potential," *Ceres* 14:13–21; S. Gregory, 1969, "Rainfall reliability," in M. Thomas and G. Whittington (eds.), 1969, *Environment and Land Use in Africa* (London: Methuen), 57–82.
48. Grigg, 1993, p. 154.
49. Grigg, 1993.
50. E. H. Hartmans, 1983, "African food production: Research against time," *Outlook on Agriculture* 12:165–171.
51. P. Harrison, 1981, "The inequities that curb potential," *Ceres* 81:22–26. In 1970, about 80 percent of Latin America's farmland was in the 8 percent of the holdings that exceeded one hundred hectares.
52. E. Ortega, 1982, "Peasant agriculture in Latin America," *Cepal Review* 16:75–111.
53. U. P. Koehn, 1982, "African approaches to environmental stress: A focus on Ethiopia and Nigeria," in R. N. Barrett (ed.), *International Dimensions of the Environmental Crisis* (Boulder, Colo.: Westview Books).
54. Bray, 1994; Shiva, 1991.
55. Much of this section is based on G. C. Daily and P. R. Ehrlich, 1995, "Socioeconomic equity, sustainability, and Earth's carrying capacity." *Ecological Applications* (in press).

56. W. Fernandes and G. Menon, 1987, *Tribal Women and Forest Economy* (New Delhi, India: Indian Social Institute).
57. V. Singh, 1987, "Hills of hardship," *The Hindustan Times Weekly,* 18 January.
58. P. Dasgupta, 1993, *An Inquiry into Well-Being and Destitution* (Oxford, Eng.: Clarendon Press).
59. Harris and Ross, 1987, pp. 157–159.
60. John M. Brewster, 1950, "The machine process in agriculture and industry," *Journal of Farm Economics* 32:69–81.
61. W. Murdoch, 1990, "World hunger and population," in C. Carroll, J. Vandermeer, and P. Rosset (eds.), *Agroecology* (New York: McGraw-Hill), 3–20.
62. Wallerstein, 1974.
63. Y. Hayami and V. W. Ruttan, 1985, *Agricultural Development: An International Perspective* (Baltimore: Johns Hopkins Press); T. Dyson, 1994, "Population growth and food production: Recent global and regional trends," *Population and Development Review* 20(2):397–411.
64. Calculated from FAO figures on production and trade in *World Resources,* 1994, pp. 291–299. Since the FAO bases its figures on unmilled rice, whereas the USDA uses milled rice, FAO's figures tend to underestimate the share of cereals produced by developed nations as compared to USDA figures (90% of the world's rice is produced and consumed in Asia).
65. Or, more correctly, the "production possibility frontier."
66. Hayami and Ruttan, 1985.
67. E.g., K. A. Dahlberg, 1979, *Beyond the Green Revolution* (New York: Plenum).
68. Bray, 1994; H. M. Cleaver, Jr., 1972, "The contradictions of the green revolution," *American Economic Review* 62:177–186; W. P. Falcon, 1970, "The green revolution: Generations of problems," *American Journal of Agricultural Economics* 52:698–710; B. F. Johnston and J. Cownie, 1969, "The seed-fertilizer revolution and labor force absorption," *American Economic Review* 59:569–582; Shiva, 1991; C. R. Wharton, 1969, "The green revolution: Cornucopia or Pandora's box?," *Foreign Affairs* 47:464–476.
69. R. Naylor, 1994, "Herbicide use in Asian rice production," *World Development* 22:55–70.
70. Y. Hayami and M. Kikuchi, 1982, *Asian Village Economy at the Crossroads: An Economic Approach to Institutional Change* (Baltimore: Johns Hopkins Press).
71. Hayami and Ruttan, 1985.
72. Videotaped interview with S. Hurst, 14 December 1994, Madras.
73. K. Griffin, 1989, *Alternative Strategies for Economic Development* (New York: St. Martin's Press), 147.
74. Shiva, 1991.
75. Madhav Gadgil, personal communication, 1994; Shiva, 1991. By mixing rice and pulses, the amino-acid deficiencies of each are com-

pensated, and the protein quality of the diet is greatly enhanced (see, e.g., F. Lappé, 1971, *Diet for a Small Planet,* rev. ed. (New York: Ballantine Books).

76. Gadgil, personal communication, 1994; Shiva, 1991.
77. R. Sharma and T. Poleman, 1993, *The New Economics of India's Green Revolution: Income and Employment Diffusion in Uttar Pradesh* (Ithaca, N.Y.: Cornell University Press).
78. P. R. Ehrlich and A. H. Ehrlich, 1970, *Population, Resources, Environment: Issues in Human Ecology* (San Francisco: W. H. Freeman & Co.); Ehrlich, Ehrlich, and Holdren, 1977, ch. 7; Shiva, 1991.
79. E.g., Swaminathan and Sinha, 1986; Dyson, 1994.
80. F. Lappé and J. Collins, 1977, *Food First* (New York: Houghton Mifflin).
81. Drèze and Sen, 1991.
82. UNICEF, 1992, *State of the World's Children 1992* (New York: United Nations).
83. International agencies differ in their estimates of the numbers of undernourished people because they use somewhat different criteria for their calculations. The Food and Agriculture Organization of the United Nations (FAO) estimated that about 786 million people in developing regions were chronically undernourished in the period 1988–1990; see FAO, 1992, *Food and Nutrition: Creating a Well-fed World* (Rome: FAO); FAO, 1992, *World Food Supplies and Prevalence of Chronic Undernutrition in Developing Regions as Assessed in 1992* (Rome: FAO). The World Bank estimated that in the mid-1980s, some 920 million people in developing regions outside China were underfed, nearly half of them getting too little food to prevent stunting of growth or threats to health; see World Resources Institute, 1988, *World Resources 1988–89* (New York: Basic Books), ch. 4. R. Kates and V. Haarmann—in 1992, "Where the poor live: Are the assumptions correct?," *Environment* 34 (May):5–11, 25–28—cite a figure of over a billion for "energy deficient for work" derived from the World Hunger Program, based on World Bank estimates.
84. P. Svedberg, 1991, "Undernutrition in Sub-Saharan Africa: A critical assessment of the evidence," in J. Drèze and A. Sen, *The Political Economy of Hunger,* vol. 3, *Endemic Hunger* (Oxford, Eng.: Oxford University Press), 155–193.
85. L. R. Brown, 1988, "The changing world food prospect: The nineties and beyond, *Worldwatch Report 85* (Worldwatch Institute, Washington, D.C.), Nov. *See also* L. R. Brown et al., *State of the World 1989, State of the World 1990,* and ch. 10 of *State of the World 1994* (New York: Norton).
86. Brown, Kane and Roodman, 1994, pp. 26–27; U.S. Department of Agriculture, Foreign Agriculture Service, 1994, *World Agricultural Production* (WAP 9-94), Sept.
87. Dyson, 1994.
88. Brown et al., 1994, ch. 10.

89. Brown, Kane, and Roodman, 1994, pp. 30–31; World Resources Institute, 1994, p. 352, About 12 to 15 mmt of the total fisheries harvest is produced in freshwater fisheries. Included in the marine and freshwater totals are some 13 mmt produced in aquaculture.

90. J. R. McGoodwin, 1990, *Crisis in the World's Fisheries* (Stanford, Calif: Stanford University Press).

91. A. Swardson, 1994, "A loss that's deeper than the ocean: Overharvesting is devastating the world's fish population," *Washington Post Weekly,* 24–30, October, 8.

92. J. H. Ryther, 1969, "Photosynthesis and fish production in the sea," *Science* 166:72–76; McGoodwin, 1990. There is considerable doubt today about whether MSY has been a useful management concept; see, e.g., D. Ludwig, R. Hilborn, and C. Walters, 1993, "Uncertainty, resource exploitation, and conservation: Lessons from history," *Science* 260:17, 36. Whether it is or not is not germane to the level of discussion here.

93. Brown, Kane, and Roodman, 1994, pp. 32–35.

94. FAO, 1993, *Marine Fisheries and the Law of the Sea: A Decade of Change,* Fisheries circular no. 853, Rome.

95. Brown, Kane, and Roodman, 1994, pp. 32–35.

96. Swardson, 1994.

97. World Resources Institute, 1994, *World Resources 1994–95* (New York: Oxford University Press), 184–185.

98. D. Ludwig, R. Hilborn, and C. Walters, 1993, pp. 17, 36.

99. FAO, 1993.

100. P. Weber, 1994, "Safeguarding oceans," in L. R. Brown et al., 1994 (New York: Norton), ch. 3, 41–60.

101. T. Goreau, personal communication; P. Weber, 1994, pp. 122–123.

102. L. R. Brown and H. Kane, 1994, *Full House* (New York: Norton).

103. J. H. Cushman Jr., 1994, "Panel recommends virtual end to fishing fleet in Georges Bank," *The New York Times,* 27 October, 1.

104. World Resources Institute, 1992.

105. N. Myers, 1983, *A Wealth of Wild Species: Storehouse for Human Welfare* (Boulder, Colo.: Westview Press).

106. E. Wilson, 1992, *The Diversity of Life* (Cambridge, Mass.: Harvard University Press).

107. FAO, 1993.

Chapter Six

1. Nutritional security is defined as "physical and economic access to balanced diets and safe drinking water to all children, women, and men at all times," after M. Swaminathan, 1987, "Building national and global nutrition security systems," in M. Swaminathan and S. Sinha (eds.), *Global Aspects of Food Production* (Oxford, Eng.: Tycooly International).

2. This chapter is based in part on P. Ehrlich, A. Ehrlich, and G. Daily,

1993, "Food security, population, and environment," *Population and Development Review* 19(1):1–32.

3. Thomas Malthus, 1798, "An essay on the principle of population as it affects the future improvement of society . . . (published anonymously). A later (1803) expanded version was reprinted in Philip Appleman (ed.), 1976, *An Essay on the Principle of Population* (New York: Norton). For an optimist's view, see the interesting article by M. K. Bennett, 1949, "Population and food supply: The current scare," *Scientific Monthly* 68 (January), reprinted in 1992, *Population and Development Review* 18(2):341–358. Bennett was totally misled by demographic projections indicating a world population of 3.3 billion in the year 2000, and was completely unaware of the ecological dimensions of the agricultural situation. But he was confident about feeding the expected population, even though he also underestimated (as did many others) the success in increasing production of what became known as the "green revolution."

4. United Nations Children's Fund (UNICEF), 1992, *State of the World's Children, 1992* (New York: United Nations).

5. FAO, 1992, *Food and Nutrition: Creating a Well-fed World* (Rome: FAO); FAO, 1992, *World Food Supplies and Prevalence of Chronic Undernutrition in Developing Regions as Assessed in 1992* (Rome: FAO); R. Kates and V. Haarmann, 1992, "Where the poor live: Are the assumptions correct?," *Environment* 34(May):5–11, 25–28.

6. FAO, 1992, *World Food Supplies;* R. Stone, 1992, "A snapshot of world hunger," *Science* 257:876.

7. L. Brown and E. Wolfe, 1984, "Soil erosion: Quiet crisis in the world economy," *Worldwatch Paper 60* (Worldwatch Institute, Washington, D.C.; A. Savory, 1994, "Will we be able to sustain civilization?," *Population and Environment* 16(2):139–147.

8. The net drain on the Ogallala Aquifer, which underlies the American Great Plains, is roughly 2 trillion gallons per year; calculation based on figures in M. Reisner, 1986, *Cadillac Desert: The American West and Its Disappearing Water* (New York: Viking). *See also* the excellent overviews of freshwater resources in S. Postel, 1992, *Last Oasis: Facing Water Scarcity* (New York: Norton) and in P. Gleick (ed.), 1993, *Water in Crisis: A Guide to the World's Fresh Water Resources* (New York: Oxford University Press).

9. P. Ehrlich and A. Ehrlich, 1981, *Extinction: The Causes and Consequences of the Disappearance of Species* (New York: Ballantine Books); E. Wilson, 1992, *The Diversity of Life* (Cambridge, Mass.: Harvard University Press).

10. P. Vitousek, P. R. Ehrlich, A. H. Ehrlich, and P. Matson, 1986, "Human appropriation of the products of photosynthesis," *BioScience* 36 (June):368–373.

11. World Resources Institute, 1992, *World Resources 1992–93* (New York: Oxford University Press); L. Oldeman, V. Van Engelen, and J. Pulles, 1990, "The extent of human-induced soil degradation,"

Annex 5 of L. Oldeman et al., *World Map of the Status of Human-Induced Soil Degradation: An Explanatory Note,* rev. 2nd ed. (Waginengen, Netherlands: International Soil Reference and Information Centre [ISRIC]); L. R. Brown et al., 1990, *State of the World 1990* (New York: Norton); Brown and Wolfe, 1984; J. Aber and J. Melillo, 1991, *Terrestrial Ecosystems* (Philadelphia: Saunders).

12. S. Postel, 1990, "Water for Agriculture: Facing the Limits," *Worldwatch Paper 93* (Worldwatch Institute, Washington, D.C.); M. Falkenmark and C. Widstrand, 1992, "Population and water resources: A delicate balance," *Population Bulletin* 47(3):1–36.

13. National Research Council, Board on Agriculture, 1991, *Managing Global Genetic Resources* (Washington, D.C.: National Academy Press); D. Plucknett, N. Smith, J. Williams, and N. Anishetty, 1987, *Gene Banks and the World's Food* (Princeton, N.J.: Princeton University Press).

14. L. R. Brown, 1991, "Fertilizer engine losing steam," *World Watch* 4(5):32–33; Brown et al., 1990; V. Smil, 1991, "Population growth and nitrogen: An exploration of a critical link," *Population and Development Review* 17(4):569–601.

15. C. Francis, C. Flora, and L. King, 1990, *Sustainable Agriculture in Temperate Zones* (New York: Wiley-Interscience); P. Ehrlich and A. Ehrlich, 1991, *Healing the Planet* (Boston: Addison-Wesley).

16. R. Worrest and L. Grant, 1989, "Effects of ultraviolet-B radiation on terrestrial plants and marine organisms," in R. Jones and T. Wigley, *Ozone Depletion: Health and Environmental Consequences* (New York: Wiley), 197–206.

17. World Resources Institute, 1986, *World Resources 1986* (New York: Basic Books), ch. 10.

18. M. Parry, 1990, *Climate Change and World Agriculture* (London: Earthscan Publications); S. Schneider, 1989, *Global Warming: Entering the Greenhouse Century* (San Francisco: Sierra Club Books).

19. Ehrlich and Ehrlich, 1991.

20. L. R. Brown, H. Kane, and D. M. Roodman, 1994. *Vital Signs* (New York: W. W. Norton). See also original analysis in L. R. Brown, 1988, "The changing world food prospect: The nineties and beyond," *Worldwatch paper 85* (Worldwatch Institute, Washington, D.C.), Oct.

21. Food and Agriculture Organization of the United Nations, 1956–91, *Production Yearbook,* vols. 10–44 (Rome: FAO).

22. T. Dyson, 1994, "Population growth and food production: Recent global and regional trends," *Population and Development Review* 20(2):397–411.

23. Conversions also could reduce nutritional security by diverting land and other resources to the production of cash crops or locally consumed luxury crops that are inaccessible to the average person in the producer country.

24. The American Farmland Trust, 1993, "Twelve tabbed as most threatened agricultural regions in U.S.," news release. For more information, contact their national office at: 1920 N Street, NW, Suite 400; Washington, DC 20036.

25. FAO estimate cited in Brown et al., 1990, p. 65. One need only visit the outskirts of New Delhi or Manila to see this loss occurring at dramatic rates.

26. H. Colby, F. Crook, and S.-E. Webb, 1992, "Agricultural Statistics of the People's Republic of China, 1949–1990," *Statistical Bulletin* no. 844 (Washington, D.C.: United States Department of Agriculture).

27. P. Shenon, 1994, "Good Earth is squandered. Who'll feed China?," *The New York Times,* 21 September.

28. M. Imhoff et al., 1986, "Monsoon flood boundary delineation and damage assessment using space-borne imaging radar and Landsat data," *Photogrammatric Engineering and Remote Sensing* 53:405–413.

29. Ibid. This section also benefited from discussions with Imhoff in 1990.

30. J. Holdren, 1991, "Population and the energy problem," *Population and Environment* 12:231–255.

31. D. O. Hall, F. Rosillo-Calle, R. H. Williams, J. Woods, 1992, "Biomass for energy: Supply prospects," in T. B. Johansson, H. Kelly, A. K. N. Reddy, and R. H. Williams (eds.), *Renewable Energy: Sources for Fuels and Electricity* (Washington, D.C.: Island Press), 593–652.

32. See H. Braunstein, P. Kanciruk, R. Roop, F. Sharples, J. Tatum, K. Oakes, F. Kornegay, and T. Pearson, 1981, *Biomass Energy Systems and the Environment* (New York: Pergamon Press); J. Cook, J. Beyea, and K. Keeler, 1991, "Potential impacts of biomass production in the United States on biological diversity," *Annual Review of Energy and the Environment* 16:401–431; D. Pimentel, C. Fried, L. Olson, S. Schmidt, K. Wagner-Johnson, A. Westman, A. Whelan, K. Foglia, P. Poole, T. Klein, R. Sobin, and A. Bochner, 1984, "Environmental and social costs of biomass energy," *BioScience* 34(2):89–94.

33. These set-asides amount to some tens of millions of hectares. For comparison, the total area planted in cereals in 1991 amounted to over 700 million hectares; see FAO, 1992, *World Food Supplies.*

34. This section is adapted from G. Daily, 1994, "Restoring productivity to the world's degraded lands," *Morrison Institute for Population and Resource Studies Working Paper no. 52.*

35. R. Houghton, 1994, "The worldwide extent of land-use change," *BioScience* 44(5):305–313. See also A. Ehrlich, 1988, "Development and agriculture," in P. R. Ehrlich and J. P. Holdren (eds.), *The Cassandra Conference* (College Station, Tex.: Texas A & M University Press); P. Sanchez, 1976, *Properties and Management of Soils in the Tropics* (New York: Wiley-Interscience); J. Tivy, 1990, *Agricultural Ecology* (Essex, Eng.: Longman, Harlow).

36. See discussion of soil organisms in P. Ehrlich, A. Ehrlich, and J. Holdren, 1977, *Ecoscience* (San Francisco: W. H. Freeman & Co.), 640–641.

37. D. Hillel, 1991, *Out of the Earth: Civilization and the Life of the Soil* (New York: Free Press); D. Pimentel, U. Stachow, D. Takacs, H. Brubaker, A. Dumas, J. Meaney, J. O'Neil, D. Onsi, and D. Corzilius, 1992, "Conserving biological diversity in agricultural/ forestry systems," *BioScience* 42(5):354–362.

38. These estimates are from Brown and Wolfe, 1984. Like many published estimates (e.g., extent of world hunger, rates of soil erosion) related to the world food situation, these numbers have an illusory precision. There is, for example, enormous controversy about UNEP estimates of desertification; see F. Pearce, 1992, "Mirage of the shifting sands," *New Scientist,* 12 December, 38–42. What is indisputable is that substantial portions of Earth's surface critical to agriculture have been degraded already, and that degradation is continuing at a time when substantially increased food production will be needed.

39. C. Riskin, 1994, "Panel on environment and poverty," paper presented at the International Symposium on Social Development, Beijing, China, October.

40. Hillel, 1991.

41. *Critias* 111B, cited in J. D. Hughes, 1975, *Ecology in Ancient Civilizations* (Albuquerque: University of New Mexico Press), 61.

42. Hillel, 1991, p. 104.

43. What follows is largely summarized from M. H. Glantz, 1994, "The West African Sahel," in M. H. Glantz (ed.), 1994 *Drought Follows the Plow* (Cambridge, Eng.: Cambridge University Press), 33–43.

44. For an early study that focused on drought as a cause more than as an effect, see Secretariat, United Nations Conference on Desertification, 1977, *Desertification: Its Causes and Consequences* (Oxford, Eng.: Pergamon Press).

45. For a detailed government interpretation, see "Desertification in the Eghazer and Azawak region: Case study presented by the government of Niger," in J. Mabbutt and C. Floret (eds.), 1980, *Case Studies of Desertification* (Paris: UNESCO).

46. A. Grainger, 1982, *Desertification* (London: International Institute for Environment and Development).

47. J. L. Cloudsley-Thompson, 1974, "The expanding Sahara," *Environmental Conservation* 1:5–13.

48. Glantz, 1994, *Drought Follows the Plow.*

49. D. J. Campbell, 1977, "Strategies for coping with drought in the Sahel: A study of recent population movements in the Department of Maradi, Niger," Ph.D. dissertation, Clark University, Worcester, Mass., 79.

50. A. Magalhães and P. Magee, 1994, "The Brazilian Nordeste (Northeast)," in Glantz (ed.), *Drought Follows the Plow,* pp. 59–76.

51. J. McCann, 1994, "Ethiopia," and W. Swearingen, 1994, "Northwest Africa," in Glantz (ed.), *Drought Follows the Plow,* pp. 103–115 and 117–133, respectively.

52. C. Vogel, "South Africa"; I. Zonn, M. Glantz, and A. Rubenstein, "The virgin lands scheme in the former Soviet Union"; and R. L. Heathcote, "Australia," all in Glantz (ed.), 1994, *Drought Follows the Plow,* pp. 151–170, 135–150, and 91–102, respectively.

53. D. Sheridan, 1981, *Desertification of the United States* (Washington, D.C.: Council on Environmental Quality), 121; *see also* Glantz (ed.), 1994, *Drought Follows the Plow,* pp. 21–30.

54. C. Maternowska, 1994, "Real lives 1: Haiti," *People and Planet* 3(4):16–19.
55. J. Terborgh, 1994, *Where Have All the Birds Gone?* (Princeton, N.J.: Princeton University Press), 165.
56. Maternowska, 1994, p. 19.
57. Grainger, 1982.
58. F. Braudel, 1972, *The Mediterranean and the Mediterranean World in the Age of Philip II*, vol. 1 (New York: Harper & Row), 243.
59. R. Repetto, 1994, *The "Second India" Revisited: Population, Poverty, and Environmental Stress over Two Decades* (Washington, D.C.: World Resources Institute), 88.
60. M. Hossain, 1994, "Asian population growth is overtaking rice output," *International Herald Tribune*, 18 March; population growth rate from C. Haub and M. Yanagishita, 1994, *World Population Data Sheet 1994* (Washington, D.C.: Population Reference Bureau). As elsewhere, this is the source of most current population data.
61. Anonymous, 1994, "Green revolution blues," *International Agricultural Development* May/June, 7–9. See other articles in this issue as well.
62. Hossain, 1994. See also J. Walsh, 1991, "Preserving the options: Food productivity and sustainability," Consultative Group on International Agricultural Research (CIGIAR), Issues in Agriculture no. 2.
63. R. Roy, 1987, "Trees: Appropriate tools for water and soil management," in B. Glaeser (ed.), *The Green Revolution Revisited* (Boston: Allen and Unwin), 124.
64. Postel, 1990.
65. National Research Council, Committee on the Role of Alternative Farming Methods in Modern Production Agriculture, Board on Agriculture, 1989, *Alternative Agriculture* (Washington, D.C.: National Academy Press); Reisner, 1986.
66. Postel, 1990.
67. Postel, 1992.
68. National Research Council, 1989.
69. Postel, 1990.
70. Interview of Helmut Hess by George Moffett, 23 March 1993, Dhaka, cited in G. Moffett, 1994, *Critical Masses* (New York: Viking), 72.
71. World Resources Institute, 1992, ch. 11.
72. C. Holdren and A. Ehrlich, 1984, "The Virunga volcanoes: Last redoubt of the mountain gorilla," *Not Man Apart*, June, 8–9; A. Ehrlich and P. Ehrlich, 1987, *Earth* (New York: Franklin Watts).
73. Postel, 1992.
74. M. Reisner, 1986, *Cadillac Desert*.
75. C. Fowler and P. Mooney, 1990, *Shattering: Food, Politics, and the Loss of Genetic Diversity* (Tucson: University of Arizona Press).
76. E. Hoyt, 1988, *Conserving the Wild Relatives of Crops* (Rome and Gland: International Board for Plant Genetic Resources, International Union for the Conservation of Nature and Natural Resources [IUCN], and World Wide Fund for Nature [WWF]); D. Vaughan and

T. Chang, 1993, "In situ conservation of rice genetic resources," *Economic Botany* 46 (in press).

77. Plucknett, et al., 1987.

78. Communicated by Norman Myers to Al Gore and reported in A. Gore, 1992, *Earth in the Balance: Ecology and the Human Spirit* (Boston: Houghton Mifflin), 133. See also N. Myers, 1983, *A Wealth of Wild Species: Storehouse for Human Welfare* (Boulder, Colo.: Westview Press).

79. T. Eisner, 1992, testimony before the Subcommittee on Environmental Protection of the U.S. Senate Committee on Environment and Public Works hearings on reauthorization of the Endangered Species Act, April 10.

80. National Research Council, 1991; J. Cohen, J. Williams, D. Plucknett, and H. Shands, 1991, "Ex-situ conservation of plant genetic resources: Global development and environmental concerns," *Science* 253:866–872; Plucknett et al., 1987.

81. World Resources Institute, 1992, ch. 11; National Research Council, 1989.

82. Smil, 1991.

83. A. Ehrlich, 1990, "Agricultural contributions to global warming," in J. Leggett (ed.), *Global Warming: The Greenpeace Report* (New York: Oxford University Press), 400–420; M. Eichner, 1990, "Nitrous oxide emissions from fertilized soils: Summary of available data," *Journal of Environmental Quality* 19:272–280; J. Houghton, G. Jenkins, and J. Ephraums (eds.), 1990, *Climate Change: The IPCC Scientific Assessment* (New York: Cambridge University Press, Intergovernmental Panel on Climate Change—IPCC).

84. Smil, 1991.

85. Council for Agricultural Science and Technology, 1975, "Potential for energy conservation in agricultural production" (Ames, Iowa: Iowa State University); L. Brown, 1975, "The world food prospect," *Science* 190:1053–1059; E. Terhune, 1977, "Prospects for increasing food production in less developed countries through efficient energy utilization," in W. Lockeretz (ed.), *Agriculture and Energy* (New York: Academic Press), 625–635.

86. J. Maass and F. Garcia-Oliva, 1992, "Erosión de suelos y conservación biológica en México y Centroamérica," in R. Dirzo, D. Pinera, and M. Kalin-Arroyo (eds.), *Conservacion y Manejo de Recursos Naturales en América Latina* (Santiago, Chile: Red Latinoamericana de Botánico).

87. *Popline*, 1991, "Nepal's forest loss linked to population," Nov./Dec.; A. Durning, 1989, "Poverty and the environment: Reversing the downward spiral," *Worldwatch Paper 92* (Worldwatch Institute, Washington, D.C.), Nov.; Kates and Haarmann, 1992; D. Norse, 1992, "A new strategy for feeding a crowded planet," *Environment* 34(5):6–11, 32–39; P. Dasgupta, 1993, "Poverty, resources, and fertility: The household as a reproductive partnership," ch. 11 in P. Dasgupta (ed.), *An Inquiry into Well-Being and Destitution* (Oxford, Eng.: Clarendon Press).

88. D. Pimentel et al., 1989, "Ecological resource management for a pro-

ductive, sustainable agriculture," in D. Pimentel and C. Hall (eds.), *Food and Natural Resources* (New York: Academic Press), 301–323.

89. An estimated 99 percent of potential insect pests are controlled by their natural enemies; see P. DeBach, 1974, *Biological Control by Natural Enemies* (Cambridge, Eng.: Cambridge University Press).

90. For more details, see P. R. Ehrlich, 1986, *The Machinery of Nature* (New York: Simon & Schuster).

91. This was the message so powerfully conveyed by Rachel Carson in *Silent Spring* (Boston: Houghton Mifflin, 1962).

92. National Research Council, 1989.

93. Various sources cited in K. Holl, G. Daily, and P. Ehrlich, 1990, "Integrated pest management in Latin America," *Environmental Conservation* 17:341–350.

94. R. Smith, 1969, "The new and the old in pest control," *Memorie della Reale Academia Nazionale dei Lincei* (Rome), mimeograph.

95. For further details on the Cañete Valley disaster and related events, see Ehrlich, Ehrlich, and Holdren, 1977, pp. 643–657.

96. K. Wirth and F. Schima, 1994, *The Pesticide PACs: Campaign Contributions and Pesticide Policy* (Washington, D.C.: The Environmental Working Group).

97. D. Weir and M. Schapiro, 1981, *Circle of Poison* (San Francisco: Institute for Food Development and Policy).

98. R. Naylor, 1994, "Herbicide use in Asian rice production," *World Development* 22:55–70. We are indebted to Dr. Naylor for much of what follows.

99. K. Ampong-Nyarko and S. K. De Datta, 1991, *A Handbook for Weed Control in Rice* (Los Banos, Philippines: International Rice Research Institute—IRRI); Moody, 1991, "Weed management in rice," in D. Pimentel (ed.), *Handbook of Pest Management in Agriculture* (Boca Raton, Fla.: CRC Press Inc.), 301–328.

100. Thirty million tons of rice would provide about a 25 percent increase in calories for 500 million people, which is about the current estimate of the number of people in Asia now significantly underfed (see Kates and Haarmann, 1992).

101. H. Shibayama, 1994, "The intensive rice-herbicide system: A case study of Japan," paper presented at Conference on Herbicide Use in Asian Rice Production, Institute for International Studies, Stanford University, Stanford, Calif., March 28–30.

102. K. Kim, 1994, "Ecological forces influencing weed competition and resistance to herbicides," paper presented at Conference on Herbicide Use in Asian Rice Production, Stanford University, Stanford, Calif., March 28–30.

103. N. Ho, K. Itoh, and A. Othman, 1992, paper given at International Rice Research Conference (IRRI), Los Banos, Philippines, April 21–25).

104. P. Pingali, C. Marquez, F. Palis, and A. Rola, 1994, "Impact of pesticides on farmer health—a medical and economic analysis," paper pre-

sented at Conference on Herbicide Use in Asian Rice Production, Stanford University, Stanford, Calif., March 28–30.

105. Much of what is known was learned after American servicemen and perhaps millions of Vietnamese people were exposed to Agent Orange in the Vietnam war. For a summary of what was known in the late 1970s, see A. W. A. Brown, 1978, *Ecology of Pesticides* (New York: Wiley-Interscience).

106. K. Li and P. Li, 1994, "The potential impacts of herbicides in aquaculture," paper presented at Conference on Herbicide Use in Asian Rice Production, Stanford University, Stanford, Calif., March 28–30.

107. D. G. Crosby, 1994, "Environmental impacts of rice herbicide use," paper presented at Conference on Herbicide Use in Asian Rice Production, Stanford University, Stanford, Calif., March 28–30.

108. E.g., V. Shiva, 1991, *The Violence of the Green Revolution* (London: Zed Books).

109. O. Loucks, 1989, "Large-scale alteration of biological productivity due to transported pollutants," in D. B. Botkin, M. F. Caswell, J. E. Estes, and A. A. Orio (eds.), *Changing the Global Environment* (London: Academic Press), 101–116.

110. J. MacKenzie and M. El-Ashry, 1988, *Ill Winds: Air Pollution's Toll on Trees and Crops* (Washington, D.C.: World Resources Institute).

111. W. Chameides, P. Kasibhatla, J. Yienger, and H. Levy II, 1994, "Growth of continental-scale metro-agro-plexes, regional ozone pollution, and world food production," *Science* 264:74–76.

112. Worrest and Grant, 1989.

113. F. Quaite, B. Sutherland, and J. Sutherland, 1992, "Action spectrum for DNA damage in alfalfa lowers predicted impact of ozone depletion," *Nature* 358:576–578.

114. R. Smith, B. Prezelin, K. Baker, R. Bidigare, N. Boucher, T. Coley, D. Karentz, S. MacIntyre, H. Matlick, D. Menzies, M. Ondrusek, Z. Wan, and K. Waters, 1992, "Ozone depletion: Ultraviolet radiation and phytoplankton biology in Antarctic waters," *Science* 255:952–959.

115. T. Appenzeller, 1991, "Ozone loss hits us where we live," *Science* 254:645; 1991, "Europe's lost ozone," *New Scientist* 27 (July):15.

116. R. Kerr, 1992, "Pinatubo fails to deepen the ozone hole," *Science* 258:395.

117. R. Kerr, 1994, "Antarctic ozone hole fails to recover," *Science* 266:217.

118. Houghton et al., 1990; S. Schneider, 1989.

119. Parry, 1990; R. Peters and T. Lovejoy (eds), 1992, *Global Warming and Biological Diversity* (New Haven, Conn.: Yale University Press).

120. G. Daily and P. Ehrlich, 1990, "An exploratory model of the impact of rapid climatic change on the world food situation," *Proceedings of the Royal Society of London* 241:232–244.

121. Houghton et al., 1990.

122. Ibid.

123. Peters and Lovejoy, 1992.

124. B. Grodzinski, 1992, "Plant nutrition and growth regulation by CO_2 enrichment," *BioScience* 42 (July/Aug.):517–525; A. Bazzaz and

E. Fajer, 1992, "Plant life in a CO_2-rich world," *Scientific American,* Jan., 68–74; C. Korner and J. Arnone III, 1992, "Responses to elevated carbon dioxide in artificial tropical ecosystems," *Science* 257: 1672–1675.

125. S. Sinha, N. Rao, and M. Swaminathan, 1990, "Food security in the changing global climate," in J. Kayser and L. Pollard (eds.), *Climate and Food Security* (Manila, Philippines: International Rice Research Institute), 594. This edited volume provides a good introduction to the complexity of crop-climate interactions.

126. P. Matson and P. Vitousek, 1990, "Ecosystem approach to a global nitrous oxide budget," *BioScience* 40:667–672; Ehrlich, 1990, Eichner, 1990.

127. Ehrlich and Ehrlich, 1981; Ehrlich and Ehrlich, 1991; D. Pimentel, U. Stachow, D. Takacs, et al., 1992, "Conserving biological diversity in agricultural/forestry systems," *BioScience* 42(May):354–362.

128. E. Wilson and F. Peter (eds.), 1988, *Biodiversity* (Washington, D.C.: National Academy Press); P. Ehrlich and E. Wilson, 1991, "Biodiversity studies: Science and policy," *Science* 253:758–762; Wilson, 1992.

129. E. Wilson, 1989, "Threats to biodiversity," *Scientific American* 261 (Sept.):108–116.

130. World Resources Institute, 1986, ch. 11.

131. E. Lefroy, R. Hobbs, and M. Scheltema, 1993, "Reconciling agriculture and nature conservation: Toward a restoration strategy for the Western Australian wheatbelt," in D. Saunders, R. Hobbs, and P. Ehrlich, *Reconstruction of Fragmented Ecosystems: Global and Regional Perspectives* (Chipping Norton, NSW, Australia: Surrey Beatty & Sons). The dollar values given here are converted from Australian to U.S. dollars. In 1991, the three of us visited the Western Australian wheatlands in connection with a joint meeting between the Commonwealth Scientific and Industrial Research Organization, Stanford University's Center for Conservation Biology, and the Tammin Land Conservation District Committee. The meeting and situation are described in Saunders et al., 1993.

132. P. Chamberlin, 1994, "Keating takes the chainsaw to pledge," *Canberra Times,* 22 December.

133. M. Schiff and A. Valdes, 1992, *The Plundering of Agriculture in Developing Countries* (Washington, D.C.: World Bank); World Bank, 1990, *World Development Report: Poverty* (Washington, D.C.: World Bank); World Bank, 1992, *World Development Report: Development and Environment* (Washington, D.C.: World Bank).

134. J. Drèze and A. Sen (eds.), *The Political Economy of Hunger:* vol 1, *Entitlement and Well-being* (1990); vol 2, *Famine Prevention* (1990); vol 3, *Endemic Hunger* (1991) (Oxford, Eng.: Oxford University Press).

135. For more detailed discussion, see G. Daily and P. Ehrlich, 1995, "Socioeconomic equity, sustainability, and carrying capacity," *Ecological Applications* (forthcoming).

136. Described in an excellent historical review by M. Gadgil and

R. Guha, 1992, *This Fissured Land: An Ecological History of India* (Berkeley: University of California Press).

137. World Bank, 1987, "The threat of protectionism," in *World Development Report* (New York: Oxford University Press), 133–153.

138. E.g., "Trade and the environment," *Economist,* 27 February 1993.

139. See P. Samuelson, 1962, "The gains from international trade once again," *Economic Journal* 72:820–829; and Summary in P. Ekins, C. Folke, and R. Costanza, 1994, "Trade, environment and development: The issues in perspective," *Ecological Economics* 9:1–12.

140. I. Røpke, 1994, "Trade, development and sustainablility—a critical assessment of the 'free trade dogma,'" *Ecological Economics* 9:13–22.

141. M. Ritchie, 1992, "Free trade versus sustainable agriculture: The implications of NAFTA," *The Ecologist* 22:221–227.

142. P. Dasgupta, 1982, *The Control of Resources* (Cambridge, Mass.: Harvard University Press); M. Glantz, 1987, *Drought and Hunger in Africa* (Cambridge, Eng.: Cambridge University Press).

143. B. DeWalt, S. Stonich, and S. Hamilton, 1993, "Honduras: Population, inequality, and resource destruction," in C. L. Jolly and B. B. Torrey (eds.), *Population and Land Use in Developing Countries* (Washington, D.C.: National Academy Press), 106–123. For an ecological characterization of the development of shrimp farming in Colombia, see J. Larsson, C. Folke, and N. Kautsky, 1994, "Ecological limitations and appropriation of ecosystem support by shrimp farming in Colombia," *Beijer Institute Discussion Paper Series* no. 29 (Stockholm: The Royal Swedish Academy of Sciences).

144. Reviewed in World Resources Institute, 1994, *World Resources 1994–95* (New York: Oxford University Press); *see also* S. Lonergan, 1993, "Impoverishment, population, and environmental degradation: The case for equity," *Environmental Conservation* 20:328–334.

145. From S. George, 1992, *The Debt Boomerang* (Boulder, Colo.: Westview Press), based on OECD data.

146. World Resources Institute, 1994.

147. Kates and Haarmann, 1992; Dasgupta, 1993, *An Inquiry into Well-Being and Destitution.*

Chapter Seven

1. M. S. Swaminathan, 1986, "Building national and global nutrition security systems," in M. S. Swaminathan and S. K. Sinha (eds.), *Global Aspects of Food Production* (Oxford, Eng.: IRRI, Tycooly International), 417–449.

2. Food and Agriculture Organization of the United Nations, 1983, *World Food Report 1983* (Rome: FAO).

3. United Nations Food and Agriculture Organization (FAO), 1950–1991, *Annual Production Yearbooks* (Rome: FAO). *See also* L. R. Brown et al., 1994, *State of the World 1994* (New York: Norton), 181. Brown uses USDA data, which differ from FAO data in some minor respects.

4. L. R. Brown et al., 1990, *State of the World 1990* (New York: Norton).
5. J. Walsh, 1991, "Preserving the options: Food productivity and sustainability," *Issues in Agriculture 2* (Consultative Group on International Agricultural Research—CGIAR).
6. B. Holmes, 1994, "Super rice extends limits to growth," *New Scientist,* 29 Oct., 4. This news was quickly picked up and misinterpreted by ecologically ignorant commentators: e.g., T. Sowell, 1994, "Unnatural resources," *Forbes,* 24 Oct., 130.
7. This is a conservative assessment of future needs based primarily on projected population growth and not considering dietary changes.
8. Walsh, 1991, p. 25.
9. Brown et al., 1990.
10. L. Brown, H. Kane, and D. Roodman, 1994, *Vital Signs* (New York: Norton), 42–43.
11. Dr. Gareth Fry, Norwegian Institute for Nature Research, personal communication, October 1992.
12. P. Webb and J. von Braun, 1994, *Famine and Food Security in Ethiopia: Lessons for Africa* (New York: Wiley). The timing and placement of fertilizers is crucial to optimize plant uptake, but may be constrained by availability of machinery or labor at the critical time. Hence losses are often large.
13. C. Gasser and R. Fraley, 1989, "Genetically engineering plants for crop improvement," *Science* 244:1293–1299.
14. A. McCalla, 1994, "Agriculture and food need to 2025: Why we should be concerned" (Washington, D.C.: Consultative Group on International Agricultural Research, CGIAR Secretariat, World Bank), 9.
15. P. R. Ehrlich and A. H. Ehrlich, 1991, *Healing the Planet* (Reading, Mass.: Addison-Wesley), 201.
16. E.g., M. Giampietro, 1994, "Sustainability and technological development in agriculture: A critical appraisal of genetic engineering," *BioScience* 44:677–689; and D. Pimentel, 1989, "Benefits and risks of genetic engineering in agriculture," *BioScience* 39:606–614.
17. Much of this section is based on G. Daily, 1995, "Restoring productivity to the world's degraded lands," *Science,* in press.
18. P. R. Ehrlich and H. A. Mooney, 1983, "Extinction, substitution, and ecosystem services," *BioScience* 33:248–254; United Nations Population Fund, 1991, *Population and the Environment: The Challenges Ahead* (New York: UNFPA); G. C. Daily and P. R. Ehrlich, 1992, "Population, sustainability, and Earth's carrying capacity," *BioScience* 42:761–771; P. R. Ehrlich, A. H. Ehrlich, and G. C. Daily, 1993, "Food security, population, and environment," *Population and Development Review* 19:1–32; D. O. Hall, F. Rosillo-Calle, and R. H. Williams, "Biomass for energy: Supply prospects," in T. B. Johansson et al. (eds.), *Renewable Energy: Sources for Fuels and Electricity* (Washington, D.C.: Island Press), 593–652.
19. E. Kessler and C. Folke (eds.), 1993, *Ambio* 22 (special issue).
20. H. A. Mooney, P. M. Vitousek, and P. A. Matson, 1987, "Exchange

of materials between terrestrial ecosystems and the atmosphere," *Science* 238:926–932; J. T. Houghton, G. J. Jenkins, and J. J. Ephraums (eds.), 1990, *Climate Change: The IPCC Scientific Assessment* (Cambridge, Mass.: Cambridge University Press); R. A. Houghton, 1991, "Releases of carbon to the atmosphere from degradation of forests in tropical Asia," *Canadian Journal of Forest Research* 21:132–142; W. H. Schlesinger, 1991, *Biogeochemistry: An Analysis of Global Change* (San Diego: Academic Press).

21. C. A. S. Hall, 1992, "Economic development or developing economics: What are our priorities?," in M. K. Wali (ed.), *Ecosystem Rehabilitation,* vol. 1 (The Hague: SPB Academic Publishing, bv), 101–126; T. N. Khoshoo, 1992, "Degraded lands for agroecosystems," in M. K. Wali (ed.), *Ecosystem Rehabilitation,* vol. 2 (The Hague: Academic Publishing, bv), 3–18; M. Gadgil, 1993, "Biodiversity and India's degraded lands," *Ambio* 22:167–172.

22. C. Uhl, 1987, "Factors controlling succession following slash-and-burn agriculture in Amazonia," *Journal of Ecology* 75:377–407.

23. E.g., C. Uhl and K. Clark, 1983, "Seed ecology of selected Amazon Basin successional species," *Botanical Gazette* 144(3):419–425.

24. W. Peters and L. Neuenschwander, 1988, *Slash and Burn: Farming in the Third World Forest.* Moscow, Idaho: University of Idaho Press.

25. D. Janos, 1980, "Mycorrhizae influence tropical succession," *Biotropica* 12:56–64; C. Uhl, 1988, "Restoration of degraded lands in the Amazon Basin," in E. O. Wilson (ed.), *Biodiversity* (Washington, D.C.: National Academy Press), 326–332; C. Uhl, H. Clark, K. Clark, and P. Maquirino, 1982, "Successional patterns associated with slash-and-burn agriculture in the Upper Rio Negro region of the Amazon Basin," *Biotropica* 14(4):249–254.

26. C. Uhl, R. Buschbacher, and E. Serrao, 1988, "Abandoned pastures in eastern Amazonia. I. Patterns of plant succession," *Journal of Ecology* 76:663–681.

27. J. Beard, 1993, "How hungry ants hinder return of the forest," *New Scientist,* 22 May, 16; C. Uhl, 1988, "Restoration of degraded lands in the Amazon Basin," in Wilson, 1988, *Biodiversity,* pp. 326–332.

28. C. Uhl, 1987; R. Buschbacher, C. Uhl, and E. Serrao, 1992, "Reforestation of degraded Amazon pasture lands," in Wali, *Ecosystem Rehabilitation,* vol. 2 of *Ecosystem Analysis and Synthesis,* 257–274.

29. Buschbacher, Uhl, and Serrao, 1992.

30. C. Uhl and R. Buschbacher, 1985, "A disturbing synergism between cattle ranch burning practices and selective tree harvesting in the Eastern Amazon," *Biotropica* 17:265.

31. C. D'Antonio and P. Vitousek, 1992, "Biological invasions by exotic grasses, the grass/fire cycle, and global change," *Annual Review of Ecology and Systematics* 23:63–87.

32. E. G. Egler, 1961, "A Zona Bragantina do Estado do Pará." *Revista Brasileira de Geografia* 23:527.

33. P. S. Ramakrishnan, 1992, "Ecology of shifting agriculture and

ecosystem restoration," in Wali, *Ecosystem Rehabilitation,* vol. 2, pp. 19–35.

34. J. Walls (ed.), 1982, *Combatting Desertification in China* (Nairobi: United Nations Environment Programme); J. Ewel, 1986, "Designing agricultural systems for the humid tropics," *Annual Review of Ecology and Systematics* 17:245–271; E. Allen (ed.), 1988, *The Reconstruction of Disturbed Arid Lands* (Boulder, Colo.: Westview Press); J. Cairns (ed.), 1988, *Rehabilitating Damaged Ecosystems* (Boca Raton, Fla.: CRC Press); D. Saunders, R. Hobbs, and P. Ehrlich (eds.), 1993, *Reconstruction of Fragmented Ecosystems,* (Chipping Norton, NSW, Australia: Surrey Beatty & Sons); B. Walker, 1993, "Rangeland ecology: Understanding and managing change," *Ambio* 22:80–87.

35. V. M. Kline and E. A. Howell, 1987, "Prairies," in W. R. Jordan III, M. E. Gilpin, and J. D. Aber (eds.), *Restoration Ecology* (Cambridge, England: Cambridge University Press), 53–74; P. Singh, 1992, "Grasslands of India—rehabilitation and management," in Wali, *Ecosystem Rehabilitation,* vol. 2, pp. 51–61.

36. P. Blaschke, N. Trustrum, and R. DeRose, 1992, "Ecosystem processes and sustainable land use in New Zealand steeplands," *Agriculture, Ecosystems, and Environment* 41:153–178.

37. United Nations Environment Programme and Commonwealth of Australia, 1987, *Drylands Dilemma: A solution to the Problem.* Canberra: Australian Government Publishing Service.

38. A. Grainger, 1982, *Desertification* (London: International Institute for Environment and Development), 67.

39. In Mali, now often spelled Tombouctou (P. R. Ehrlich and A. H. Ehrlich, personal observation, 1991).

40. P. Warshall, personal communication, October 1994.

41. Ibid. Many African cultures exhibit a blend of traditional religious beliefs with those that have been more recently superimposed.

42. Information on game ranching is from P. Ehrlich, 1986, *The Machinery of Nature* (New York: Simon & Schuster), 262–267.

43. Saunders, Hobbes, and Ehrlich, 1993.

44. Ibid.

45. National Research Council, 1989, *Alternative Agriculture* (Washington, D.C.: National Academy Press).

46. R. Harwood, 1990, "A history of sustainable agriculture," in C. Edwards et al. (eds), *Sustainable Agricultural Systems* (Ankeny, Iowa: Soil and Water Conservation Society).

47. R. Lowrance, B. R. Stinner, and G. J. House (eds.), 1984, *Agricultural Ecosystems: Unifying Concepts* (New York: Wiley); J. Tivy, 1990, *Agricultural Ecology* (London: Longman; New York: Wiley).

48. M. Altieri, 1983, *Agroecology: The Scientific Basis of Alternative Agriculture* (Berkeley: University of California, Division of Biological Control); M. Dover and L. M. Talbot, 1987, *To Feed the Earth: Agro-Ecology for Sustainable Development* (Washington, D.C.: World Resources Institute).

49. E. Hartmans, 1983, "African food production: Research against time," *Outlook on Agriculture* 12:165–171.

50. R. Tourte and J. Moomaw, 1977, "Traditional African systems of agriculture and their improvement," in C. Leakey and J. Wills (eds.), *Food Crops of the Lowland Tropics* (Oxford: Oxford University Press), 195–311.

51. Consultative Group on International Agricultural Research, 1994, *CGIAR Annual Report 1993–94* (Washington, D.C.).

52. Cole wrote about the problems of tropical agriculture in the 1960s: D. G. Cole, 1968, "The myth of fertility dooms development plans," *National Observer,* 22 April.

53. Martha Rosemeyer, personal communications, March 1991, 1992, 1993, and 1994.

54. R. Chambers, A. Pacey, and L.-A. Thrupp, 1989, *Farmers First: Farmer Innovation and Agricultural Research* (London: Intermediate Technology Publications).

55. J. Fairhead and M. Leach, 1994, "Declarations of difference," in I. Scoones and J. Thompson, 1994, *Beyond Farmer First* (London: Intermediate Technology Publications), 76–77.

56. W. de Boef, K. Amanor, K. Wellard, and A. Bebbington, 1993, *Cultivating Knowledge: Genetic Diversity, Farmer Experimentation and Crop Research* (London: Intermediate Technology Publications).

57. P. Sikana, 1994, "Indigenous soil characterization in northern Zambia," in Scoones and Thompson, pp. 80–82.

58. S. Fujisaka, 1994, "Will farmer participatory research survive in the International Agricultural Research Centres?," in Scoones and Thompson, pp. 227–235.

59. J. Pretty and R. Chambers, 1994, "Towards a learning paradigm: New professionalism and institutions for a sustainable agriculture," in Scoones and Thompson, pp. 182–202.

60. Fujisaka, 1994.

61. Some studies suggest that the green revolution technologies eventually have very positive effects on employment among poor people: e.g., R. Sharma and T. Poleman, 1993, *New Economics of India's Green Revolution: Income and Employment Diffusion in Uttar Pradesh* (Ithaca, N.Y.: Cornell University Press).

62. National Research Council, 1993, *Sustainable Agriculture and the Environment in the Humid Tropics* (Washington, D.C.: National Academy Press).

63. Dover and Talbot, 1987.

64. National Academy of Sciences, 1975, *Underexploited Tropical Plants of Promising Economic Value* (Washington, D.C.: National Academy of Sciences); P. R. Ehrlich and A. H. Ehrlich, 1981, *Extinction: The Causes and Consequences of the Disappearance of Species* (New York: Random House), 63; National Academy of Sciences, 1984, *Amaranth: Modern Prospects for an Ancient Crop* (Washington, D.C.: National Academy of Sciences); C. Tudge, 1988, *Food Crops for the Future* (Oxford, Eng.: Basil Blackwell).

65. Tudge, 1988, gives an excellent summary along with an overview of future crop improvement.

66. P. R. Ehrlich, A. H. Ehrlich, and J. P. Holdren, 1977, *Ecoscience; Population, Resources, Environment* (San Francisco: W. H. Freeman & Co.), 343–345.

67. S. Agboola, 1977, "The spatial dimension in Nigerian agricultural development," *Inaugural Lecture* no. 22, University of Ife, Ife, Nigeria; B. Glaeser and K. Phillips-Howard, 1987, "Low-energy farming systems in Nigeria," in B. Glaeser (ed.), *The Green Revolution Revisited* (Boston: Allyn and Unwin), 126–149.

68. Chambers, Pacey, and Thrupp, 1989.

69. C. Wille, 1994, "The birds and the beans," *Audubon,* Nov./Dec., 58–64.

70. R. J. Marquis, and C. J. Whelan, 1994. "Insectivorous birds increase growth of white oak through consumption of leaf-chewing insects," *Ecology* 75:2007–2014. C. Yoon, 1994, "More than decoration, songbirds are essential to forests' health," *The New York Times,* 8 November.

71. World Resources Institute, 1994, *World Resources 1994–95* (New York: Oxford University Press), 192.

72. World Resources Institute, 1994, pp. 190–191.

73. World Resources Institute, 1993, *The 1994 Information Please Environmental Almanac* (Boston: Houghton Mifflin), 320.

74. The following material on the Calcutta marshes is based on an account in C. Pye-Smith and G. Feyerabend, 1994, *The Wealth of Communities* (London: Earthscan).

75. D. Pimentel, 1978, "Socioeconomic and legal aspects of pest control," in E. Smith and D. Pimentel (eds.), *Pest Control Strategies* (New York: Academic Press), 55–71.

76. P. Ehrlich, 1991, "Pest control and the human predicament," in J. Menn and A. Steinhauer (eds.), *Progress and Perspectives for the 21st Century* (Lanham, Md.: Entomological Society of America), 119–126.

77. Here again, estimates are very approximate, and this one may be high (although it may be low for certain easily spoiled foods such as fruits).

78. Ehrlich, Ehrlich, and Daily, 1993.

79. See Ehrlich, Ehrlich, and Holdren, 1977, for a summary of early literature and references. For more recent treatments, see D. Horn, 1988, *Ecological Approach to Pest Management* (London: Elsevier); and D. Pimentel and H. Lehman (eds.), 1993, *The Pesticide Question* (New York: Chapman and Hall).

80. R. Frisbie and P. Adkisson (eds.), 1985, *Integrated Pest Management in Major Agricultural Systems* (College Station, Tex.: Texas A&M University Press).

81. K. Holl, G. Daily, and P. Ehrlich, 1990, "Integrated pest management in Latin America," *Environmental Conservation* 17:341–350.

82. M. Altieri et al., 1989, "Classical biological control in Latin America: Past, present, and future," in T. Rusher (ed.), *Principles and Application of Biological Control* (Berkeley: University of California Press), 30–38.

83. R. Chambers and B. Ghildyal, 1985, "Agricultural research for resource-poor farmers: The farmer's first and last model," *Agricultural Administration* 20:1–30.

84. Y. Winarto, 1994, "Encouraging knowledge exchange: Integrated pest management in Indonesia," in Scoones and Thompson, *Beyond Farmer First*, p. 151.

85. E.g., R. Sinha and F. Watters, 1985, "Insect pests of flour mills, grain elevators, and feed mills and their control," *Agriculture Canada* (Ottawa).

86. P. Warshall, 1989, *Mali: Biological Diversity Assessment* (US AID, Bureau of Africa, project no. 698-0467), and Warshall, personal communication, December 1994.

87. L. R. Brown et al., 1989, *State of the World 1989* (New York: Norton); L. R. Brown et al., 1994, *State of the World 1994* (New York: Norton), ch. 10; D. Pimentel and C. Hall (eds.), 1989, *Food and Natural Resources* (San Diego: Academic Press).

88. Conversion efficiencies differ from animal to animal. In a feedlot situation, it takes about 7 kg of grain to produce 1 kg of beef, 4 kg to produce 1 kg of pork, and 2 kg to produce 1 kg of poultry or fish; see L. Brown and H. Kane, 1994, *Full House* (New York: Norton), 66–67.

89. Quoted in A. B. Durning, 1989, "Poverty and the environment: Reversing the downward spiral," *Worldwatch Paper 92* (Worldwatch Institute, Washington, D.C.), Nov.

90. United Nations Development Program (UNDP), 1993, *Human Development Report 1993* (New York: Oxford University Press); Durning, 1989; World Bank, 1992, *World Development Report 1992* (New York: Oxford University Press).

91. Food and Agriculture Organization (FAO), 1992, *World Food Supplies and Prevalence of Chronic Undernutrition in Developing Regions as Assessed in 1992* (Rome: FAO); UN Population Fund (UNFPA), *State of the World's Population 1992* (New York: United Nations).

92. UNFPA, 1993, *State of the World's Population, 1993* (New York: United Nations).

93. World Bank, 1992; Durning, 1989; FAO, 1992.

94. R. Summers and A. Heston, 1988, "A new set of international comparisons of real product and price level estimates for 130 countries, 1950–1985," *Review of Income and Wealth,* March, cited in Durning, 1989.

95. UNDP, 1993.

96. UNDP, 1992, *Human Development Report 1992* (New York: Oxford University Press).

97. S. Postel, 1994, "Carrying capacity: Earth's bottom line," in Brown et al., p. 5.

98. Elizabeth Dowdeswell, 1993, editorial, *Our Planet* (UNEP), 5:2; reprinted in The International Society of Naturalists, 1993, *Environment Awareness* (Baroda, India) 16 (April/June):2

99. A. T. Durning, "Redesigning the forest economy," ch. 2 in Brown et al.

100. R. Repetto and T. Holmes, 1983, "The role of population in resource depletion in developing countries," *Population and Development Review* 9 (Dec.):609–632; R. Repetto et al., 1989, *Wasting Assets: Natural*

Resources in the National Income Accounts (Washington, D.C.: World Resources Institute).

101. This section is based on G. Daily and P. Ehrlich, 1995, "Socioeconomic equity, sustainability, and carrying capacity," *Ecological Applications* (forthcoming).

102. Women's Environment and Development Organization (WEDO), 1994, "African women fight desert encroachment," *News and Views* 7:11; World Resources Institute, 1994; J. Pointing and S. Jockes, 1990, "Women in pastoral societies in east and west Africa," unpublished manuscript, Institute of Development Studies, Brighton, Eng., Nov.

103. P. Dasgupta, 1993, *An Inquiry into Well-Being and Destitution* (Oxford, Eng.: Oxford University Press); World Resources Institute, 1994, ch. 3.

104. J. Holmberg, 1991, *Poverty, Environment and Development: Proposals for Action* (Stockholm: Swedish International Development Agency); S. C. Lonergan, 1993, "Impoverishment, population, and environmental degradation: The case for equity," *Environmental Conservation* 20 (Winter):328–334.

105. World Resources Institute, 1994, p. 51.

106. C. Udry, 1994, "Gender and agricultural productivity," *Food Policy* (forthcoming).

107. M. Gadgil and R. Guha, 1992, *"This Fissured Land: An Ecological History of India* (Berkeley: University of California Press); Anil Agarwal, 1993, "The land of milk from water," *Our Planet* 5(2):8–9; O. Sattaur, 1990, "The green solution for India's poor," *New Scientist,* 15 Sept., pp. 28–29; A. Agarwal and S. Narain, 1990, *Towards Green Villages: A Strategy for Environmentally Sound and Participatory Rural Development* (New Delhi: Center for Science and Environment).

108. Dasgupta, 1993.

109. Personal communication from a Mexican friend who wishes not to be identified.

110. A. Reding, 1994, "Chiapas is Mexico: The imperative of political reform," *World Policy Journal* 11(1):11–25.

111. Pye-Smith and Feyerabend, 1994.

112. Most poor nations, consciously or unconsciously, have heeded a message in economist Ester Boserup's influential book, *The Conditions of Agricultural Growth* (Chicago: Aldine, 1965). She asserted that urban industrialization necessarily preceded the adoption of "scientific" and "industrial" agriculture. Unhappily, she had it just about backward. Scientific and industrial agriculture is a mixed blessing from the standpoint of sustainability. More important, a strong agricultural economy with relatively equitable distribution of land is the best foundation for national modernization, as the history of the United States and many other nations clearly illustrates. And modernization, fortunately, does not necessarily involve large-scale industrialization. It is impossible for all nations to be net exporters of manufactured goods, and industrialization of all nations in the Western model would destroy Earth's capacity to support civilization.

113. M. Glantz, 1987, *Drought and Hunger in Africa* (Cambridge: Eng.: Cambridge University Press).
114. Reported in Dasgupta, 1993; M. Schiff and A. Valdéz, 1992, *The Plundering of Agriculture in Developing Countries* (Washington, D.C.: World Bank).
115. According to UNDP, 1993.
116. UNDP, 1993.
117. Even China, formerly immune to this phenomenon, is now undergoing a flood of rural migrants into its cities. Complicating the process is a huge and growing "floating population" (of as many as 150 million by 1994) in precarious circumstances, who move back and forth from rural to urban areas as a result of the economic liberalization of the 1980s; see K. Wakabayasho, 1990, "Migration from rural to urban China," *The Developing Economies* 28 (Dec.); Nicholas D. Kristof, 1993, "China sees 'market-Leninism' as way to future," and "Riddle of China: Repression as standard of living soars," *The New York Times*, 6, 7 September.
118. M. Lofchie, 1987, "The decline of African agriculture," in Glantz, *Drought and Hunger in Africa*, pp. 85–109.
119. R. Baker, 1987, "Linking and sinking: Economic externalities and the persistence of destitution and famine in Africa," in Glantz, *Drought and Hunger in Africa*, pp. 149–168; Schiff and Valdéz, 1992.
120. R. Naylor, 1994, "Herbicide use in Asian rice production," *World Development* 22:55–70.
121. Lofchie, 1987.
122. Madhav Gadgil, personal communication, 1994; T. N. Srinivasan, 1988, "Population growth and food: An assessment of issues, models, and projections," in R. D. Lee, W. B. Arthur, A. C. Kelley, G. Rodgers, and T. N. Srinivasan (eds.), *Population, Food, and Rural Development* (Oxford, Eng.: Clarendon Press), 11–39.
123. E.g., Arjun Makhijani, 1992, *From Global Capitalism to Economic Justice* (New York: Apex Press); *see also* Sylvia Nasar, 1994, "Economics of equality: A new view," *The New York Times* (Business Day section), 8 January.
124. C. Timmer, W. Falcon, and S. Pearson, 1983, *Food Policy Analysis* (Baltimore: Johns Hopkins Press).
125. UNICEF, 1989, *State of the World's Children, 1989* (New York: United Nations), quoted in Durning, 1989.
126. U.S. Congress, Office of Technology Assessment, 1994, *Perspectives on the Role of Science and Technology in Sustainable Development*, OTA-ENV-609 Sept., (Washington, D.C.: U.S. Government Printing Office).
127. Office of Technology Assessment, 1994, based in part on K. Hansen-Kuhn, 1993, "Sapping the economy: Structural adjustment policies in Costa Rica," *The Ecologist* 23:179–184.
128. Ehrlich, Ehrlich, and Holdren, 1977, ch. 7.
129. Ibid.

130. Repetto et al., 1989. The study focused on two very different developing nations: Indonesia and Costa Rica.
131. C. A. Meyer, 1993, "Environmental and natural resource accounting: Where to begin?," *Issues in Development* (World Resources Institute, Washington, D.C.), Nov.
132. They do not, for example, consider the contributions of deforestation to global warming or of soil erosion to the damaging of fisheries.
133. See J. Holdren, 1992, "The transition to costlier energy," in L. Schipper and S. Meyers (eds.), *Energy Transitions* (Stockholm: Stockholm Environmental Institute), a version of which appeared as "Energy in transition," in *Scientific American,* Sept. 1990.
134. This is the number Holdren originally used in his scenario, which assumed strenuous efforts at population limitation. United Nations projections now suggest that number may be passed in the middle of the century. Precise numbers in scenarios depend heavily on how nations are classified into rich and poor, but the basic thrust of Holdren's argument holds with any reasonable division.
135. Indeed, depending on assumptions made about total energy use in the future—which technologies supply what fraction and for how long—efficiency itself could make available 10 to 40 TW by 2050. Holdren's scenario depends on increased efficiency "supplying" some 45 TW by 2100—the difference between the scenario's 30 TW and the 75 TW that would be required to give 10 billion people a lifestyle resembling that of the rich in the 1990s, fueled by 1990 technologies requiring 7.5 kW per capita.
136. For instance, T. Johansson, H. Kelly, A. Reddy, and R. Williams, 1993, *Renewable Energy* (Washington, D.C.: Island Press).
137. See the summary in A. McCalla, 1994, "Agriculture and food needs to 2025: Why we should be concerned," *Consultative Group on International Agricultural Research* (Washington, D.C.: CGIAR Secretariat, World Bank).
138. Ehrlich, Ehrlich, and Holdren, 1977, p. 343.
139. Brown, Kane, and Roodman, 1994, pp. 36–37.
140. E.g., G. Daily and P. Ehrlich, 1990, "An exploratory model of the impact of rapid climate change on the world food situation," *Proceedings of the Royal Society of London,* B 241:232–244.

Chapter Eight

1. D. Pirages and P. Ehrlich, 1974, *Ark II: Social Response to Environmental Imperatives* (New York: Viking), v.
2. The problem of growing antibiotic resistance has finally been widely recognized by the popular press, e.g., M. Lemonick, 1994, "The killers," *Time,* 12 Sept.
3. For reports on bodies, see *Illustrated London News,* Jan. 1847. The quote is from C. Woodham-Smith, 1962, *The Great Hunger, Ireland 1845–9* (London: Hamish-Hamilton), 18. For more details and refer-

ences on the famine, see P. Ehrlich, L. Bilderback, and A. Ehrlich, 1979, *The Golden Door: International Migration, Mexico, and the United States* (New York: Ballantine Books; rev. ed., New York: Wideview Books, 1981).

4. E.g., O. L. Bettmann, 1974, *The Good Old Days—They Were Terrible!* (New York: Random House). This is the source for the following statistics and anecdotes.

5. R. Hunter, 1965, *Poverty* (New York: Harper Torchbooks).

6. For a technical discussion of the myth of the noble savage, see R. B. Edgerton, 1992, *Sick Societies: Challenging the Myth of Primitive Harmony* (New York: Free Press).

7. T. Pakenham, 1991, *The Scramble for Africa, 1876–1912* (New York, Random House).

8. M. Harner, 1977, "The ecological basis for Aztec sacrifice," *American Ethnologist* 4:117–135; M. Harris, 1977, *Cannibals and Kings: The Origins of Cultures* (New York: Random House).

9. A. Balikci, 1970, *The Netsilik Eskimo* (Garden City, N.Y.: Natural History Press).

10. M. Harris, 1989, *Our Kind* (New York: Harper & Row).

11. D. Grigg, 1993, *The World Food Problem* (Oxford, Eng.: Blackwell), 54.

12. R. Douthwaite, 1993, *The Growth Illusion* (Tulsa, Okla.: Council Oak Books).

13. R. Barnet and J. Cavanagh, 1994, *Global Dreams: Imperial Corporations and the New World Order* (New York: Simon & Schuster), 17.

14. Li Qiang, 1994, "Social indicators and social structural changes in China today," paper presented at International Symposium on Social Development, Beijing, China, Oct. See also N. Kristof and S. WuDunn, 1994, *China Wakes* (New York: Random House).

15. Peng Xhizhe, 1994, *Recent Trends in China's Population and Their Implications* (London: London School of Economics, Research Programme on the Chinese Economy, CP/30, STICERD); Li Qiang, 1994.

16. Li Qiang, 1994, p. 8.

17. C. Riskin, 1994, "Panel on environment and poverty," paper presented at International Symposium on Social Development, Beijing, China, Oct.

18. V. K. Bhargava and C. N. Lucas, 1994, "Advancing social development in China," paper presented at International Symposium on Social Development, Beijing, China, Oct.

19. All personal material in this section is based on visits to the Beijing area in September 1991 and December 1994.

20. Up to 1979, China manufactured only some 160,000 vehicles each year, with buses and trucks making up over 90 percent of the production. In 1993, production reached 1.25 million, and the target for 2000 is 3 million. The numbers given in the text are for cars only. M. J. Dunne, 1994, "The race is on," *The China Business Review* (March/April): 16. The official newspaper, *China Daily,* described the Chinese market as consisting of "about 300 million potential car own-

ers," in Jeffrey Parker, 1994, "Mercedes, Porsche down-shift for vast China market," Reuters, Beijing, 15 Nov.

21. J. Parker, 1994, "Corporate giants vie to produce China's family car," Reuters, Beijing, 14 Nov.

22. J. Kahn, 1994, "China's next great leap: The family car," *The Wall Street Journal,* 16 Nov.

23. An Argentine-style restaurant specializing in meat (especially beef) dishes.

24. Ye Ruqui, 1994, "Break the vicious cycle of poverty and environmental deterioration," paper presented at International Symposium on Social Development, Beijing, China, Oct.

25. Li Qiang, 1994.

26. J. Cooley, 1984, "The war over water," *Foreign Policy* 54:3–26. For more on water shortages as a population-related source of conflict, see P. H. Gleick, 1994, "Water, war and peace in the Middle East," *Environment* 36(3):6–15.

27. T. Homer-Dixon, J. Boutwell, and G. Rathjens, 1993, "Environmental change and violent conflict," *Scientific American,* Feb., 38–45.

28. N. Keyfitz, 1991, "Population and development within the ecosphere: One view of the literature," *Population Index* 57(1):5–22.

29. The current situation is well summarized in the fascinating book by Barnet and Cavanagh, 1994.

30. *The Economist,* 1993, "A survey of multinationals," 27 March, 5–6.

31. Barnet and Cavanagh, 1994, p. 14.

32. Ibid., p. 13.

33. Ibid., pp. 347–348.

34. United Nations estimates cited in R. Barnet, 1993, "The end of jobs," *Harper's Magazine,* Sept., 47–52.

35. M. S. Swaminathan, 1994, interview, 13 Dec. (M. S. Swaminathan Research Foundation, Taramani Institutional Area, Madras 600 113, India).

36. H. M. Enzensberger, 1990, *Civil War* (London: Granta Books), 36.

37. Much of the material in this section is adapted from G. Daily and P. Ehrlich, 1995, "Socioeconomic equity, sustainability, and carrying capacity," *Ecological Applications* (forthcoming). References to statements can be found there. See also the insightful article by B. Hannon, 1987, "The discounting of concern," *Environmental Economics,* G. Pillet and T. Mucota, eds. (Geneva: R. Leingruber), 227–281.

38. R. Ornstein and P. Ehrlich, 1989, *New World/New Mind* (New York: Doubleday).

39. See, e.g., H. Albers, A. Fisher, and W. Hardmann, 1993, "Valuation and management of tropical forests: A theoretical and empirical analysis," *Giannini Foundation of Agricultural Economics Working Paper no. 681* (University of California, Berkeley), Aug.; H. Albers, 1995, "Modeling ecological constraints on tropical forest management: Spatial interdependence, irreversibility, and uncertainty," *Journal of Environmental Economics and Management* (in press).

40. Proposed in Pirages and Ehrlich, 1974.

41. For instance, the current energy system rests on a capital investment of about $8 trillion to $9 trillion. The system of industrial energy now depends about 85 percent on fossil fuels; if a significant portion of that is to be replaced by 2040 by other energy sources, the changes in delivery infrastructure should be on the drawing boards today (J. P. Holdren, 1994, personal communication).

42. For interesting recent examples of problems in this area, see Barnet and Cavanagh, 1994, especially pp. 250–256.

43. Solar cookers can help solve fuelwood problems, but they sometimes are too expensive for poor farmers, cannot be used for cooking in the evening, and often cannot maintain the varied temperatures required for different dishes. They have been used communally in West Africa with considerable success (Patrick Gonzalez, personal communication, 15 November 1994). For a discussion of substitution of other fuels for dwindling firewood supplies, see G. Leach and R. Mearns, 1988, *Beyond the Woodfuel Crisis: People, Land and Trees in Africa* (London: Earthscan). For more information on increasing the efficiency of cookstoves, see K. Smith, 1993, "One hundred million improved cookstoves in China: How was it done?," *World Development* 21:941–961.

44. R. Falk, 1993, "The making of global citizenship," in J. Brecher, J. Childs, and J. Cutler (eds.), *Global Visions: Beyond the New World Order* (Boston: South End Press), 39.

45. In a classic paper, ecologist Garrett Hardin described the problems that can arise from the exploitation of common property resources (1968, "The tragedy of the commons," *Science* 162:1243–1248). Where no rent is (or can be) charged for a scarce resource like wood from a communal woodlot, there will be a tendency for each community member to take what he or she desires, regardless of whether this leads to exploitation. The philosophy is "I might as well use it; if I don't, someone else will—it won't be preserved in any case." In Kenya, private rather than communal ownership has been given much of the credit for increases in woody biomass on farmland in the face of rapid population growth; see P. Holmgren, E. Masakha, and H. Sjöholm, 1994, "Not all African land is being degraded: A recent survey of trees on farms in Kenya reveals rapidly increasing forest resources," *Ambio* 23:390–395. The communal land tenure system (the "ejido" system) that is used on about 50 percent of the productive land in Mexico has been described as "an environmental disaster," due to the problems of the communal use of the marginal rangelands (Peter Warshall, personal communication, December 1994).

46. P. Ehrlich and A. Ehrlich, 1970, *Population, Resources, Environment: Issues in Human Ecology* (San Francisco: W. H. Freeman & Co.), 304. In this book, we also used the terms *de-development* and *semi-development* for what rich and poor nations should do, but neither the terms nor the concepts were paid any attention. *Sustainable development* has caught on as a slogan, but certainly rich nations show no signs of wanting to do it themselves.

47. Barnet and Cavanagh, 1994, p. 246.
48. See R. Costanza (ed.), 1991, *Ecological Economics* (New York: Columbia University Press); P. Ehrlich, 1994, "Ecological economics and the carrying capacity of Earth," in A.-M. Jansson et al., *Investing in Natural Capital: The Ecological Economics Approach to Sustainability* (Washington, D.C.: Island Press), 38–56; J. Gowdy, 1994, *Coevolutionary Economics: The Economy, Society and the Environment* (Boston: Kluwer Academic Publishers).
49. C. Folke, 1994, "Askö meeting of ecologists and economists" (submitted to *BioScience* and *Journal of Economic History*). Economist signatories were: Partha Dasgupta; Frank Ramsey (professor of economics, University of Cambridge); Geoffrey M. Heal (associate dean and professor of economics, Graduate School of Business, Columbia University); Karl-Goran Mäler (professor of economics, Stockholm School of Economics); Charles Perrings (Professor of environmental economics and management, University of York); and David A. Starrett (chairman and professor of economics, Stanford University). A second meeting resulted in a discussion paper: K. Arrow et al., 1995, "Economic growth, carrying capacity, and the environment," *Science* 268:520–521.
50. The ecologists present at the first meeting were Robert Costanza (professor at the Maryland International Institute for Ecological Economics); P. R. Ehrlich (professor of biological sciences, Stanford University); Carl Folke (deputy director of the Beijer Institute); C. S. Holling (professor of zoology, University of Florida); A.-M. Jansson (associate professor of ecology, University of Stockholm); B.-O. Jansson (professor of marine ecology, University of Stockholm); and Jonathan Roughgarden (professor of biological sciences and earth sciences, Stanford University).
51. P. Ehrlich and A. Ehrlich, 1991, *Healing the Planet* (New York: Addison-Wesley); G. Daily, A. Ehrlich, and P. Ehrlich, 1994, "Optimum human population size," *Population and Environment* 15:469–475.
52. J. Fisher, 1994, "Third World NGOs: A missing piece to the population puzzle," *Environment* 36(7):6–11, 37–41.
53. One guide to wise-use NGOs is C. Deal, 1993, *The Greenpeace Guide to Anti-environmental Organizations* (Berkeley, Calif.: Odonian Press).
54. See G. Hardin, 1993, *Living within Limits* (New York: Oxford University Press).
55. R. Benedict, 1934, *Patterns of Culture* (Boston: Houghton Mifflin), 278. For an early criticism of the relativist position, see C. Kluckhohn, 1955, "Ethical relativity: Sic et non," *Journal of Philosophy* 52:663–677.
56. Regulation can be accomplished through various mechanisms. There is a large and rapidly expanding literature on this topic, especially now in ecological economics.
57. R. Benedick, 1991, *Ozone Diplomacy* (Cambridge, Mass.: Harvard University Press).
58. Anonymous, 1994, "Ozone depletion: Holes galore," *The Economist,* 15 October, 114.
59. That would result in, roughly, a 75 TW world. Leveling up to the

consumption of the average U.S. citizen would yield a 110 TW world; leveling up to twice the average consumption of today's rich nations would produce a 150 TW world. The latter assumes both that equity would be achieved *and* consumption of the rich would continue to grow, as recommended by the Brundtland Report.

60. World Commission on Environment and Development, G. Brundtland (chair), 1987, *Our Common Future* (Oxford, Eng.: Oxford University Press). The report envisioned economic activity multiplying "five- or ten-fold in the coming half-century" (p. 4).

61. World Resources Institute, 1994, *World Resources 1994–95* (New York: Oxford University Press), 260.

62. A. Ehrlich and J. Birks (eds.), 1990, *Hidden Dangers: Environmental Consequences of Preparing for War* (San Francisco: Sierra Club Books).

63. Brown, Kane, and Roodman, 1994, pp. 110–111.

64. Latest figures (1992) on non-military assistance are from *1994 Information Please Almanac,* (New York: Houghton Mifflin).

65. The Clinton administration, to its credit, has established a Council on Sustainable Development, including environmentalists, business leaders, and government officials, to thrash out these issues. But as of mid-1995, its deliberations have remained outside the public eye.

66. For a view of the human predicament from a religious perspective by one of the world's greatest ecologists, see C. Birch, 1993, *Regaining Compassion* (Kensington, NSW, Australia: New South Wales University Press).

INDEX

government policies, 103–37
  health care and, 63, 66, 72
  reduction of, 69–72, 265
  regulation of, 37, 52–55, 60–64
  women's education and, 72–76
  women's health and, 76–80
  See also Contraception; Fertility; Total
    Fertility Rate
Bongaarts, John, 56
Borlaug, Norman, 161
Botswana, 18
Brandt, Willy, 238
Braudel, Fernand, 178
Brazil, 67, 82
Breastfeeding, 41, 78
  wet nurses, 33, 46–47
Brown, Lester, 15, 162
Brundtland Report (1987), 5–6, 279
Bucharest Conference (1974), 68
Burkina Faso, 232
Bush administration, 103, 105, 106
Bush, George, 105,106

Cain, Mead, 89
Cairo Conference. *See* International
  Conference on Population and
  Development
California, 181
Canada, 109, 122, 192, 219, 220
Carbon dioxide emissions, 278
Carr-Saunders, A. M., 38, 53
Carrying capacity, 3–5, 6, 154–61, 261
Cash-crops, 198–99
Catholicism. *See* Roman Catholic Church
Catholic Relief Services (CRS), 127–28
Cattle. *See* Livestock
Cavanagh, John, 250
Center for the Improvement of Maize and
  Wheat (CIMMYT), 150
Cervical cancer, 78
Cervical cap, 79
CFC's, 279
CGIAR. *See* Consultative Group on
  International Agriculture Research
Chatfield, Joselyn, 212
Chemical pollutants, 25–26, 205
Chen, Robert, 23, 227
Chengalpattu health clinic (India), 85–86

Childbearing
  maternal mortality, 77–78
  pronatalism, 36, 54, 109, 123, 131
  right to, 117, 276–77
  societal expectations for, 14
  STDs and, 79
  unmarried, 108–9
  See also Abortion; Birth rates;
    Contraception; Fertility; Government
    fertility policies
Child care, 87–91, 109–10
Child family policy, 90,103–4, 115–119
Children
  abuse of, 34, 43, 47–48
  AIDS and, 16
  death rates. *See* Death rates, children;
    Infanticide
  health care, 76–78
  as laborers, 232, 233
  malnutrition, 22, 162
  value of, 87–91, 156
Child support, 14
China
  agriculture, 90, 152, 190, 218
  automobile production, 29, 119,
    252–55
  birth rate, 67–68
  farmland loss, 172–73
  female infanticide, 48, 116–17
  fertility policy, 19–20, 51, 57, 68, 69,
    90, 103–4, 106, 113–19, 115
  grain needs in, 15
  growing socioeconomic inequity,
    251–55
  historical famines, 114–15
Climate change, 192–94
Clinton administration, 103, 106
Cole, Darryl, 214
Colombo Plan, 149
Comoros, 11, 12, 13–14, 18
Condoms
  as AIDS prevention, 16, 56, 57, 125,
    128
  availability in fertility-control programs,
    84, 85, 94, 96, 117
  as contraception, 36, 41, 55, 79
  female, 57
  Vatican opposition to, 125, 126–27, 128